Nanoelectronic
Devices

Byung-Gook Park
Sung Woo Hwang
Young June Park

Nanoelectronic
Devices

PAN STANFORD PUBLISHING

Published by

Pan Stanford Publishing Pte. Ltd.
Penthouse Level, Suntec Tower 3
8 Temasek Boulevard
Singapore 038988

Email: editorial@panstanford.com
Web: www.panstanford.com

British Library Cataloguing-in-Publication Data
A catalogue record for this book is available from the British Library.

ISBN 978-981-4364-00-3 (Hardcover)
ISBN 978-981-4364-01-0 (eBook)

Contents

Preface

The transition from microelectronics to nanoelectronics occurred at the dawn of the new millennium. It was the production of CMOS devices with a 100 nm feature size that delineated the beginning of the nanoelectronics era. Microelectronics was developed for more than 50 years since the bipolar transistor was invented in 1947. During this time, there was an enormous progress in the development of integrated circuit (IC) technologies. The invention of MOSFET in 1960 and of the CMOS circuit in 1963 ignited the exponential shrinkage of devices, called scaling, and device scaling has been going on for more than 48 years. Every three years, the average design rule has shrunk to the 70 percent of the previous generation. Such a rate of shrinkage brings roughly an order of magnitude reduction every 20 years. In addition, there was a short period of accelerated scaling around the microelectronics-to-nanoelectronics transition point. Thus, in about 50 years, some devices have been shrunk down by almost three orders of magnitude in size, so that the leading-edge IC products have a minimum feature size of 20–30 nm in 2011.

The transition between microelectronics and nanoelectronics was smooth and took an evolutionary path rather than a revolutionary one. The basic device structures and operating principles remain more or less the same. However, device scaling has been challenging us with many difficult issues. There are several traditional issues that have been troubling device engineers in the microelectronics era. Short-channel effects (threshold voltage roll-off and drain-induced barrier lowering), punch-through, and velocity saturation are in this category. To cope with these issues, MOSFET scaling theories have been developed and successfully applied down to 100 nm channel devices. When we move into the nanoelectronics era, however, not only the traditional issues become extremely severe, but also a few new issues start to challenge us. One of them is the fluctuation in various device parameters due to the granularity of the microscopic world, and the other is tunneling and other quantum phenomena due to the wave nature of electrons. To overcome the problems in the nanoelectronics era, new device structures, operating principles, and materials are introduced.

The purpose of this book is to provide the readers with the fundamentals of nanoelectronic devices. The book starts with a brief review of quantum mechanics and solid-state physics that can form the basis of semiconductor physics. The basic physics of electron transport and p–n junctions, as they relate to the fundamental principles of MOSFET and other nanoelectronic devices, are covered. From there, the basic operations of MOS capacitor and MOSFET are developed and some basic CMOS circuits are introduced. The nanoelectronic devices are categorized into three types: quantum well, quantum wire, and quantum dot devices depending on the dimensionality of their active region (in most cases, the channel). Three-dimensional device structures, new materials, and new operating principles based on quantum mechanics or discreteness of charge are emphasized. The last chapter is devoted to the nanotechnology application of field-effect transistors, focusing on the chemical and biochemical sensors.

This book is suitable for use as a textbook by senior under-graduate or graduate students in nanotechnology, nanoscience, and electrical engineering. We wanted the book to be self-contained, considering its introductory nature. For practicing engineers and scientists involved in research and development in the IC industry, this book may serve as an introductory reference for them to stay up to date in this field. Nanoelectronic devices are too huge a subject to be covered in one book. We have chosen to cover only the fundamental aspects here.

We would like to thank the Ministry of Education, Science and Technology, and the Korea Nanotechnology Research Soci-ety for their support and funding. Without their initiation and encouragement, the writing of this book could not have even started. Young June Park would like to acknowledge the secretarial support of Yejin Choi. Sung Woo Hwang would like to thank

Prof. Yun Seop Yu for providing contents on single-electron circuits. Finally, we would like to give special thanks to our families — Jeehee Lee, Albert, Paul (B.G.P.); Keummi Jang, Jeffrey, Sandy (S.W.H.); and Sukhee Kim, Dongwoo, Hyori (Y.J.P.) — for their warm support and understanding during this task.

Byung-Gook Park
Sung Woo Hwang
Young June Park

Chapter 1

Quantum Mechanics for Nanoelectronic Devices

1.1 Fundamental Concepts of Quantum Mechanics

1.1.1 *Wave Nature of Particles*

In the late 19th century, scientists started to notice that there were phenomena that could not be explained by the classical theory of mechanics. One of them was the spectrum of absorption and emission by atoms. According to classical mechanics, the spectrum should be continuous, i.e., the emitted or absorbed light could have any frequency (wavelength). The experimental results, however, were surprising. The atoms emitted and absorbed light with discrete values of frequency only.

Let us examine the situation in more detail. Figure 1.1(a) shows an experimental setup to measure the emission spectrum of a hydrogen discharge lamp. If we pump out the air in a vacuum tube and fill it with a small amount of hydrogen (H_2) gas, an arc discharge can be initiated by supplying sufficient bias voltage to the electrodes in the tube. The electrons in the hydrogen atoms absorb energy from the arc discharge and emit light with various frequencies. One startling feature of this process is that the emitted light has certain discrete frequencies only, as shown in Fig. 1.1(b).

Nanoelectronic Devices
Byung-Gook Park, Sung Woo Hwang, and Young June Park
Copyright © 2012 Pan Stanford Publishing Pte. Ltd.
www.panstanford.com

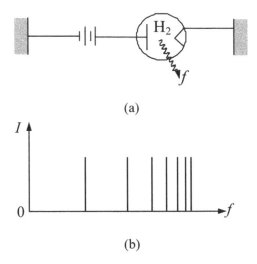

(a)

(b)

Figure 1.1. Spectrum of hydrogen discharge lamp: (a) experimental setup, (b) emission spectrum.

Many people studied this phenomenon and some came up with the following formula for the observed frequencies.

$$f = \text{const.} \left(\frac{1}{n_1^2} - \frac{1}{n_2^2} \right), \tag{1.1}$$

where n_1 and n_2 are integers ($n_1 < n_2$). Depending on the value of n_1, the discrete frequencies form distinctive groups, called Lyman ($n_1 = 1$), Balmer ($n_1 = 2$), and Paschen ($n_1 = 3$) series after their early investigators.

This phenomenon, however, cannot be understood at all in terms of classical mechanics. Maxwell's theory of electromagnetic radiation predicts that there should be an emission of radiation since the electron is accelerated toward the positively charged nucleus. The electron loses its energy due to the emission of radiation and forms a continuously shrinking orbit. The continuous motion of the electron should generate a continuous emission spectrum as shown in Fig. 1.2(a). There is no way of explaining the discrete spectrum in the frame of classical mechanics. Furthermore, the electron should eventually settle down on the surface of nucleus (Fig. 1.2(b)). This conclusion again contradicts Rutherford's well-known experiment

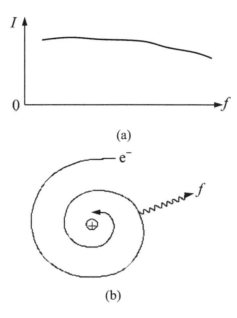

(a)

(b)

Figure 1.2. Classical explanation of the emission spectrum of atoms: (a) emission intensity as a function of frequency, (b) mechanism of photo emission. Since the electron is accelerated toward the positively charged nucleus, it emits electromagnetic radiation and loses energy. The radius of electron orbit should shrink rapidly and the electron will eventually come in contact with the nucleus.

that showed that there is a vast space between the electron and the nucleus.

In order to explain the emission spectrum of atoms, Niels Bohr proposed a model for the hydrogen atom, based on the motion of the planetary system. His main hypothesis concerning the electron motion in an atom was the existence of quantized stable orbits. The condition for a stable orbit is that the phase integral of the electron angular momentum, P_φ, should be an integer multiple of Planck's constant, h.

$$\oint P_\varphi d\varphi = nh \tag{1.2}$$

In the case of a circular orbit with radius r, mass m_e, and velocity v, we can easily evaluate the phase integral, using the definition of angular momentum ($P_\varphi = |\vec{r} \times m_e\vec{v}| = rm_ev$).

$$rm_ev \cdot 2\pi = nh \tag{1.3}$$

For a circular motion, the Coulomb attraction between the electron and the nucleus should be balanced by the centrifugal force of the electron:

$$\frac{e^2}{4\pi \varepsilon_0 r^2} = \frac{m_e v^2}{r},$$ (1.4)

where e is the charge of an electron and ε_0 is the permittivity of vacuum. Combining Eqs. (1.3) and (1.4), we can obtain the velocity of the electron.

$$v = \frac{e^2}{2nh\varepsilon_o}$$ (1.5)

Thus, the velocity of the electron in a stable orbit is quantized and the corresponding energy is also quantized as

$$E_n = \frac{1}{2}m_e v^2 - \frac{e^2}{4\pi \varepsilon_o r} = -\frac{1}{2}m_e v^2 = -\left(\frac{m_e e^4}{8h^2 \varepsilon_0^2}\right)\frac{1}{n^2} = -\frac{13.6}{n^2}[\text{eV}].$$ (1.6)

By the time Bohr proposed his model, it was well established that light is emitted as an energy quantum ($E = hf$) called the photon, and he postulated that a photon is emitted when an electron jumps from one stable orbit (with higher energy E_{n_2}) to another (with lower energy E_{n_1}).

$$hf = E_{n_2} - E_{n_1} = 13.6\left(\frac{1}{n_1^2} - \frac{1}{n_2^2}\right)[\text{eV}]$$ (1.7)

Now this equation is the same as Eq. (1.1) and the atomic spectrum is explained finally.

There was one major deficiency in Bohr's theory, however, since he could not provide any reason why the angular momentum should be quantized. It was Louis de Broglie who provided an explanation for the quantization of angular momentum by recognizing the wave nature of particles. If a wave is associated with an electron, which is a well-known particle, the wave should form a standing wave in a given orbit: otherwise, it cannot form a stable orbit since it will interfere destructively with itself and disappear. Figure 1.3 shows one of the standing waves in a circular orbit ($2\pi r = 3\lambda$, where λ is the wavelength).

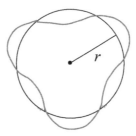

Figure 1.3. Standing wave formed in a circular orbit. de Broglie hypothesized that there was a wave associated with a particle (electron), and that it should form a standing wave to make a stable orbit.

Following de Broglie's argument, we can obtain the relationship between the wavelength of the associated wave and the momentum of the electron. In order for the wave to form a standing wave, the circumference of the orbit should be an integer multiple of its wavelength.

$$n\lambda = 2\pi r \tag{1.8}$$

The angular momentum, P_φ, of the electron can be expressed as

$$P_\varphi = rp, \tag{1.9}$$

where p is the linear momentum of the electron ($p = m_e v$). To explain the atomic spectrum, the angular momentum should satisfy Bohr's condition:

$$\oint P_\varphi d\varphi = \oint rp\, d\varphi = 2\pi rp = nh. \tag{1.10}$$

From Eqs. (1.8) and (1.10), we obtain

$$\lambda = h/p. \tag{1.11}$$

This deceptively simple equation has a tremendous significance in the history of quantum mechanics. The wave nature of particles was recognized for the first time and the wavelength of a particle was expressed in terms of the momentum, which used to be regarded as purely a property of particles. The concept that a particle has an associated wave appeared so strange and unrealistic that nobody seriously tried to seek other evidence of the associated wave until Erwin Schrödinger set up his wave equation and obtained a complete solution for the hydrogen atom.

1.1.2 *Wave Functions and Operators*

Before we delve into Schrödinger's great achievement, let us explore the world of waves for a short while. The simplest wave is a plane wave. A plane wave moving in the x-direction can be expressed as

$$\Psi = A \exp[j(kx - \omega t)], \qquad (1.12)$$

where j is the unit of imaginary number ($\sqrt{-1}$), k is the wave number, and ω is the angular frequency. k and ω are defined from the wavelength, λ, and frequency, f, as

$$k = \frac{2\pi}{\lambda} \qquad (1.13)$$

and

$$\omega = 2\pi f, \qquad (1.14)$$

respectively. The wave function, Ψ, can be a complex number, since it is not a directly measurable quantity such as field intensity or density. Using $E = hf = \hbar\omega$ ($\hbar = \frac{h}{2\pi}$ is the reduced Planck constant) and Eq. (1.11), the wave function can be rewritten as

$$\Psi = A \exp\left[\frac{j}{\hbar}(px - Et)\right] \qquad (1.15)$$

If we have a plane wave in a three-dimensional space, we can write

$$\Psi = A \exp\left[\frac{j}{\hbar}(\vec{p} \cdot \vec{r} - Et)\right], \qquad (1.16)$$

using a momentum vector, $\vec{p}$, and a position vector, $\vec{r}$.

In classical mechanics, a wave function directly represents a physical quantity such as field intensity or density. A physical quantity can be calculated arithmetically from the value of the wave function. For example, the value of an electric field wave function is the field intensity itself, and the energy density stored in the electrical field can be evaluated by squaring the field intensity and multiplying it by an appropriate constant. In quantum mechanics, however, the wave function is just related to the existence of a particle and the physical quantities such as momentum and energy are embedded in the wave function, so that the wave function does not yield any physical quantity under arithmetic operations. Then, how can

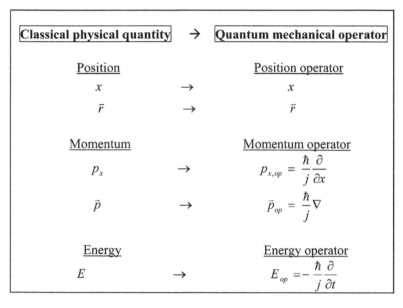

Figure 1.4. Classical physical quantities and corresponding quantum mechanical operators. The position operator is just a real number, but the momentum and energy operators are differentiation operators. ∇ is defined as $\hat{x}\frac{\partial}{\partial x} + \hat{y}\frac{\partial}{\partial y} + \hat{z}\frac{\partial}{\partial z}$ in Cartesian coordinate system.

we extract a physical quantity from a wave function in quantum mechanics? The answer lies in the concept of quantum mechanical operators.

The correspondence between the classical physical quantity and the quantum mechanical operator is shown in Fig. 1.4. The position operator and its functions are just real numbers. Most of the other operators are expressed as differentiation operators in the scheme of wave function representation. Let us examine how we can apply a quantum mechanical operator to a wave function for the extraction of the classical physical quantity. For simplicity, we will use a plane wave for the wave function. When the momentum operator in the x-direction is applied to the wave function, we obtain the x-component of the momentum as an eigenvalue.

$$p_{x,op}\Psi = \frac{\hbar}{j}\frac{\partial}{\partial x}\Psi = \frac{\hbar}{j}\frac{\partial}{\partial x}\left\{A\exp\left[\frac{j}{\hbar}(p_x x - Et)\right]\right\} = p_x\Psi \quad (1.17)$$

Thus, the plane wave in Eq. (1.12) is an eigenfunction of the momentum operator in the x-direction. If we apply the energy operator to the same wave function, we can confirm that the plane wave is also an eigenfunction of the energy operator.

$$E_{op}\Psi = -\frac{\hbar}{j}\frac{\partial}{\partial t}\Psi = -\frac{\hbar}{j}\frac{\partial}{\partial t}\left\{A\exp\left[\frac{j}{\hbar}(p_x x - Et)\right]\right\} = E\Psi \quad (1.18)$$

1.1.3 Schrödinger's Wave Equation

In classical mechanics, the energy of a particle is the sum of kinetic energy and potential energy.

$$\begin{array}{ccccc}
\text{kinetic energy} & + & \text{potential energy} & = & \text{total energy} \\
\frac{\vec{p}^2}{2m_e} & + & V(\vec{v}, t) & = & E
\end{array} \quad (1.19)$$

Schrödinger formulated his famous wave equation on the basis of this fundamental relationship. The first step was to transform the classical physical quantities into the quantum mechanical operators,

$$\frac{\vec{p}^2}{2m_e} \rightarrow -\frac{\hbar^2}{2m_e}\nabla^2, V(\vec{r}, t) \rightarrow V(\vec{r}, t), E \rightarrow -\frac{\hbar}{j}\frac{\partial}{\partial t}, \quad (1.20)$$

where ∇^2 is $\frac{\partial^2}{\partial x^2} + \frac{\partial^2}{\partial y^2} + \frac{\partial^2}{\partial z^2}$ in Cartesian coordinate system. The next natural step is to apply the operator to a wave function $\Psi(\vec{r}, t)$, and we obtain

$$\left[-\frac{\hbar^2}{2m_e}\nabla^2 + V(\vec{v}, t)\right]\Psi(\vec{r}, t) = -\frac{\hbar}{j}\frac{\partial}{\partial t}\Psi(\vec{r}, t). \quad (1.21)$$

Schrödinger applied this equation to the hydrogen atom and obtained a complete set of solutions, succeeding in fully explaining atomic spectra and, eventually, the chemical properties of atoms. We will come back to this point in Section 1.2.

When the potential energy is not a function of time ($V(\vec{r}, t) = V(\vec{r})$), the total energy is constant. For such a time-independent potential, we can use the method of separation of variables. Let us assume that $\Psi(\vec{r}, t)$ can be expressed as a product, $\psi(\vec{r})\phi(t)$. We can rewrite Schrödinger's equation as

$$\left[-\frac{\hbar^2}{2m_e}\nabla^2 + V(\vec{r})\right]\psi(\vec{r})\phi(t) = -\frac{\hbar}{j}\frac{\partial}{\partial t}[\psi(\vec{r})\phi(t)]. \quad (1.22)$$

Carrying out the differentiation operation and dividing the result by $\psi(\vec{r})\phi(t)$, Eq. (1.21) becomes

$$\left\{\left[-\frac{\hbar^2}{2m_e}\nabla^2 + V(\vec{r})\right]\psi(\vec{r})\right\}\bigg/\psi(\vec{r}) = \left\{-\frac{\hbar}{j}\frac{\partial}{\partial t}\phi(t)\right\}\bigg/\phi(t) = E.$$
(1.23)

Here we have intentionally used the symbol E as a constant because we know that the eigenvalue of the energy operator should be the constant energy E. Finally, we end up with two independent differential equations.

$$-\frac{\hbar}{j}\frac{\partial}{\partial t}\phi(t) = E\phi(t)$$
(1.24)

$$\left[-\frac{\hbar^2}{2m_e}\nabla^2 + V(\vec{r})\right]\psi(\vec{r}) = E\psi(\vec{r})$$
(1.25)

Equation (1.24) has a simple solution that is common to all time-independent potential cases. It can be explicitly written as

$$\phi(t) = A_1\exp\left(-j\frac{E}{\hbar}t\right) = A_1\exp(-j\omega t),$$
(1.26)

where A_1 is a constant. Equation (1.25) is called the time-independent Schrödinger equation. Often, the technique to solve it is quite involved mathematically. In fact, there are analytical solutions for only a few cases.

The simplest case is a particle in free space where the potential energy is zero throughout the whole space ($V(\vec{r}) = 0$ for all $\vec{r}$). Equation (1.25) becomes

$$-\frac{\hbar^2}{2m_e}\nabla^2\psi(\vec{r}) = E\psi(\vec{r}).$$
(1.27)

The solution to this differential equation is again an exponential function

$$\psi(\vec{r}) = A_2\exp(j\vec{k}\cdot\vec{r}),$$
(1.28)

where A_2 is a constant and $\vec{k}$ should satisfy the following relationship:

$$\frac{\hbar^2|\vec{k}|^2}{2m_e} = E$$
(1.29)

If we combine Eqs. (1.26) and (1.28),

$$\Psi(\vec{r}, t) = A_1 \exp(j\vec{k}\cdot\vec{r})\cdot A_2 \exp(-j\omega t) = A \exp(j\vec{k}\cdot\vec{r} - j\omega t) \quad (1.30)$$

We can see that this is exactly the same plane wave that we had in Eq. (1.16) if we recognize that

$$\vec{k} = \frac{\vec{p}}{\hbar}, \omega = \frac{E}{\hbar} \quad (1.31)$$

1.1.4 *Meaning of the Wave Function*

Even though the Schrödinger equation provided the wave function with a strong foothold in quantum mechanics, the physical meaning of the wave function still defied the grasp of Schrödinger and other physicists. They recognized that the wave function is related to the existence of a particle. If they interpret it as a density function, however, they were confronted with a serious problem. When a particle is in its free state, it should form a plane wave, and the density function should be spread out through the whole universe. If we capture the particle, the density function should be localized instantly, which means that the material should move with a speed faster than that of light. This is clearly incompatible with Einstein's relativity theory.

In 1926, Max Born proposed a radically different interpretation of the wave function. He associated the wave function with the probability of finding a particle at a given position and time. The wave function is not the probability itself but the probability amplitude, which can be used to calculate the probability in the following way:

$$\Psi(\vec{r}) = \text{probability amplitude} \quad (1.32)$$

$|\Psi(\vec{r})|^2 d\vec{r}$ = probability of finding a particle in the volume element

$$d\vec{r} \text{ at position } \vec{r} \quad (1.33)$$

This interpretation is radical since it abandons the classical concept of the position of a particle. In classical mechanics, a particle has a definite position at any moment. In the probabilistic interpretation, however, only the probability of finding the particle can be determined. As an extreme example, we can take the case of a plane wave. Since the absolute value of the probability amplitude is the same everywhere, the probability of finding the particle is the same at any place. The position of the particle is completely unknown. Even though quite a number of physicists, including Einstein, have been against it, the probabilistic interpretation is still regarded as the standard interpretation of wave function.

If we accept the probabilistic interpretation of wave function, one of its natural consequences is the condition of normalization, which states that the probability of finding the particle in all space is 1.

$$\int_{all\ space} |\Psi(\vec{r}, t)|^2 d\vec{r} = 1 \tag{1.34}$$

Another important concept is the expectation value of a physical quantity. An expectation value is the average of all values measured in independent and identical systems. Let q be a physical quantity that is a function of $\vec{r}$ and t. For a time-independent potential, the expectation value of q is defined as

$$<q> = \int_{all\ space} \psi^*(\vec{r}) q_{op} \psi(\vec{r}) d\vec{r}. \tag{1.35}$$

For example, the expectation value of the momentum can be calculated as

$$<\vec{p}> = \int_{all\ space} \psi^*(\vec{r}) \frac{\hbar}{j} \nabla \psi(\vec{r}) d\vec{r}. \tag{1.36}$$

Let us calculate the expectation value of the momentum for a plane wave ($\psi(\vec{r}) = A \exp(j\vec{k} \cdot \vec{r})$) in a free space.

$$<\vec{p}> = \int \psi^*(\vec{r}) \frac{\hbar}{j} \nabla \psi(\vec{r}) d\vec{r} = \int \psi^*(\vec{r}) \hbar \vec{k} \psi(\vec{r}) d\vec{r}$$

$$= \hbar k \int \psi^*(\vec{r}) \psi(\vec{r}) d\vec{r} = \hbar \vec{k} = \vec{p} \tag{1.37}$$

We obtain this simple result, since the wave function of a plane wave is an eigenfunction of the momentum operator ($\vec{p}_{op} \psi = \vec{p} \psi$), i.e., $\vec{p}$ is a constant of motion.

1.1.5 *Electron in a Box*

One of the solvable problems that have practical importance is the "particle in a box" problem. Almost any material or device can be approximated as a box that contains electrons in it. For simplicity,

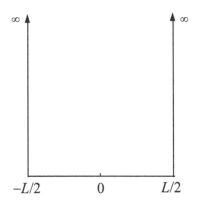

Figure 1.5. Potential energy for the particle-in-a-box problem. The potential energy is zero inside the box $(-L/2 < x < L/2)$ and infinity outside it $(x < -L/2$ or $x > L/2)$.

the potential energy for the box is assumed to have a form shown in Fig. 1.5. The potential energy is zero inside the box $(-L/2 < x < L/2)$ and infinity outside it $(x < -L/2$ or $x > L/2)$. In real materials, the potential energy barrier outside the box is finite. As long as the barrier height is high enough, however, the infinite barrier assumption in Eq. (1.38) is a good approximation.

$$V(x) = \begin{cases} 0, & |x| < L/2 \\ \infty, & |x| > L/2 \end{cases} \tag{1.38}$$

Since the potential energy is time independent and one dimensional, we can use the one-dimensional (1D) time-independent Schrödinger equation.

$$\left[-\frac{\hbar^2}{2m_e} \frac{\partial^2}{\partial x^2} + V(x) \right] \psi(x) = E\psi(x) \tag{1.39}$$

Because of the infinite potential barrier, the wave function $\psi(x)$ should be zero for $|x| \geq L/2$. This boundary condition allows only sinusoidal functions as valid solutions. We have an infinite number of solutions with discrete energy levels E_n. The wave functions have the form of a standing wave the same as the standing waves generated in the string of a guitar.

$$\psi_n(x) = \begin{cases} \sqrt{2/L} \cos a_n x, \ n = 1, 3 \dots \\ \sqrt{2/L} \sin a_n x, \ n = 2, 4, \dots \end{cases} \quad \text{(1.40a)}$$

$$a_n = \sqrt{\frac{2m_e E_n}{\hbar^2}} = \frac{n\pi}{L} \quad \text{(1.40b)}$$

$$E_n = \frac{n^2 \pi^2 \hbar^2}{2m_e L^2} \quad \text{(1.40c)}$$

Example 1.1: Quantized energy of an electron in a box

Calculate the quantized energy of the ground state for an electron in a box with the following dimension:

(a) $L = 50$ nm
(b) $L = 1$ nm

Solution:

(a) $E_1 = \dfrac{\pi^2 \hbar^2}{2m_e L^2} = \dfrac{3.14^2 \times (1.06 \times 10^{-34})^2}{2 \times 9.1 \times 10^{-31} \times (50 \times 10^{-9})^2}$
$= 2.43 \times 10^{-23} (J)$

$E_1 = 0.15$ meV
This energy is much smaller than the thermal energy $(k_B T = 26$ meV) at room temperature.

(b) $E_1 = \dfrac{3.14^2 \times (1.06 \times 10^{-34})^2}{2 \times 9.1 \times 10^{-31} \times (1 \times 10^{-9})^2} = 6.09 \times 10^{-20} (J)$

$E_1 = 380$ meV
This energy is much larger than the thermal energy $(k_B T = 26$ meV) at room temperature.

To examine the characteristics of these wave functions, let us try to extract the momentum in the x-direction. We need only to apply the momentum operator to the wave function.

$$\frac{\hbar}{j} \frac{\partial}{\partial x} \psi(x) = \frac{\hbar}{j} \frac{\partial}{\partial x} \left(\sqrt{2/L} \cos a_n x \right)$$
$$= -\frac{\hbar}{j} a_n \sqrt{2/L} \sin a_n x \neq (\text{const.}) \cdot \psi \quad \text{(1.41)}$$

It is clear that none of the wave functions is an eigenfunction of the momentum operator. The momentum is not a constant of motion for the electron in a box. This result can be understood by realizing that a sinusoidal function is a combination of two exponential functions, one of which is the complex conjugate of the other.

$$\cos a_n x = \frac{1}{2}\left[\exp(ja_n x) + \exp(-ja_n x)\right] \qquad (1.42)$$

Even though the exponential functions are eigenfunctions of the momentum operator, their sum is not an eigenfunction of the momentum operator, since the sum contains two plane waves with different momenta. The standing wave in the box is the superposition of two plane waves with opposite direction of propagation, and it has two momentum values, $\hbar a_n$ and $-\hbar a_n$. Thus, the momentum cannot be determined, and it is not a constant of motion.

If we evaluate the expectation value of the momentum, it turns out to be zero as shown in the following equation:

$$\int \psi^* p_{op} \psi \, dx = \frac{\hbar}{j} \frac{2}{L} \int_{-L/2}^{L/2} (\cos a_n x) \frac{\partial}{\partial x} (\cos a_n x) \, dx = 0 \qquad (1.43)$$

This result agrees with our intuition, which says that the standing wave is not moving in any direction.

1.2 Solid State and Energy Bands

1.2.1 *Electron Waves and Energy Levels in Atoms*

After formulating his wave equation, Schrödinger applied it to the hydrogen atom. The first step was to set it up in the spherical coordinate system (Fig. 1.6), since the spherical symmetry of the potential energy made the equation simpler, facilitating the solution of the differential equation. The potential energy is time independent, so we can use the time-independent Schrödinger equation. From Eq. (1.25),

$$\nabla^2 \psi + \frac{2m_e}{\hbar^2}(E - V)\psi = 0, \qquad (1.44)$$

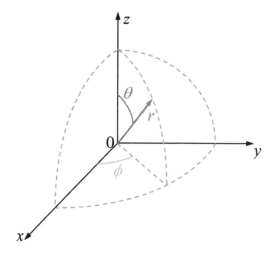

Figure 1.6. Spherical coordinate system. Three variables, r, θ, and ϕ, are defined.

where

$$V = -\frac{e^2}{4\pi\varepsilon_0 r}, r = \sqrt{x^2 + y^2 + z^2}. \tag{1.45}$$

The final form of the equation is

$$\nabla^2\psi + \frac{2m_e}{\hbar^2}\left(E + \frac{e^2}{4\pi\varepsilon_0 r}\right)\psi = 0. \tag{1.46}$$

In this equation, the eigenvalue E should be determined by an appropriate boundary condition, i.e., the condition that the wave function ψ should be finite and unique.

The next step is to find solutions to Eq. (1.46) using appropriate boundary conditions. We can put $\psi(r, \theta, \phi) = R(r)\Theta(\theta)\Phi(\phi)$ and use the technique of separation of variables. The Laplacian, ∇^2, in the spherical coordinate systems is

$$\nabla^2 = \frac{1}{r^2}\frac{\partial}{\partial r}\left(r^2\frac{\partial}{\partial r}\right) + \frac{1}{r^2\sin\theta}\frac{\partial}{\partial\theta}\left(\sin\theta\frac{\partial}{\partial\theta}\right) + \frac{1}{r^2\sin^2\theta}\frac{\partial^2}{\partial\phi^2}. \tag{1.47}$$

After some manipulations, we obtain the following three separated equations:

$$\frac{\partial^2 \Phi}{\partial \phi^2} = -m^2 \Phi \tag{1.48}$$

$$\frac{1}{\sin \theta} \frac{\partial}{\partial \theta} \left(\sin \theta \frac{\partial \Theta}{\partial \theta} \right) + \left[l(l+1) - \frac{m^2}{\sin^2 \theta} \right] \Theta = 0 \tag{1.49}$$

$$\frac{1}{r^2} \frac{\partial}{\partial r} \left(r^2 \frac{\partial R}{\partial r} \right) + \left\{ \frac{2m_e}{\hbar^2} [E - V(r)] - \frac{l(l+1)}{r^2} \right\} R = 0 \tag{1.50}$$

Equation (1.48) can be easily solved, and the result is

$$\Phi(\phi) = A_\Phi \exp [jm\phi]. \tag{1.51}$$

For uniqueness, Φ should have the same value when ϕ changes by 2π. This condition requires m to be an integer.

The techniques to solve Eqs. (1.49) and (1.50) are quite involved mathematically, and we will not go into the details of the calculations here. The solution to Eq. (1.49) can be expressed in the following form:

$$\Theta(\theta) = A_\Theta P_l^m (\cos \theta), \tag{1.52}$$

where P_l^m is an associated Legendre polynomial. The solution for Eq. (1.50) can be written as

$$R(r) = A_R \left(\frac{r}{na_0} \right)^l \exp \left(-\frac{r}{na_0} \right) L_{n+l}^{2l+1}, \tag{1.53}$$

where L_{n+l}^{2l+1} is a Laguerre polynomial and a_0 is the Bohr radius.

$$a_0 = \frac{4\pi \varepsilon_0 \hbar^2}{m_e e^2} \tag{1.54}$$

If we apply boundary conditions to $R(r)\Theta(\theta)$, conditions for the numbers n and l are obtained: n should be a positive integer and l should be a non-negative integer. In addition, the range of integers m is constrained by the value of l, and the range of l is in turn constrained by the value of n. The integers n, l, and m are called quantum numbers. For each set of quantum numbers n, l, and m, a unique wave function is defined. Let us examine the characteristics of these quantum numbers.

(1) Principal quantum number (n): A quantum number that determines the radial characteristics of the wave function. It is the same as the electron-orbital quantum number in Bohr's theory and determines the energy of the wave function as follows:

$$E_n = -\frac{m_e e^4}{8\varepsilon_0^2 h^2 n^2} \tag{1.55}$$

(2) Angular momentum quantum number (l): A quantum number determined by the condition for finite magnitudes of the wave function. For a given principal quantum number, n, the angular momentum quantum number can take only the following values:

$$l = 0, 1, 2, \ldots, n-1 \tag{1.56}$$

The angular momentum quantum number determines the angular momentum, L, of the electron.

$$L = \sqrt{l(l+1)}\hbar \tag{1.57}$$

(3) Magnetic quantum number (m): A quantum number determined by the uniqueness of the wave function Φ. For a given angular momentum quantum number l, m can have only the following values:

$$m = -l, -l+1, \ldots, 0, \ldots, l-1, l \tag{1.58}$$

The magnetic quantum number determines the angular momentum component L_z in the direction of the magnetic field.

$$L_z = m\hbar \tag{1.59}$$

Hence, the total number of quantum states for a given principal quantum number, n, is

$$\sum_{l=0}^{n-1}(2l+1) = n^2 \tag{1.60}$$

Since a set of quantum numbers n, l, and m determines a unique wave function, we can use this set of quantum numbers as a name for the wave function as follows:

$$\psi_{nlm}(r, \theta, \phi) = R_{nl}(r)\Theta_{lm}(\theta)\Phi_m(\phi) \tag{1.61}$$

Table 1.1. Five wave functions for electrons in a hydrogen atom

n	l	m	state	wave function
1	0	0	$1s$	$\psi_{100} = \frac{1}{\sqrt{\pi a_0^3}} \exp\left(-\frac{r}{a_0}\right)$
2	0	0	$2s$	$\psi_{200} = \frac{1}{4\sqrt{2\pi a_0^3}} \left(2 - \frac{r}{a_0}\right) \exp\left(-\frac{r}{2a_0}\right)$
2	1	0	$2p$	$\psi_{210} = \frac{1}{4\sqrt{2\pi a_0^3}} \frac{r}{a_0} \exp\left(-\frac{r}{2a_0}\right) \cos\theta$
2	1	1	$2p$	$\psi_{211} = \frac{1}{8\sqrt{\pi a_0^3}} \frac{r}{a_0} \exp\left(-\frac{r}{2a_0}\right) \sin\theta \exp(j\phi)$
2	1	−1	$2p$	$\psi_{21\bar{1}} = \frac{1}{8\sqrt{\pi a_0^3}} \frac{r}{a_0} \exp\left(-\frac{r}{2a_0}\right) \sin\theta \exp(-j\phi)$

Table 1.1 shows five normalized wave functions with the principal quantum number, n, up to 2.

The field of spectrum analysis has names for the various numerical values of angular momentum quantum number, l, and this naming scheme is often more popular than the method of using the numerical value of l. The first several names are provided in Table 1.2. For angular momentum quantum number larger than 3, we simply use letters from the alphabet in their usual order: g, h, i, ...

As we have already seen in Table 1.1, to characterize a wave function we put the character for the angular momentum quantum number after the principal quantum number. Following Bohr's orbit model, we often refer to a set of wave functions with the same principal and angular momentum quantum numbers ($1s$, $2s$, ...) as an "orbital."

As can be noticed in Eq. (1.61) and Table 1.1, the radial function depends on n and l, the θ function on l and m, and the ϕ function on m. Based on the wave functions in Table 1.1, we can sketch the shape

Table 1.2. Alternative names for angular momentum quantum numbers

angular momentum	alternative name
$l = 0$	s (sharp)
$l = 1$	p (principal)
$l = 2$	d (diffuse)
$l = 3$	f (fundamental)
$l = 4$	g
⋮	⋮

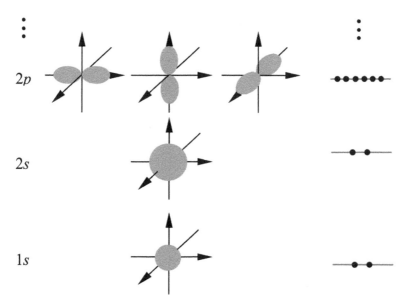

Figure 1.7. Shape of the probability density functions for the electrons in a hydrogen atom.

of the probability density functions. These shapes are illustrated in Fig. 1.7. In this figure, we have used sinusoidal functions instead of the exponential functions for the wave functions with $n = 2, l = 1, m = \pm 1$.

Attentive readers may have noticed that there are peculiar symbols (dots on a bar) on the rightmost side of Fig. 1.7. They represent the number of allowable states for the given type of orbital (1s, 2s, ...) shown on the leftmost side of the figure. In order to understand these numbers, we have to realize that there is another quantum number related to the electron "spin" (in fact, all elementary particles such as proton and neutron possess spin). Spin is the intrinsic angular momentum of an electron, and its projection to the axis of magnetic field is given as

$$L_s = \pm \frac{\hbar}{2}. \tag{1.62}$$

An electron can have a spin of 1/2 in units of $\hbar$, and the numbers $\pm(1/2)$ are called spin angular momentum quantum numbers (s). The spin angular momentum is positive (spin up) or negative (spin down) depending on the direction of the spin. Now, each allowable state of the electron should have four quantum numbers, n, l, m, and s. In order to obtain the total number of allowable states, we need to multiply the number of possible wave functions by 2. Thus, we have two allowable states for the $1s$ orbital and six allowable states for the $2p$ orbitals.

The Schrödinger equation is set up for the hydrogen atom, and all the solutions are obtained for one electron. Using appropriate approximations, however, we can extend its applicability to the case of atoms with multiple electrons. The procedure can be described as follows. Let us assume that there are N electrons in a given atom. The wave functions can be calculated by using a potential that includes the screening effect from all $(N - 1)$ electrons other than the one under consideration. The wave functions maintain the general shape of a hydrogen atom, so that the quantum numbers, n, l, m, and s for the hydrogen atom can be used. One difference from the hydrogen atom is that the energy level depends not only on n but also on l for atoms with multiple electrons. As l increases, the probability density of electrons near the nucleus decreases, and the screening effect from other electrons increases. Hence, the energy increases as the angular momentum quantum number increases. In multi-electron atoms, $2p$ orbitals have higher energy than $2s$ orbitals.

Figure 1.8 shows the energy levels of electron orbitals in a multi-electron atom. Each box represents the state defined by n, l, and m,

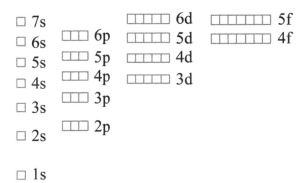

Figure 1.8.   Energy levels of electron orbitals in a multiple-electron atom.

and its position in the vertical direction reflects the energy. For each box, there are two allowed states for spin up and down. Due to the Pauli exclusion principle, only one electron can occupy a quantum state. The electrons fill the orbitals starting from the lowest energy and moving upward in energy. For example, a carbon atom has two electrons in the $1s$ orbital, two electrons in the $2s$ orbital, and two electrons in the $2p$ orbitals.

1.2.2 Numerical Experiments with Quantum Well Model

What happens to the electron wave functions when the atoms are bound together and form a molecule or a solid? The wave function of an electron should be recalculated considering the potential energy produced by all the atoms (nucleus and electrons). Let us do some numerical experiments to find out the shape of wave functions. For simplicity, we confine ourselves to one dimension and approximate the potential as a square well (box) with a finite barrier height as shown in Fig. 1.9.

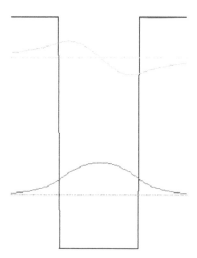

Figure 1.9.   Bound state energy levels and the shape of the wave functions for one-dimensional square well potential. The well depth (barrier height) is 3 eV and the well width is 0.5 nm. The calculated energy levels are marked by dotted lines and the corresponding wave functions are drawn.

The well depth (barrier height) is 3 eV and the well width is 0.5 nm. The time-independent Schrödinger equation is solved with the boundary conditions that the wave functions should be zero at infinity. In Fig. 1.9, the calculated energy levels are marked by dotted lines, and the corresponding wave functions are drawn at the position of the energy. The dotted lines for the energy levels also serve as the abscissa for the graph of the corresponding wave function. We can see that there are two energy levels in this single square well, and the wave function corresponding to the lower energy level is symmetric, while the one corresponding to the upper energy level is anti-symmetric. The anti-symmetric wave function corresponds to the higher energy level, since it has one node (meaning higher frequency) in the well while the other has none.

If we have two square wells, a systematic change occurs in the energy levels and wave functions as shown in Fig. 1.10(a). The energy level in a single square well splits into two, one section of which has a slightly higher energy than the single well case, while the other has a slightly lower energy than that for the single well. The amount of the split is larger for the upper two levels. The wave functions appear to be a linear combination of the single-well wave functions. For symmetric wave functions with lower energies, their additive combination has a lower energy than their subtractive combination,

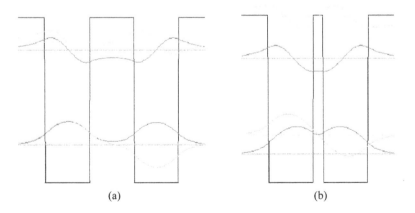

(a)   (b)

Figure 1.10. Bound state energy levels and the shape of the wave function for two square wells. The well depth (barrier height) is 3 eV and the well width is 0.5 nm. The spacing between wells is (a) 0.5 nm and (b) 0.1 nm.

since the former has a smaller number of nodes than the latter. For anti-symmetric wave functions with higher energies, however, their additive combination has a higher energy than their subtractive combination, since the former has larger number of nodes than the latter.

If we decrease the spacing between the wells, the amount of level splitting increases as shown in Fig. 1.10(b). It is clear that the degree of wave function overlap plays an important role in determining the amount of level splitting. Let us examine the wave functions corresponding to the lowest two levels. The wave function with the lower energy appears to be obtained by an additive combination of the single-well wave functions. Due to the large extent of overlap, the combined wave function has a large magnitude even at the center region where there is an energy barrier. It looks as though the wave functions are bonded together and the corresponding energy level is decreased. Thus, this type of wave function combination is called "bonding." The wave function with the higher energy appears to be obtained by subtracting different single-well wave functions. In this case, the combined wave function has a node at the center region, so that the wave function is clearly divided into two. This type of wave function combination is called "anti-bonding."

If we increase the number of wells to three, we can observe that the trend seen with the two wells continues. Each single well energy level splits into three and the corresponding wave function appears to be a linear combination of the original wave functions. The total amount of energy split (the difference between the lowest energy and the highest energy of the triplet) is somewhat larger than that of the two wells. The center level of the triplet has almost the same energy as the single well level. The wave functions have more nodes as the corresponding energy level moves upward.

Figure 1.12 shows the bound state energy levels and the shape of the wave functions for eight square wells. A similar trend is observed in the energy level split and the shape of wave functions. The total amount of energy split (the difference between the lowest energy and the highest energy of the octet) is larger than that of the three wells. The separation between two energy levels in one octet group, however, is smaller than that in the triplet group shown in Fig. 1.11. Clearly, the amount of total energy split in a group is

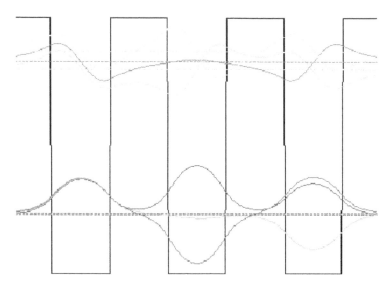

Figure 1.11. Bound state energy levels and the shape of the wave function for three square wells. The well depth (barrier height) is 3 eV, and the well width is 0.5 nm. The spacing between wells is 0.5 nm. See also Color Insert.

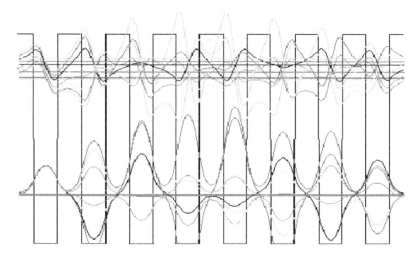

Figure 1.12. Bound state energy levels and the shape of the wave function for eight square wells. The well depth (barrier height) is 3 eV, and the well width is 0.5 nm. The spacing between wells is 0.5 nm. See also Color Insert.

saturating. Each group of energy levels originating from the same energy level of the single well tends to form a distinctive band. Even though it is difficult to distinguish each wave function in Fig. 1.12, a careful examination reveals that the wave functions have more nodes as the corresponding energy increases.

Let us summarize the trends we have observed in this numerical experiment.

(1) Each atomic level tends to form a distinctive band of energy levels in a multi-well system.
(2) For n wells, the number of split levels in a band is n.
(3) The energy band width increases with the number of wells, but it saturates eventually.
(4) The energy band with a lower average energy shows a smaller band width.
(5) Within an energy band, the energy level separation becomes smaller near the band edges.

Since the multi-well system models the array of atoms in a solid, we can draw important conclusions on the electron wave functions. Each energy level of the electron in an atom tends to form a distinctive band of energy levels in a solid. For tightly bound states, the energy band width is narrower because there is less overlap of the atomic wave functions.

1.2.3 *Formation of Solids and Energy Bands*

In the previous section, we have found that each energy level of the electron in an atom tends to form a band in a solid composed of the same atoms. This band-forming tendency originates from the property of atomic wave functions, which generate a global wave function by combining linearly with each other. The individual atomic orbitals interact with each other and form a global orbital when the atoms are brought together in a solid. Figure 1.10 shows that the interaction becomes stronger when the distance between atoms becomes smaller. The stronger interaction results in a larger energy split.

Based on these arguments, we can perform a thought experiment in which we visualize the band formation process. Let us consider a carbon atom as a specific example. We start with an arrangement of

N carbon atoms separated by an infinite distance. Even though the carbon atoms are separated by an infinite distance, we still assume that their arrangement maintains the diamond structure. At this initial state, the electrons in each atom retain their atomic energy levels and the wave functions, since there is no interaction between atomic orbitals. There are $2N$ degenerate states that share the energy of carbon $1s$ orbital, $2N$ degenerate states with the energy of carbon $2s$ orbital, $6N$ states with the energy of carbon $2p$ orbital, and so on. Since all the states originating from the same atomic orbital share the same energy, we can see only discrete lines in this configuration (Fig. 1.13).

As we bring the carbon atoms closer together without changing the structure of arrangement, the atomic wave functions begin to interact and exhibit energy splitting. The width of the energy bands, especially those originating from the $2p$ and $2s$ orbitals, increases continuously, while the increase in the width of the energy band originating from the $1s$ orbital is negligibly small. The wave functions of the $2p$ and $2s$ orbitals interact more than those of the $1s$ orbital, since the average radius of the former is larger than that of the latter.

This process of band broadening continues until the $2p$ and $2s$ bands merge. At this point, an interesting phenomenon called hybridization occurs. Up to this point, the $2p$ orbitals have maintained higher energy than those of the $2s$ orbitals. At this distance, however, the $2p$ and $2s$ orbitals lose their separate identity and are combined to form four sp^3 hybrid orbitals. The probability density functions of the sp^3 hybrid orbitals look like a lopsided dumbbell, as shown in Fig. 1.14. By reconfiguring the $2p$ and $2s$ orbitals into the sp^3 hybrid orbitals, the carbon atoms gain a chance to lower their energy by forming a bonding orbital with their nearest neighbor.

Once this hybridization occurs, the sp^3 hybrid orbitals of the nearest neighbor atoms begin to form bonding and anti-bonding orbitals through the linear combinations of wave functions. In this process, the overlap of atomic wave functions is quite significant and the interaction is very strong, so the energy difference between the bonding and anti-bonding levels should be fairly large. Since there are four sp^3 hybrid orbitals in a carbon atom, $4N$ bonding and $4N$ anti-bonding orbitals are generated in a solid with N carbon

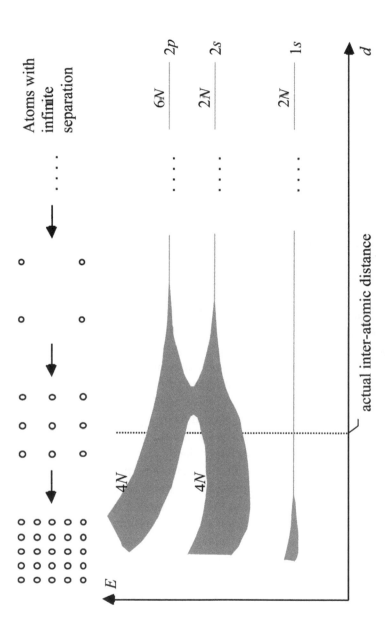

Figure 1.13. Formation of a solid and its energy bands. In this thought experiment, we assume that we can build a solid by bringing the atoms together from an infinite distance and that the inter-atomic distance can be adjusted at our will.

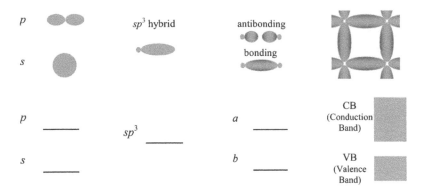

Figure 1.14. Explanation of the formation of energy bands in a solid after hybridization.

atoms, and they will interact with each other (bonding with bonding, anti-bonding with anti-bonding orbitals), resulting in bonding and anti-bonding energy bands. The band originating from the bonding orbitals is called a **valence band** and the one originating from the anti-bonding orbitals is called a **conduction band**. The reason behind these names will be explained in the next paragraph.

There are six electrons in a carbon atom. A solid (diamond) with N carbon atoms should have $6N$ electrons, which will completely fill the $1s$ and bonding sp^3 hybrid band if they occupy the allowed states starting from the lowest-energy state. Thus, all of the allowed states in the bonding sp^3 hybrid band should be occupied by the valence electrons. This is why the bonding sp^3 hybrid band is called a "valence band." A band that is completely filled with electrons, however, cannot flow any current, since all the electron momentum values in one direction are exactly compensated by those in the other direction. Another way to understand this phenomenon is to realize the fact that all the electrons in the valence band are involved in chemical bonding, so there is no net current flow. The current flow can occur when some of the electrons in the valence band acquire sufficient energy to overcome the energy gap between the bonding and anti-bonding energy bands. The electrons that occupy the allowed energy states in the anti-bonding band can move around and produce a net current flow under the influence of an electric field. Due to this reason, the anti-bonding band is called a "conduction band."

As mentioned in the previous paragraph, the space between two energy bands is called an energy gap. There are no allowed energy states in the energy gap, and the energy value in this gap is "forbidden" to electrons. In Fig. 1.13, we can find two forbidden energy gaps: one between the conduction band and the valence band, and the other between the valence band and the 1s band. The energy gap between the conduction band and the valence band has a very important role in the determination of the electrical properties of a solid. In order to get into the conduction band, electrons need to jump up above this energy gap. Without any external energy supply such as strong electric field or light, the electrons have to acquire the necessary amount of energy from the thermal energy in their environment. Since the thermal energy that can be provided by the environment decreases exponentially as the energy gap increases, the amount of electrons in the conduction band critically depends on the value of the energy gap. Therefore, the electrical conductivity of a solid is a sensitive function of the energy gap.

1.3 Energy Bands and Electrical Properties

1.3.1 *Wave Function Analysis and Envelope Functions*

In Section 1.2.2, we found that the wave function of a multiple quantum well can be approximated as a linear combination of single-well wave functions. There is another way to approximate the wave function of a multiple quantum well system. As shown in Fig. 1.15, we can approximate the two-well wave functions by the product of single-well wave functions and an **envelop function** (shown as a dotted line in Fig. 1.15). The envelope function is a slowly varying function that can approximate a multi-well wave function when it is multiplied by a periodic function composed of single-well wave functions. The shape of an envelope function is determined by the global boundary condition, which, in this case, is a flat-top barrier with infinite thickness. Often, envelope functions can appear repeatedly in a set of multi-well wave functions.

For the readers who may not be familiar with the concept of an envelope function, we would like to take an analogy shown in

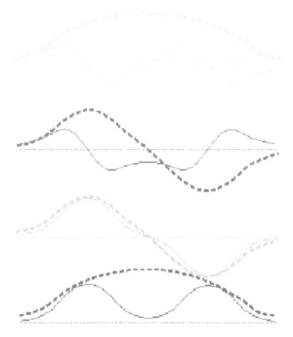

Figure 1.15. Analysis of the wave functions. Roughly, a wave function can be approximated by the product of atomic wave functions and an envelope function.

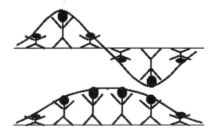

Figure 1.16. Concept of an envelope function. In this analogy, the envelope function is shown to modulate each figure's height, implementing a wave.

Fig. 1.16. Stick figures are performing a mass calisthenics. Their heights are modulated by a well-defined function shown as a solid line. This function is called an envelope function. The envelope function modulates the heights of the figures and implements a wave. The wave can be a standing wave or a propagating wave. Each stick figure corresponds to a single-well wave function and the complete group of stick figures forms a periodic function.

The observation that the wave functions in a multiple quantum well system can be approximated by the product of an envelope function and a periodic function is useful in understanding the relationship between the single-well wave functions and the multi-well wave functions, but it is not a mathematically rigorous result. In fact, we can find an exception to this simple approach. In Fig. 1.15, the second wave function from the top cannot be well approximated well by a product of an envelope function and a periodic function. Such a product should have a node at the center, while the actual wave function does not. Obviously, there is a qualitative difference. The wave function may be better approximated by a linear combination (in this case, a subtraction) of the single-well wave functions. This type of exception occurs because the multi-well potential is not a well-defined periodic potential. The double well potential has just two identical wells, and the number of wells is too small to define a periodicity. In order to deal with the periodic potential in a solid, we need a theorem that is formal and mathematically rigorous. In 1928, Felix Bloch established his theorem on the quantum mechanics of electrons in a crystal lattice.

1.3.2 Bloch Theorem

For a periodic potential with period $\vec{R}$ ($V(\vec{r} + \vec{R}) = V(\vec{r})$), the solutions (wave functions) to the Schrödinger wave equation can be expressed as the product of a plane wave and a periodic function with the same period as the potential.

$$\psi(\vec{r}) = \exp(j\vec{k} \cdot \vec{r})u(\vec{r}) \qquad (1.63)$$
$$u(\vec{r} + \vec{R}) = u(\vec{r}) \qquad (1.64)$$

The plane wave ($\exp(j\vec{k} \cdot \vec{r})$) can be considered an envelope function and the periodic function ($u(\vec{r})$) as an array of atomic wave functions.

The Bloch theorem can be proven mathematically with full rigor, but we will not prove it here. Instead, we would like to provide a heuristic argument to help the readers understand the meaning of the Bloch theorem. Let us start with a simplest periodic potential where the potential energy is constant everywhere. Then the solution will be just a plane wave, $\psi(\vec{r}) = \exp(j\vec{k} \cdot \vec{r})$, and the periodic function is unity ($u(\vec{r}) = 1$). If we introduce a non-zero potential gradually, the periodic potential will modulate the wave function periodically. This periodic modulation can be represented by the periodic function, $u(\vec{r})$. When the periodic potential is weak, the periodic modulation would be weak and $u(\vec{r})$ will maintain its magnitude close to one. For a strong periodic potential, however, $u(\vec{r})$ will resemble the superposition of bound-state wave functions in potential wells.

1.3.3 *Wave Vectors and **k**-Space*

In the Bloch theorem, the plane wave plays the role of an envelope function. It determines the overall behavior of the wave function in a solid, so that it should be constrained by the boundary conditions of the solid. Figure 1.17 shows the shape of the potential in a solid. We can see that a solid can be approximated as a box that can contain electrons in it. However, infinite potential barriers, which are often used as boundary conditions for an electron in a box, generate only standing waves. Since standing waves have zero average momentum, it is inconvenient when we deal with the movement of electrons. Besides, details of boundary condition may not be important in a sample with a size larger than a few tens of nanometers, since scattering generally prevents electrons from feeling the effect of the boundary. Scattering can force an electron to change its direction of motion long before it reaches the boundary of the sample. Thus,

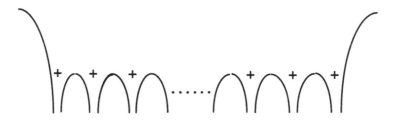

Figure 1.17. Shape of the potential in a solid. The + symbols represent the positive ions and they determine the shape and the periodicity of the potential.

we often use periodic boundary conditions following Born and von Karman.

Let us find the allowed wave vectors in a cubic solid using periodic (Born–von Karman) boundary conditions. If one side of the cube is L, the boundary conditions are

$$\psi(\vec{r}) = \psi(\vec{r} + L\hat{x}) \tag{1.65a}$$

$$\psi(\vec{r}) = \psi(\vec{r} + L\hat{y}) \tag{1.65b}$$

$$\psi(\vec{r}) = \psi(\vec{r} + L\hat{z}), \tag{1.65c}$$

where $\hat{x}$, $\hat{y}$, and $\hat{z}$ are the unit vectors in the x-, y-, and z-directions, respectively. In order for the Bloch wave function (Eqs. (1.63) and (1.64)) to satisfy these boundary conditions, the wave vector, $\vec{k}$, whose x-, y-, and z-components are k_x, k_y, and k_z, respectively, should satisfy the following condition:

$$k_x = \frac{2\pi n_x}{L}, \, k_y = \frac{2\pi n_y}{L}, \, k_z = \frac{2\pi n_z}{L}, \tag{1.66}$$

where n_x, n_y, and n_z are integers. The boundary conditions require the constraints only on the plane wave ($\exp(j\vec{k}\cdot\vec{r})$) of the Bloch wave function, since $u(\vec{r}) = u(\vec{r} + L\hat{x})$, $u(\vec{r}) = u(\vec{r} + L\hat{y})$, and $u(\vec{r}) = u(\vec{r} + L\hat{z})$ as long as the cube contains integer number of atoms on each side (Problem 4). Equation (1.66) tells us that only discrete values are allowed for the components of wave vectors.

Figure 1.18 shows that the endpoints of the allowed wave vectors form a three-dimensional lattice in k-space. Since the typical size of the solid is much larger than the inter-atomic distance, the spacing between discrete k values is extremely small compared with

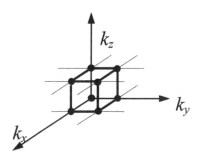

Figure 1.18. Allowed k vectors in k-space determined by periodic boundary conditions.

the value of the zone boundary (π/a, where a is the lattice constant of the solid) determined by the Bragg reflection condition. Bragg reflection will be explained in the following section. From the viewpoint of the overall structure of the k-space, the allowed k values are densely packed, and they can often be treated as a continuous variable. The cube shown in Fig. 1.18 is a unit cell of the Bravais lattice in k-space, and its volume is $(2\pi/L)^3$. Thus, the density of the allowed wave vectors in k-space is $(L/2\pi)^3$, and we may consider the k vector a continuous variable spread in k-space with such a density.

1.3.4 *Bragg Reflection and Brillouin Zones*

One of the main effects of periodic structure on the propagation of wave is the Bragg reflection. When an X-ray is incident on a crystalline solid with a lattice constant similar to the wavelength of the X-ray, the partial waves scattered by the atoms will interfere with each other. If all the scattered partial waves interfere constructively, there will be a strong reflection. In 1913, Bragg proposed an explanation of such reflections and derived a condition for the constructive interference. He modeled the crystal as a set of parallel lattice planes separated by a constant parameter d. Instead of considering the effect of scattering from all atoms or lattice points, he assumed that the X-ray is specularly reflected by the parallel lattice planes. In Fig. 1.19, the path length difference between the two parallel rays reflected from the adjacent lattice planes is $2d \cos\theta$. For constructive

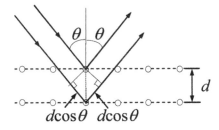

Figure 1.19. Reflection of X-ray by a crystal with a lattice constant d. The X-ray will be strongly reflected when the scattered partial waves interfere constructively. The dotted lines represent the parallel lattice planes of Bragg.

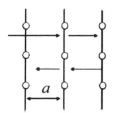

Figure 1.20. Reflection of electron wave inside a crystal with a lattice constant *a*. When the Bragg reflection condition is satisfied, a standing wave will be formed inside the solid due to the multiple reflections and constructive interference in the forward and backward directions.

interference, it should be an integer multiple of the wavelength:

$$2d \cos\theta = n\lambda \tag{1.67}$$

This is the well-known Bragg condition. A more general analysis including the effect of scattering from all lattice points can be done, but the condition for the constructive interference is exactly the same as the Bragg condition.

Since the Bragg theory can be applied to any wave incident on a periodic structure, we can employ it in the analysis of electron waves in a solid. In this case, the concept corresponding to the incident wave function is the envelop function of the Bloch wave in a solid. The focus of the analysis is on the behavior of electron wave inside the solid, which is different from the case of X-ray. For simplicity, let us consider the case of normal incidence to lattice planes shown in Fig. 1.20. The condition for constructive interference is

$$2a = n\lambda, \tag{1.68}$$

where *a* is the lattice constant of the solid. While a wave incident to the solid will experience a strong reflection outside the solid, a standing wave is formed inside the solid due to the multiple reflections and constructive interference in the forward and backward directions.

Using the relationship between the wavelength and the wave number, we can obtain the wave number *k* at which the Bragg condition is satisfied:

$$k = \frac{2\pi}{\lambda} = \frac{n\pi}{a} \tag{1.69}$$

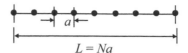

$$L = Na$$

Figure 1.21. One-dimensional lattice structure with lattice constant a and length L. If there are N lattice points, L is equal to Na.

The formation of a standing wave has a special implication on the behavior of electrons with the wave number given in Eq. (1.69). Since the standing wave cannot propagate, the electrons with such a wave number cannot move around inside a solid. In addition, standing waves have nodes (zero displacement) and anti-nodes (maximum displacement), and there will be a corresponding variation in the probability density of finding an electron. Such a variation in the probability density and the immobility of the electron can generate an energy difference for the same wave number, resulting in the energy gap.

In order to illustrate the relationship between the lattice structure in the real space and the discrete k values in the k-space, let us consider a one-dimensional lattice shown in Fig. 1.21. Any lattice point can be expressed as $x = ma$, where a is the lattice constant and m is an integer. If there are N lattice points, L is equal to Na.

In the k-space, the allowed k values are determined by the periodic boundary condition and given as

$$k = \frac{2\pi n}{L} = \frac{2\pi n}{Na}, \tag{1.70}$$

where n is an integer. The k values satisfying the Bragg condition is given in Eq. (1.69). Figure 1.22 shows the allowed k values and the k values for the Bragg reflection. The continuous regions in the k-space that are divided by the points satisfying the Bragg condition are called Brillouin zones. For example, the region with $-(\pi/a) < k < (\pi/a)$ is called the first Brillouin zone, and the regions with $\pi/a < |k| < 2\pi/a$ are called the second Brillouin zone. As can be seen in the next section, the importance of the Brillouin zone stems from the fact that the Bloch wave functions can be completely characterized by their behavior in a single Brillouin zone. Each Brillouin zone contains exactly N allowed k points.

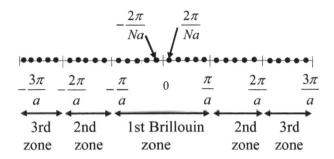

Figure 1.22. Structure of k-space. All allowed k-values are represented as discrete points. The points that satisfy the Bragg condition are marked by short lines. Brillouin zones are defined as the continuous regions in the k-space that are divided by the points satisfying the Bragg condition. Exactly N allowed k points are contained in each Brillouin zone.

Example 1.2: Size of a Brillouin zone and k spacing

Calculate the following sizes using the given values of the lattice constant (a) and the size of a solid (L).

(a) Size of the Brillouin zone for $a = 0.2$ nm
(b) k spacing for $L = 1$ mm

Solution:

(a) $\dfrac{2\pi}{a} = \dfrac{2 \times 3.14}{0.2 \times 10^{-7}} = 3.14 \times 10^8 (\text{cm}^{-1})$.

(b) $\dfrac{2\pi}{L} = \dfrac{2 \times 3.14}{1 \times 10^{-1}} = 62.8 (\text{cm}^{-1})$.

1.3.5 *Energy-versus-k Diagram in One-Dimensional Space*

For free electrons with a constant potential energy, the periodic function, $u(x)$, is a constant and the energy has a simple parabolic relationship with the wave number k:

$$E = \frac{\hbar^2 k^2}{2m_e} \tag{1.71}$$

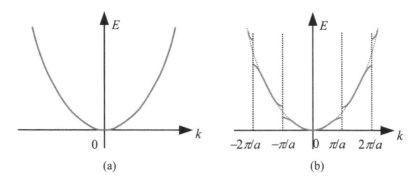

Figure 1.23. 1D energy-versus-k diagram for (a) free electrons and (b) Bloch electrons.

Bloch electrons in a periodic potential, however, have different energy from that of free electrons near the Brillouin zone boundary, $k = n\pi/a$. Figure 1.23 shows energy-versus-k diagram for free electrons and Bloch electrons.

Let us examine the reason for the deviation in the energy near the zone boundaries. At a zone boundary, the periodicity of the envelop function coincides with the periodicity of the potential. Bragg reflection results in the formation of standing waves, and the distance between the nodes (anti-nodes) is the same as the lattice constant of the solid. Moreover, either anti-nodes or nodes are aligned with the position of the positive ions. Figure 1.24 shows the average probability density of finding electrons calculated from the two standing waves. When the anti-nodes are aligned with the positions

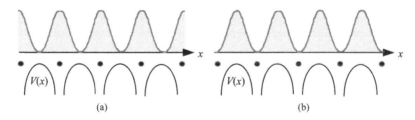

Figure 1.24. Average probability density of finding electrons calculated from the position of the two standing waves compared with the position of the ions in a solid: (a) positions of the anti-nodes coincide with those of positive ions and (b) positions of the nodes coincide with those of ions.

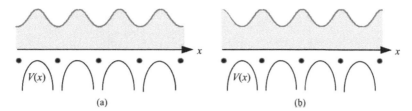

Figure 1.25. Average probability density of finding electrons calculated from the position of the two partial standing waves compared with the position of the ions in a solid: (a) positions of the anti-nodes coincide with those of the positive ions and (b) positions of the nodes coincide with those of the ions.

of the positive ions, the higher average probability density of electrons near the ion decreases the electrostatic potential energy, so that the total energy decreases below the value of free electrons. On the other hand, if the nodes are aligned with the positions of the ions, the total energy increases. Thus, a significant energy splitting can occur at the zone boundary.

Such an energy splitting is not confined to the zone boundary, but it also occurs near the zone boundary. If the wave number is close to the value of the Bragg condition, there is a partial Bragg reflection. The partial Bragg reflection generates partial standing waves as shown in Fig. 1.25, which result in the energy difference. The magnitude of the partial standing waves decreases when we move away from the zone boundary. As we can see in Fig. 1.23(b), the energy of free electrons is restored at the center of the Brillouin zones.

As mentioned in the previous section, a single Brillouin zone is enough to describe all the Bloch wave functions and the complete energy-versus-k relationship. Let us consider a Bloch wave function with a wave number $k' = k + (2\pi n)/a$. We can transform the envelope function from $\exp(jk'x)$ to $\exp(jkx)$ with the following manipulation:

$$\exp(jk'x)u(x) = \exp\left[j\left(k + \frac{2\pi n}{a}\right)x\right]u(x)$$

$$= \exp(jkx)\exp\left(j\frac{2\pi n}{a}x\right)u(x)$$

$$= \exp(jkx)u'(x) \tag{1.72}$$

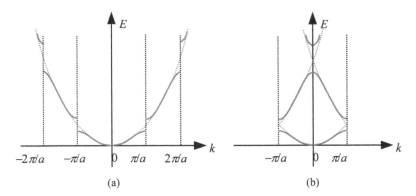

Figure 1.26. Two schemes of representing the E versus k relationship: (a) extended zone scheme and (b) reduced zone scheme.

$u'(x)$ is a periodic function with the periodicity of the lattice since

$$u'(x + a) = \exp\left[j\frac{2\pi n}{a}(x + a)\right]u(x + a)$$

$$= \exp\left(j\frac{2\pi n}{a}x\right)u(x) = u'(x). \qquad (1.73)$$

Thus, we can shift the k value by $(2\pi n)/a$ by choosing a new periodic function $u'(x)$. By selecting an appropriate periodic function, we can describe all the Bloch wave functions and their energies within a single Brillouin zone. This method is called a reduced zone scheme, while the original method is called an extended zone scheme. The two schemes shown in Fig. 1.26 are equivalent.

In the reduced zone representation of an energy-versus-k diagram, it is clear that there are allowed and forbidden regions of energy. These regions are the energy bands and gaps that we already confronted in Section 1.2.3. There, starting from the discrete energy levels of atomic states, we showed that the energy bands can be formed by the interaction between the atomic wave functions. Since the broadening of energy bands eventually saturates as the number of atoms increases, there must be an energy gap between the energy bands. Now, starting from the opposite end, i.e., an envelope function that resembles the wave function of a free electron, we arrived at the same conclusion. The Bragg reflection generates standing or

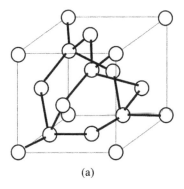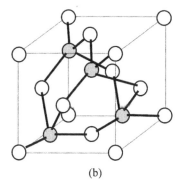

(a)   (b)

Figure 1.27.   Crystal structure of semiconductors: (a) diamond structure for group IV semiconductors (Si, Ge,...), (b) zinc blende structure for III-V or II-VI compound semiconductors (GaAs, InP, ZnS,...).

partial standing waves, and the alignment (or misalignment) of the high-probability density region and the ions produce the energy gap.

1.3.6 *Energy-versus-k Diagram in Three-Dimensional Space*

In three dimensions, the energy-versus-k relationship is much more complicated, since the lattice structures and the corresponding Brillouin zones are complex. Let us take the crystal structure of semiconductors as an example. Many semiconductor materials such as silicon (Si) and gallium arsenide (GaAs) are based on face-centered cubic (fcc) lattice structure. In more detail, the group IV semiconductors such as Si and germanium (Ge) have a diamond structure (Fig. 1.27(a)) and the III-V or II-VI compound semiconductors such as GaAs, indium phosphide (InP), and zinc sulphide (ZnS) have a zinc blende structure (Fig. 1.27(b)). By grouping two adjacent atoms as a unit called a basis, however, the two crystal structures can be reduced to the same fcc lattice structure. Since the Brillouin zones are defined from a periodic lattice structure in real space, Si and GaAs have the Brillouin zones with the same shape.

Figure 1.28 shows the cubical unit of the fcc lattice structure and its first Brillouin zone in k-space. Besides the lattice points at eight vertices of a cube, the fcc lattice has one lattice point at the center

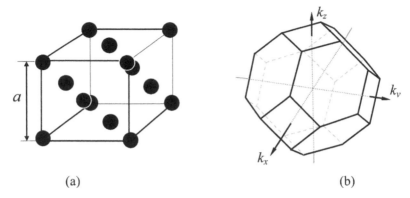

(a)   (b)

Figure 1.28. Face-centered cubic (fcc) lattice and the corresponding Brillouin zone: (a) fcc lattice structure, (b) first Brillouin zone for the fcc lattice.

of each of the six faces of the cube (Fig. 1.28(a)). Since the lattice points at the vertices are shared by eight cubical units and the lattice points at the face centers are shared by two cubical units, four lattice points are included in one cubical unit. The primitive unit cell containing only one lattice point has one-fourth of the volume of the cubical unit cell in real space. Brillouin zone boundaries can be determined by applying the Bragg reflection condition in k-space, and Fig. 1.28(b) shows the shape of the first Brillouin zone. It can be described as a truncated octahedron, i.e., we can construct this structure by truncating the regions near the six vertices of a regular octahedron.

The energy-versus-k relationship for three-dimensional lattice structures can be calculated and measured using various methods. Figure 1.29 shows the energy-versus-k relationship for GaAs and Si. In order to show the energy-versus-k relationship in a two-dimensional graph, we select a certain direction and plot the energy as a function of k values on a straight line in that direction. To define a direction, we often use three integers h, k, and l in a brace (i.e., [hkl]). For example, [100] designates the direction of the vector $(1, 0, 0)$ and [111] designates the direction of the vector $(1, 1, 1)$. These directions are marked underneath the abscissa in Fig. 1.29. We can see that the energy-versus-k relationship depends on the material and the direction of the k vector. In GaAs, the global

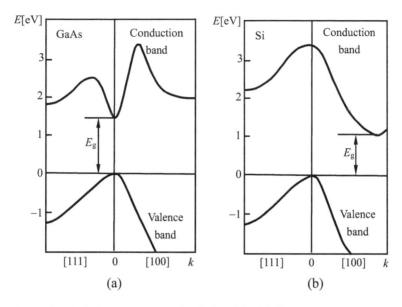

Figure 1.29. 3D energy-versus-*k* relationship: (a) direct energy gap material (GaAs) and (b) indirect energy gap material (Si).

minimum energy of the conduction band (the energy band above the energy gap) is located at the origin of *k*-space. Since the global maximum energy of the valence band (the energy band below the energy gap) is also located at the origin, an electron at the valence band edge can jump into the conduction band edge without changing its *k* value. A material that allows such a transition is called a **direct energy gap** material. In Si, the global minimum energy of the conduction band is located somewhat away from the origin, while the global maximum energy of the valence band is located at the origin. An electron at the valence band edge cannot jump into the conduction band edge unless the difference in the *k* vector is provided by a certain mechanism. This type of material is called an **indirect energy gap** material. Since the energy quanta of light (photons) have a very small *k* value, requiring the conduction band minimum and the valence band maximum to have almost the same *k* value the property of having a direct energy gap is essential to the optical emission of semiconductors. The efficiency of light emission in GaAs is very high, while hardly any light can be emitted from Si.

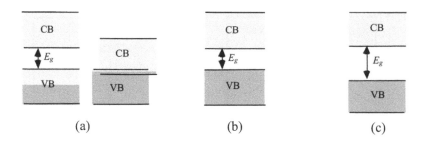

Figure 1.30. Electrical properties determined by the band structure: (a) metal, (b) semiconductor, and (c) insulator.

1.3.7 Energy Band and Classification of Solids by Electrical Conduction

As mentioned in Section 1.2.3, the electrical conductivity of a solid critically depends on the size of the energy gap and the filling status of the energy bands. If there is a partially filled band or a band overlap as shown in Fig. 1.30(a), all the electrons in the band can participate in conduction. A material with such a band structure and such a filling status is a metal. If the valence band is completely filled at 0 K and the size of the energy gap is moderate (from a few tenths of an eV to a few eV), there will be some electrons in the conduction band at room temperature. The conductivity of such a material will be several orders of magnitude lower than that of a metal, but the conductivity can be changed in a wide range by optical excitation or doping (intentional introduction of impurities that can easily donate or accept electrons). This type of material is a semiconductor. If the energy gap is large, the valence band is completely filled and the conduction band is almost empty. The electrical conduction is negligible (usually the conductivity is 20 orders of magnitude smaller than that of typical metal) and this type of material is an insulator.

1.3.8 Semiclassical Equations of Motion for Electrons

The Bloch wave functions that we have dealt with up to now have a clearly defined k value and energy, but the position of the electron is completely unknown in the crystal. In order to use the concept of position, we have to form a wave packet with a large number of Bloch

Figure 1.31. Semiclassical model of an electron. A wave packet is formed with a large number of Bloch wave functions. The wave packet should be spread in real space over many atoms, and the spread in k space should be sufficiently smaller than the scale of the Brillouin zone.

wave functions. The wave packet should be spread in real space over many atoms, and the spread in k space should be sufficiently smaller than the scale of the Brillouin zone. Such a wave packet is called a semiclassical model of an electron and is graphically shown in Fig. 1.31. The black dots represent the atoms arranged in a one-dimensional array.

The velocity of the electron is the group velocity of the wave packet. In one-dimensional space, it is

$$v = \frac{d\omega}{dk} = \frac{1}{\hbar}\frac{dE}{dk}. \tag{1.74}$$

In three-dimensional space, it is given as

$$\vec{v} = \nabla_k \omega = \frac{1}{\hbar}\nabla_k E. \tag{1.75}$$

Let us derive the semiclassical equations of motion for an electron. When a force $\vec{F}$ is applied to an electron with velocity $\vec{v}$, the absorption rate of energy is

$$\frac{dE(\vec{k})}{dt} = \vec{F}\cdot\vec{v}. \tag{1.76}$$

If we consider the movement in k-space,

$$\frac{dE(\vec{k})}{dt} = \nabla_k E(\vec{k}) \cdot \frac{d\vec{k}}{dt} = \nabla_k(\hbar\omega) \cdot \frac{d\vec{k}}{dt} = \hbar\vec{v}\cdot\frac{d\vec{k}}{dt}. \tag{1.77}$$

In order for these equations to hold for any $\vec{v}$,

$$\hbar\frac{d\vec{k}}{dt} = \vec{F}. \tag{1.78}$$

Here the quantity $\hbar\vec{k}$ plays the role of the momentum in classical mechanics. In general, $\hbar\vec{k}$ is different from the classical momentum

$m_e\vec{v}$ and is called a crystal momentum. Hence, the force is equal to the time rate of change in the crystal momentum.

1.3.9 *Effective Mass*

We use the semiclassical approach to the acceleration of an electron in one-dimensional space. The acceleration is defined as

$$a = \frac{dv}{dt}. \tag{1.79}$$

Since the velocity is a function of k ($v = v(k)$) and since k is a function of time, we obtain

$$a = \frac{dv}{dk}\frac{dk}{dt} = \frac{1}{\hbar}\frac{d^2E}{dk^2}\frac{F}{\hbar}. \tag{1.80}$$

Using an analogy with Newton's law, we can define an effective mass m^* as follows:

$$m^* = \frac{\hbar^2}{d^2E/dk^2}. \tag{1.81}$$

In this definition, the effective mass is inversely proportional to the curvature of energy band. This observation is depicted in Fig. 1.32. When the curvature is large, the effective mass is small. If the energy-versus-k curve is parabolic, the effective mass is a constant.

Let us consider a more realistic energy-versus-k curve shown in Fig. 1.33(a). Starting from a positive value at the origin, the

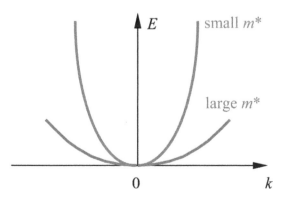

Figure 1.32. Dependence of effective mass on the curvature of energy-versus-k curve.

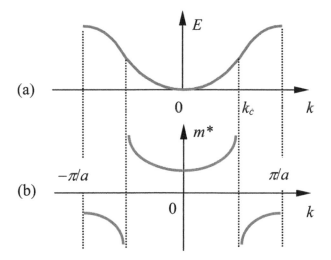

Figure 1.33. Realistic energy-versus-k relationship (a) and the effective mass (b).

curvature decreases gradually to zero (inflection point, $k = k_c$) and eventually becomes negative as the k value approaches the Brillouin zone boundary. The effective mass becomes infinity at the inflection point and turns to be negative for $k > k_c$.

The negative effective mass means that the electron is decelerated in the direction of force. The velocity in the direction of force decreases as the electron moves. Since the concept of negative effective mass is strange and awkward in the classical world, we will discuss a method of dealing with the negative effective mass more conveniently in the next section.

1.3.10 *Concept of Holes*

If a valence band is completely filled, the electrons in it do not contribute to the current at all, since any contribution of a state with a certain k is compensated for by the contribution of a state with $-k$. When there are some empty states in the valence band, however, there can be a net current contributed by the motion of electrons with uncompensated k values.

The current density due to the electrons in the valence band is expressed as

$$J = -e \sum_{i(\text{filled})} v_i = -e \sum_{i(\text{total})} v_i + e \sum_{i(\text{empty})} v_i. \qquad (1.82)$$

The valence band is an even function of k. Thus, for every v, there is a corresponding $-v$.

$$-e \sum_{i(\text{total})} v_i = 0 \qquad (1.83)$$

Therefore,

$$J = +e \sum_{i(\text{empty})} v_i. \qquad (1.84)$$

The total current density can be expressed as the sum of the product of the empty state velocity and the positive charge $+e$. We can treat the empty state as a quasi-particle with a positive charge $+e$. Such a quasi-particle is called a **hole**.

Now let us consider the effective mass of a hole. When an electric field is applied to a solid, all the electrons in the valence band change their position in k-space at the same time. The empty state changes its position in the same direction. Since our goal is to describe the behavior of all the electrons in the valence band, we have to evaluate the total energy and k values of electrons in the valence band. Compared with the total energy of electrons in Fig. 1.34(a), the total energy of electrons in Fig. 1.34(b) is higher, since an electron with a lower energy is missing in Fig. 1.34(b). We can conclude that the hole in Fig. 1.34(b) has a higher energy than the hole in Fig. 1.34(a). The total k value of electrons in Fig. 1.34(b) is more negative than that in Fig. 1.34(a), as can be seen in the figure. The hole in Fig. 1.34(b) has a more negative k value. Thus, the behavior of holes can be described more easily by introducing a hole band in k space. The hole band is drawn in Fig. 1.34. From the curvature of the hole band, we can conclude that the effective mass of a hole is a positive number.

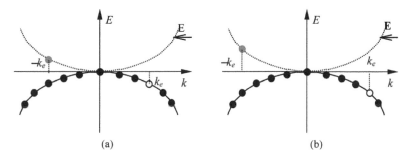

Figure 1.34. Behavior of an empty state in the valence band and the concept of a hole band: (a) $t = 0$, and (b) $t > 0$.

1.4 Nanostructure Applications of Quantum Mechanics

1.4.1 *Density of States for Electrons in a Macroscopic Box*

Let us consider a macroscopic box with electrons in it. For simplicity, we assume that the box has a cubical shape with a side L. Since the potential inside the box is constant, the wave functions should be plane waves or standing waves (superposition of plane waves). Without losing generality, we can apply periodic boundary conditions to the plane waves in this box. Then, the wave vectors are quantized due to the periodic boundary conditions. Since the wave functions are plane waves, the electrons will behave just like free particles except that their wave vectors are quantized. We would like to calculate the number of quantum states with energy ranging from E to $E + dE$. Such a number per unit volume is called a density of states.

The quantum states are evenly distributed in k space with one state taking the volume, $(2\pi/L)^3$ as shown in Fig. 1.35. The number of states in a volume element, $d\Omega_{\vec{k}}$, is

$$\frac{d\Omega_{\vec{k}}}{(2\pi/L)^3}. \tag{1.85}$$

Defining $|\vec{k}| = k$, the energy of a free electron can be expressed as

$$E = \frac{\hbar^2}{2m_e} \left(k_x^2 + k_y^2 + k_z^2 \right) = \frac{\hbar^2 k^2}{2m_e}, \tag{1.86}$$

and the volume for states with energy between E and $E + dE$ is

$$d\Omega_{\vec{k}} = 4\pi |\vec{k}|^2 d|\vec{k}| = 4\pi k^2 dk. \tag{1.87}$$

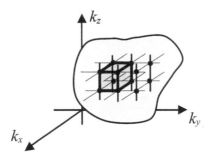

Figure 1.35. Method of calculating the number of states within a given volume element in k space. It can be obtained by dividing the volume, $d\Omega$, of the element by the volume occupied by one state.

This is the volume of a spherical shell. Now, we need to find the relationship between dE and dk. From Eq. (1.86), we can obtain k:

$$k = \sqrt{\frac{2m_e E}{\hbar^2}} \tag{1.88}$$

By differentiating this equation with respect to E, we obtain

$$dk = \sqrt{\frac{m_e}{2}} \frac{1}{\hbar\sqrt{E}} dE. \tag{1.89}$$

If we combine Eqs. (1.87), (1.88), and (1.89),

$$d\Omega_{\vec{k}} = 4\pi \frac{2m_e E}{\hbar^2} \frac{\sqrt{m_e/2}}{\hbar\sqrt{E}} dE = \frac{4\sqrt{2}\pi m_e^{3/2}}{\hbar^3} E^{1/2} dE. \tag{1.90}$$

Since two electrons with different spin states (spin up (↑) and spin down (↓)) can be accommodated in each state, we need to multiply the answer by two when we evaluate the number of states for the given energy interval:

$$N(E)dE = \frac{2d\Omega_{\vec{k}}}{(2\pi/L)^3} = \frac{8\sqrt{2}\pi L^3 m_e^{3/2}}{h^3} E^{1/2} dE. \tag{1.91}$$

Finally, if we calculate the number of states per unit volume, we have to divide $N(E)$ by L^3:

$$g(E) = \frac{N(E)}{L^3} = \frac{4\pi (2m_e)^{3/2}}{h^3} E^{1/2}. \tag{1.92}$$

Figure 1.36 shows the density of states as a function of energy. We can see that the density of states increases monotonically as the

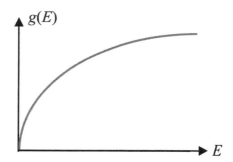

Figure 1.36. Density of states as a function of energy.

energy increases. The rate of increase in the density of states, however, is reduced at higher energies because of the density's square root dependence on the energy.

1.4.2 *Confinement and Density of States in Lower Dimensions*

If we reduce one side of the macroscopic box in the previous subsection to the point where the final length of that side is nanoscale, we have a one-dimensionally confined structure shown in Fig. 1.37. In this structure, the electrons are confined in one dimension but can move around freely in the other two dimensions. In the direction of confinement, the wave functions become standing waves and the energy level spacing increases drastically. Such a structure is

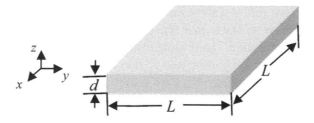

Figure 1.37. Quantum well: a two-dimensional electron system (2DES) with one-dimensional confinement. The length (d) of one side is nanoscale, while that (L) of the other two is macroscopic ($d \ll L$).

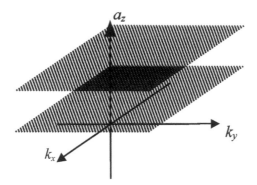

Figure 1.38. Distribution of quantum states for a 2DES in k space.

called a **quantum well** and the system of electrons is called a **two-dimensional electron system (2DES)**.

We can calculate the density of states for 2DES using the procedure described in the previous subsection. If the confinement occurs in the z-direction, the energy of an electron is given as

$$E = \frac{\hbar^2}{2m_e}\left(k_x^2 + k_y^2\right) + E_{n_z}. \tag{1.93}$$

If we assume infinite potential barriers for the quantum well in the z-direction, the z-component of energy, E_{n_z}, is given as

$$E_{n_z} = \frac{\hbar^2 a_z^2}{2m_e} = \frac{\hbar^2 n_z^2 \pi^2}{2m_e d^2}, \tag{1.94}$$

where n_z is a positive integer.

In Eq. (1.94), we used the symbol a_z for $n_z\pi/d$ instead of k_z, since a_z is not exactly a wave number in the z-direction, even though it plays a similar role to a wave number. Figure 1.38 shows the distribution of quantum states in k space. We can see that the states are not evenly distributed. The spacing in the a_z-direction is much larger than that in the k_x- or k_y-direction, so that the states with a certain a_z value form a distinctive group. Each group contains exactly the same set of k_x and k_y. The only difference between these groups is the a_z value, or the subband index, n_z. Such groups are called **subbands**.

Once we know that the quantum states can be grouped into distinct subbands, we can obtain the density of states for a 2DES by calculating the density of states for one subband and adding them

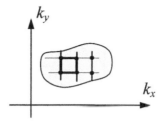

Figure 1.39. Method of calculating the number of states within a given area element in two-dimensional *k* space. It can be obtained by dividing the area, $dA_{\vec{k}}$, of the element by the area $(2\pi/L)^2$ occupied by one state.

for all subbands. Let E' be the sum of the *x*- and *y*-components of energy for a subband with index n_z:

$$E' = \frac{\hbar^2}{2m_e}(k_x^2 + k_y^2) \qquad (1.95)$$

With this energy, E', we can treat each subband as a two-dimensional (2D) *k* space. We can calculate the density of states in a 2D *k* space first and then simply shift the energy by E_{n_z}, since E_{n_z} remains as a constant throughout the calculation.

Let us calculate the density of states in a 2D *k* space. Figure 1.39 shows the method of calculating the number of states within a given area element in 2D *k* space. It can be obtained by dividing the area, $dA_{\vec{k}}$ of the element by the area taken by one state, $(2\pi/L)^2$.

The number of states in an area element $dA_{\vec{k}}$ is

$$\frac{dA_{\vec{k}}}{(2\pi/L)^2}. \qquad (1.96)$$

The area for states with energy between E' and $E' + dE'$ is the area of the ring shown in Fig. 1.40.

$$dA_{\vec{k}} = 2\pi |\vec{k}| d|\vec{k}| \\ = 2\pi k dk \qquad , \qquad (1.97)$$

where $k = \sqrt{k_x^2 + k_y^2}$.

Using the relationship $dE' = \frac{\hbar^2 k}{m_e} dk$, we obtain

$$dA_{\vec{k}} = 2\pi k dk = \frac{2\pi m_e}{\hbar^2} dE'. \qquad (1.98)$$

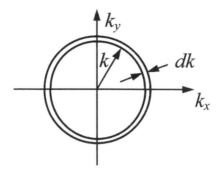

Figure 1.40. Area for states with energy between E' and $E' + dE'$.

Since two electrons (spin ↑, spin ↓) can be accommodated in each state,

$$N(E')dE' = \frac{2d A_{\bar{k}}}{(2\pi/L)^2} = \frac{L^2 m_e}{\pi \hbar^2} dE'.$$

(1.99)

Thus, per unit area, the density of states for a subband with index n_z is

$$g_{n_z}(E') = \frac{N(E')}{L^2} = \frac{m_e}{\pi \hbar^2}.$$

(1.100)

This equation is valid only for $E' > 0$. Figure 1.41 shows the 2D density of states as a function of E'.

The density of states for the complete 2DES is

$$g(E) = \sum_{n_z=1}^{\infty} g_{n_z}(E') = \sum_{n_z=1}^{\infty} g_{n_z}(E - E_{n_z}).$$

(1.101)

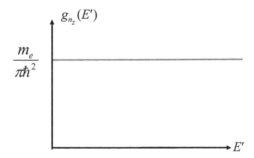

Figure 1.41. Density of states as a function of E' for a subband of 2DES.

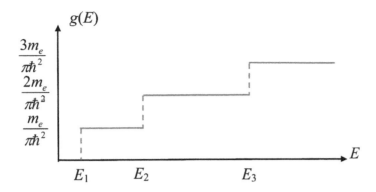

Figure 1.42. Density of states for the complete 2DES.

Figure 1.42 shows the density of states for the complete 2DES. Note that this is still a 2D density of states. If we want to convert it to a 3D density of states, we simply need to divide it by d.

Now, if electrons are confined in two dimensions as shown in Fig. 1.43(a), the system is called a **one-dimensional electron system (1DES)**. The confinement structure is called a **quantum wire**. If electrons are confined in all three dimensions as shown in Fig. 1.43(b), they lose all degrees of freedom and become a **zero-dimensional electron system (0DES)**. This confinement structure is called a **quantum dot**.

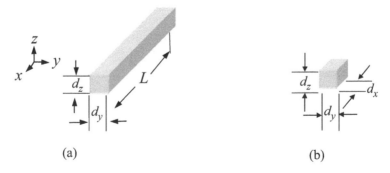

Figure 1.43. Quantum wire and quantum dot: (a) one-dimensional electron system (1DES) with two-dimensional confinement and (b) zero-dimensional electron system (0DES) with three-dimensional confinement.

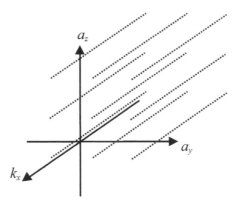

Figure 1.44. Distribution of quantum states for 1DES in k space.

We can calculate the density of states for a 1DES using a method similar to that of the 2DES. If the confinement occurs in the y- and z-directions, the energy of the electron is given as

$$E = \frac{\hbar^2 k_x^2}{2m_e} + E_{n_y} + E_{n_z} \tag{1.102}$$

If we assume infinite potential barriers in the y- and z-directions, the y- and z-components of energy are given as

$$E_{n_y} = \frac{\hbar^2 a_y^2}{2m_e} = \frac{\hbar^2 n_y^2 \pi^2}{2m_e d_y^2} \tag{1.103a}$$

$$E_{n_z} = \frac{\hbar^2 a_z^2}{2m_e} = \frac{\hbar^2 n_z^2 \pi^2}{2m_e d_z^2}, \tag{1.103b}$$

where n_y and n_z are positive integers. Figure 1.44 shows the distribution of quantum states in k space. We can see that the quantum states can be grouped into wire-like subbands and each subband can be distinguished by indices n_y and n_z.

Now, we can obtain the density of states for a 1DES by calculating the density of states for one subband and adding them for all subbands. Let E' be the x-component of energy for a subband with indices n_y and n_z.

$$E' = \frac{\hbar^2}{2m_e} k_x^2 \tag{1.104}$$

With this energy, E', we can treat each subband as a one-dimensional (1D) k space. We can calculate the density of states in

Figure 1.45. Allowed k-values in the k_x axis and the length element for states with energy between E and $E + dE$.

the 1D k space first and simply shift the energy by $E_{n_y n_z}(= E_{n_y} + E_{n_z})$, since $E_{n_y n_z}$ remains a constant throughout the calculation.

The density of states for a subband can be calculated by considering a regular array of points on the k_x axis as shown in Fig. 1.45. The array of points has a spacing of $2\pi/L$, if the length of the quantum wire is L. Thus, the number of states contained in the length element dL_k is

$$\frac{dL_k}{2\pi/L}. \tag{1.105}$$

The region for states with energy between E' and $E'+dE'$ is the two line segments shown in Fig. 1.45.

$$dL_k = 2dk. \tag{1.106}$$

Using the relationship $dk = \frac{1}{\hbar}\sqrt{\frac{m_e}{2E'}}dE'$, we obtain

$$dL_k = 2dk = \frac{1}{\hbar}\sqrt{\frac{2m_e}{E'}}dE'. \tag{1.107}$$

Since two electrons (spin ↑, spin ↓) can be accommodated in each state,

$$N(E')dE' = \frac{2dL_k}{2\pi/L} = \frac{L(2m_e)^{1/2}}{\pi\hbar}\frac{1}{\sqrt{E'}}dE'. \tag{1.108}$$

Thus, per unit length, the density of states for a subband with indices n_y and n_z is

$$g_{n_y n_z}(E') = \frac{N(E')}{L} = \frac{(2m_e)^{1/2}}{\pi\hbar}\frac{1}{\sqrt{E'}}. \tag{1.109}$$

Figure 1.46 shows the 1D density of states as a function of E'.

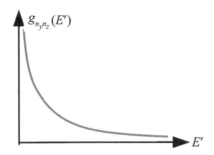

Figure 1.46. Density of states as a function of E' for a subband of 1DES.

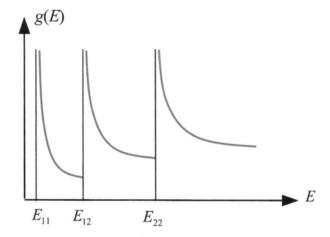

Figure 1.47. Density of states for the complete 1DES.

The density of states for the complete 1DES is

$$g(E) = \sum_{n_y=1}^{\infty} \sum_{n_z=1}^{\infty} g_{n_y n_z}(E') = \sum_{n_y=1}^{\infty} \sum_{n_z=1}^{\infty} g_{n_y n_z}(E - E_{n_y n_z}), \quad (1.110)$$

where $E_{n_y n_z} = E_{n_y} + E_{n_z}$. Figure 1.47 shows the density of states for the complete 1DES.

For a 0DES confined in a quantum dot, an electron is confined in all directions, and the energy of electron is given as

$$E = E_{n_x} + E_{n_y} + E_{n_z}. \quad (1.111)$$

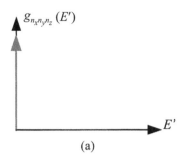

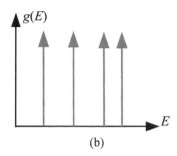

(a) (b)

Figure 1.48. Density of states: (a) for a state of 0DES, (b) for the complete 0DES.

If we assume infinite potential barriers in all directions, the x-, y- and z-components of energy are given as

$$E_{n_x} = \frac{\hbar^2 a_x^2}{2m_e} = \frac{\hbar^2 n_x^2 \pi^2}{2m_e d_x^2} \qquad (1.112a)$$

$$E_{n_y} = \frac{\hbar^2 a_y^2}{2m_e} = \frac{\hbar^2 n_y^2 \pi^2}{2m_e d_y^2} \qquad (1.112b)$$

$$E_{n_z} = \frac{\hbar^2 a_z^2}{2m_e} = \frac{\hbar^2 n_z^2 \pi^2}{2m_e d_z^2}, \qquad (1.112c)$$

where n_x, n_y, and n_z are positive integers. Each confined state forms a discrete point widely separated from the other points and the density of states for each state becomes a delta function.

$$g_{n_x n_y n_z}(E') = \delta(E') \qquad (1.113)$$

The density of states for the complete 0DES is

$$g(E) = \sum_{n_x=1}^{\infty} \sum_{n_y=1}^{\infty} \sum_{n_z=1}^{\infty} g_{n_x n_y n_z}(E') = \sum_{n_x=1}^{\infty} \sum_{n_y=1}^{\infty} \sum_{n_z=1}^{\infty} \delta(E - E_{n_x n_y n_z}),$$

$$(1.114)$$

where $E = E_{n_x} + E_{n_y} + E_{n_z}$. Figure 1.48(b) shows the density of states for the complete 0DES.

1.4.3 *Tunneling*

Tunneling is a manifestation of the wave nature of particles in quantum mechanics. When a propagating wave is confronted with a

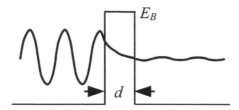

Figure 1.49. Tunneling through a potential barrier.

potential barrier, it forms an evanescent wave inside the potential barrier as long as the barrier height is finite. If the thickness of the potential barrier is also finite, there will be some non-zero wave function remaining on the other side of the potential barrier. Thus, we can have a propagating wave on the other side of the barrier, even though the amplitude might be very small. This phenomenon is called tunneling. Figure 1.49 illustrates such a phenomenon.

Now let us calculate the tunneling coefficient using a delta function potential. Figure 1.50 shows the delta function potential barrier and the propagating waves. The strength of the delta function, $E_B d$, can be considered as the product of the barrier height and the barrier thickness.

In order to set up the boundary conditions at $x = 0$, we start with the Schrödinger equation:

$$-\frac{\hbar^2}{2m_e}\frac{d^2\psi}{dx^2} + (E_B d)\delta(x)\psi = E\psi \qquad (1.115)$$

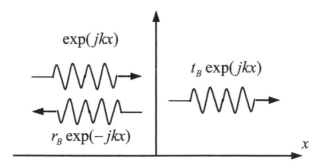

Figure 1.50. Delta function potential barrier and the propagating waves.

Integrating this equation from $-\varepsilon$ to ε (ε is a positive real number) and taking the limit where ε is indefinitely small, we obtain

$$-\frac{\hbar^2}{2m_e}\lim_{\varepsilon \to 0}\left(\frac{d\psi}{dx}\bigg|_{\varepsilon} - \frac{d\psi}{dx}\bigg|_{-\varepsilon}\right) + (E_B d)\psi(0) = 0. \qquad (1.116)$$

We can simplify this equation, if we use a notation 0^+ to represent ε that approaches zero from the positive value and 0^- to represent $-\varepsilon$ that approaches zero from the negative value.

$$-\frac{\hbar^2}{2m_e}\left(\frac{d\psi}{dx}\bigg|_{0^+} - \frac{d\psi}{dx}\bigg|_{0^-}\right) + (E_B d)\psi(0) = 0. \qquad (1.117)$$

Because of the Schrödinger equation, the wave function should be continuous at $x = 0$, and the second boundary condition is given by Eq. (1.118).

$$\psi(0^+) = \psi(0^-) \qquad (1.118)$$

Using these boundary conditions, we can calculate the transmission and reflection coefficient.

$$t_B = \frac{1}{1 - \frac{m_e E_B d}{ik\hbar^2}} = \frac{\frac{ik\hbar^2}{m_e E_B d}}{-1 + \frac{ik\hbar^2}{m_e E_B d}} \qquad (1.119a)$$

$$r_B = \frac{1}{-1 + \frac{ik\hbar^2}{m_e E_B d}} \qquad (1.119b)$$

Taking the absolute square value of Eq. (1.119b), we can see that the tunneling probability is a monotonically increasing function of the wave vector k.

PROBLEMS

1. **Propagation of electron wave:**

   (a) Consider an electron wave that is incident on a potential energy step, as shown in the following figure. Calculate the wave number k_t of the transmitted wave. The wave number of the incident wave is k_i and the electron mass is m_e. Assume that $\frac{\hbar^2 k_i^2}{2m_e} > E_B$.

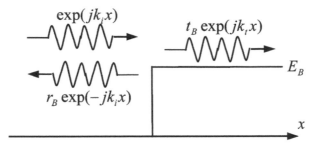

(b) Now consider an electron wave that is incident on a potential energy step in a two-dimensional space, as shown in the following figure. Calculate the wave numbers, k_{tx} and k_{ty}, of the transmitted wave. The potential energy is 0 for $y < 0$ and E_B for $y > 0$. Assume that $\frac{\hbar^2 k_{ix}^2}{2m_e} > E_B$.

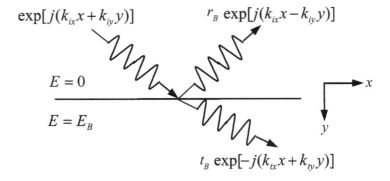

(c) Using the result of (b), we can design a lens for an electron wave. Explain how we would be able to do it.

2. **Uncertainty principle for an electron in a box:**
   In Section 1.1.5, we calculated the wave functions and momenta for an electron in a box. Based on those results, let us calculate the standard deviations of the position and momentum of the electron and show that the uncertainty principle is upheld in this case.

   (a) Show that the expectation value of the position is zero $(< x > = 0)$. Then, calculate the uncertainty, Δx, in position by evaluating the standard deviation in position, $\Delta x = \sqrt{< (x - < x >)^2 >} = \sqrt{\int_{-\infty}^{\infty} \psi^*(x) x^2 \psi(x) dx}$.

(b) We have already seen that the expectation value of the momentum is zero $(< p_{op} > = 0)$. Calculate the uncertainty, Δp, in momentum by evaluating the standard deviation in momentum, $\Delta p = \sqrt{< (p_{op} - < p_{op} >)^2 >} = \sqrt{\int_{-\infty}^{\infty} \psi^*(x) p_{op}^2 \psi(x) dx}$.

(c) Using the results of (a) and (b), calculate $\Delta x \Delta p$. The uncertainty principle discovered by Werner Heisenberg states that $\Delta x \Delta p \geq \frac{\hbar}{2}$.

3. **Atomic orbitals in many-electron atoms:**

   (a) In many-electron atoms, the energy level of an electron with a given principal quantum number, n, increases as the angular momentum quantum number, l, increases. Comparing the radial component of the wave functions for the 2s and 2p orbitals of hydrogen atom, explain why this is the case.

   (b) How does the peak probability position of an electron with a given principal quantum number, n, change as the angular momentum quantum number increases?

4. **Born–von Karman boundary conditions for a Bloch wave function:**
   In Section 1.3.3, we used the equations, $u(\vec{r}) = u(\vec{r} + L\hat{x})$, $u(\vec{r}) = u(\vec{r} + L\hat{y})$, and $u(\vec{r}) = u(\vec{r} + L\hat{z})$ without any proof. Using the fact that each side of the cube should contain integer number of atoms (lattice points), show that these equations hold.

5. **Properties of holes:**

   (a) In k-space, what would be the wave vector of the hole if there is only one empty state with a wave vector $\vec{k}_e$ in the valence band? Explain the reason briefly.

   (b) Would the site of a missing electron in the covalent bonding behave the same way as the hole defined in class? Explain your answer briefly.

6. **Calculation of density of states with hard-wall boundary conditions:**

In Section 1.4.1, we calculated the density of states assuming that the spacing between the allowed quantum states was $2\pi/L$ in each direction. This value was obtained by solving the Schrödinger equation with periodic boundary conditions for a three-dimensional potential well that had a side L. If the 3D potential well is surrounded by infinite potential barriers, we cannot use the periodic boundary conditions, so we instead solve the problem as follows.

(a) Let us assume that $V(\vec{v}) = 0$ for $0 < x < L, 0 < y < L, 0 < z < L$ and $V(x) = \infty$ otherwise. Solve the Schrödinger equation and find the spacing between the allowed k values.

(b) Calculate the density of states using the result of (a).

7. **Density of states for electrons in a quantum well:**
Consider a quantum well with infinite energy barriers at $z = 0$ and $z = 5$ nm. Assume no boundaries in the x- and y-directions. (Effective mass of electron $= 9.11 \times 10^{-31}$kg)

(a) Find three quantized energy levels (from the lowest to the third) due to the confinement in the z-direction.

(b) Calculate and plot the density of states for this system. You need to specify all the critical points and values. Your plot should include all three energy levels calculated in (a).

8. **Tunneling probability through a square energy barrier:**
Consider a square energy barrier with a finite height, E_B and thickness, d, as shown in Fig. 1.49. The effective mass of electron is m_e.

(a) By solving the Schrödinger equation and applying boundary conditions, express the tunneling probability as a function of the wave number, k, of the incident wave.

(b) Plot the tunneling probability as a function of k for $0 < k < \sqrt{\frac{2m_e E_B}{\hbar^2}}$, where $E_B = 3$ eV and $m_e = 9.11 \times 10^{-31}$kg; d is 1 nm.

(c) Plot the tunneling probability, $|t_B|^2$, calculated by Eq. (1.116b). Compare this plot with the accurate result in (b) and discuss the validity of the delta function approximation in Section 1.4.3.

Bibliography

1. L. Pauling and E., Wilson, *Introduction to Quantum Mechanics with Applications to Chemistry*, McGraw-Hill, New York (1935).

2. N. Bohr "On the Constitution of Atoms and Molecules, Part I," *Philos. Mag.* **26**, 1–24; "On the Constitution of Atoms and Molecules, Part II Systems Containing Only a Single Nucleus," *Philos. Mag.* **26**, 476–502; "On the Constitution of Atoms and Molecules, Part III Systems Containing Several Nuclei," *Philos. Mag.* **26**, 857–875 (1913).

3. L. de Broglie, *Recherches sur la théorie des quanta* (Researches on the quantum theory), Thesis, Paris (1924).

4. E. Schrödinger, "An Undulatory Theory of the Mechanics of Atoms and Molecules," *Phys. Rev.* **28**(6), 1049–1070 (1926).

5. M. Born, "Zur Quantenmechanik der Stoßvorgänge," *Zeitschrift für Physik* **37**, 863–867 (1926).

6. F. Bloch, "Über die Quantenmechanik der Elektronen in Kristallgittern," *Zeitschrift für Physik* **52**, 555-600 (1928).

7. W.L. Bragg, "The Diffraction of Short Electromagnetic Waves by a Crystal," *Proc. Camb. Philos. Soc.* 17, 43–57 (1913).

8. L. Brillouin, "Les électrons dans les métaux et le classement des ondes de de Broglie correspondantes," Comptes Rendus Hebdomadaires des Séances de l'Académie des Sciences, **191**, 292 (1930).

9. C. Kittel, *Introduction to Solid State Physics*, Wiley, New York (1996).

10. N.W. Ashcroft and N. David Mermin, *Solid State Physics*, Harcourt, Orlando (1976).

Chapter 2

Electron Transport and Device Physics

2.1 Semiconductors and Carriers

2.1.1 Strategy for the Calculation of Carrier Concentration

In a semiconductor, it is important to calculate the concentration of carriers in the conduction and valence bands. The number of carriers per unit volume (1 cm^3) is huge and we only know the occupation probability of states with a given energy. In this case, we need to calculate the number of states for a given energy first, multiply it with the occupation probability, and then add all of the products.

To understand such a procedure clearly, let us consider a method of counting the total number of people in a building. We know the occupation probability of a room and it depends on which floor the room is located. Then, we have to count the number of rooms on each floor and multiply that by the occupation probability. This will give us the number of people on each floor and we can obtain the total number of people by adding the numbers for all floors. Figure 2.1 illustrates the concept. In this analogy, the floor level corresponds to the energy level in the carrier concentration problem, the number of

Nanoelectronic Devices
Byung-Gook Park, Sung Woo Hwang, and Young June Park
Copyright © 2012 Pan Stanford Publishing Pte. Ltd.
www.panstanford.com

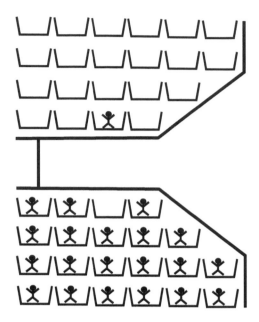

Figure 2.1. Concept of density of states and distribution function.

rooms in a floor to the density of states, and the occupation probability to the distribution function.

2.1.2 Density of States in Semiconductors

In general, the shape of the energy-versus-k diagram is not parabolic. Near the band edge, however, we can use the parabolic approximation. That is, the bottom edge of the conduction band and the top edge of the valence band can be approximated by parabolas, as shown in Fig. 2.2. The parabolic approximation of a differentiable function is possible through the Maclaurin series expansion of the function, since the first derivative is zero at an extremum (minimum or maximum) of the function.

For the conduction band edge, the energy-versus-k relationship can be approximated as

$$E = E_c + \frac{\hbar^2 k^2}{2m_n^*} \tag{2.1}$$

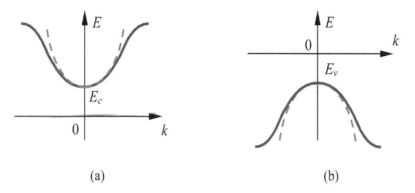

(a)   (b)

Figure 2.2. General shape of the energy versus k diagrams near the band edge and the parabolic approximation (dotted lines) for (a) conduction band minimum and (b) valence band maximum.

where m_n^* is the effective mass of an electron near the conduction band minimum. For holes near the valence band edge, we can write

$$E = E_v - \frac{\hbar^2 k^2}{2m_p^*},$$ (2.2)

where m_p^* is the effective mass of a hole near the valence band maximum.

Using Eq. (2.1) and following the procedure described in Section 1.4.1, we can obtain the density of states near the conduction band minimum.

$$g_c(E) = \frac{4\pi (2m_n^*)^{3/2}}{h^3} \sqrt{E - E_c}$$ (2.3)

Near the valence band maximum, Eq. (2.2) yields the following density of states for holes:

$$g_v(E) = \frac{4\pi (2m_p^*)^{3/2}}{h^3} \sqrt{E_v - E}$$ (2.4)

2.1.3 *Fermi–Dirac Distribution*

According to quantum mechanics, electrons are indistinguishable particles that are subjected to Pauli's exclusion principle. The exclusion principle states that only one electron can occupy each quantum state. Using this property of electrons, we can derive the

E_r ▣▢▣▣▢▢▣▢▣▢▢▢▢▢▣▢▢ g_r states, n_r particles

Figure 2.3. Occupation of g_r states with energy E_r by n_r indistinguishable Fermi particles.

Fermi–Dirac distribution function. Let us consider a system with g_r states with energy E_r. We would like to count the number of ways to put n_r electrons in this system (Fig. 2.3).

This problem can be recast to a simpler form where we evaluate the number of ways to select n_r states out of g_r states. Once n_r states are selected, we just need to put one electron in each selected state. There is only one way of doing this, since electrons are indistinguishable from each other, and only one electron can be accommodated in a state due to the exclusion principle. The number of ways to select n_r states out of g_r states is

$$\frac{g_r!}{(g_r - n_r)!\, n_r!}. \tag{2.5}$$

If we want to calculate the number of all possible arrangement of electrons, we have to evaluate the product of these numbers for all energies.

$$W = \prod_r \frac{g_r!}{(g_r - n_r)!\, n_r!} \tag{2.6}$$

Taking $\ln W$ and using Stirling's formula ($n_r \gg 1$), we can obtain the following approximation:

$$\begin{aligned}
\ln W &= \sum_r [\ln g_r! - \ln(g_r - n_r)! - \ln n_r!] \\
&\approx \sum_r [g_r \ln g_r - (g_r - n_r)\ln(g_r - n_r) - n_r \ln n_r]
\end{aligned} \tag{2.7}$$

Now we can obtain the most probable distribution by maximizing $\ln W$ under the constraint of particle and energy conservation.

$$\sum_r n_r = N, \quad \sum_r n_r E_r = E_t \tag{2.8}$$

Let us use the Lagrange multiplier method to maximize $\ln W$ with the constraints given in Eq. (2.8). Introducing the Lagrange multipliers, α and β, we define a function Φ of multiple variables (n_r).

$$\Phi(n_r) = \ln W + \alpha \left(N - \sum_r n_r \right) + \beta \left(E_t - \sum_r n_r E_r \right) \tag{2.9}$$

Equation (2.8) implies that Φ is the same as $\ln W$, so that we can maximize Φ instead of $\ln W$. At maximum, Φ should not change for any infinitesimal change in n_r's. That is, for all n_r,

$$\frac{\partial \Phi}{\partial n_r} = 0 \qquad (2.10)$$

Combining Eqs. (2.7), (2.9), and (2.10), we can find a condition for the maximum Φ.

$$n_r = \frac{g_r}{\exp(\alpha + \beta E_r) + 1} \qquad (2.11)$$

The occupation probability, n_r/g_r, of states with energy E_r should be

$$f(E_r) = \frac{n_r}{g_r} = \frac{1}{\exp(\alpha + \beta E_r) + 1} \qquad (2.12)$$

If we consider the classical limit where the occupation probability becomes very small ($f(E_r) \ll 1$), the exponential term in the denominator will dominate the other term.

$$f(E_r) \approx \exp(-\alpha - \beta E_r) \qquad (2.13)$$

This probability should be the same as the classical Maxwell–Boltzmann distribution given as

$$f_{MB}(E_r) = A \exp(-E_r/k_B T). \qquad (2.14)$$

By equating Eq. (2.13) with Eq. (2.14), we obtain

$$\beta = 1/k_B T. \qquad (2.15)$$

In addition, we define a new parameter E_F as

$$E_F = -\alpha k_B T. \qquad (2.16)$$

Using Eqs. (2.15) and (2.16), we can rewrite Eq. (2.12) as follows:

$$f(E) = \frac{1}{\exp[(E - E_F)/k_B T] + 1} \qquad (2.17)$$

This function is known as the Fermi–Dirac distribution, and it is obeyed by all particles that are subject to Pauli's exclusion principle. Figure 2.4 shows the shape of the function.

The energy, E_F, is the "Fermi energy" where the Fermi–Dirac distribution function is 1/2. In general, it depends on both the electron concentration and the temperature. At 0 K, it is the energy that delineates the filled and empty quantum states: that is, at

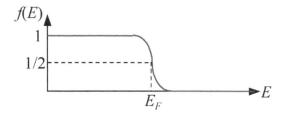

Figure 2.4. Fermi–Dirac distribution function versus energy.

the temperature of absolute zero, all states with an energy lower than E_F are filled with an electron, while all states with an energy higher than E_F are empty. It is important to realize that the Fermi–Dirac distribution function determines only the probability of occupation and that the Fermi energy should be adjusted to a proper value in order to obtain the correct electron and hole concentration. There do not have to be quantum states at the Fermi energy. Often, the Fermi energy is located inside the energy gap of a semiconductor.

2.1.4 *Electron and Hole Concentrations*

Now we can calculate the electron concentration using Eqs. (2.3) and (2.17). As described in Section 2.1.1, we need to multiply the density of states with the Fermi–Dirac distribution function for a given energy and integrate it for all possible energy values (Fig. (2.5)).

$$
\begin{aligned}
n_0 &= \int_{E_c}^{\infty} g_c(E) f(E) dE \\
&= \frac{1}{2\pi^2} \left(\frac{2m_n^*}{\hbar^2} \right)^{3/2} \int_{E_c}^{\infty} \frac{(E - E_c)^{1/2} \, dE}{1 + \exp[(E - E_F)/k_B T]}
\end{aligned}
\tag{2.18}
$$

For simplicity, let us define two dimensionless quantities by normalizing the energies with the thermal energy, $k_B T$.

$$
\eta = \frac{E - E_c}{k_B T}
\tag{2.19a}
$$

$$
\eta_F = \frac{E_F - E_c}{k_B T}
\tag{2.19b}
$$

With these definitions, Eq. (2.18) can be written as

$$
n_0 = \frac{1}{4} \left(\frac{2m_n^* k_B T}{\pi \hbar^2} \right)^{3/2} \frac{2}{\sqrt{\pi}} \int_0^{\infty} \frac{\eta^{1/2} d\eta}{1 + \exp(\eta - \eta_F)}.
\tag{2.20}
$$

The integral in Eq. (2.20) is one of the Fermi integrals, which are generally defined as

$$F_v(x) = \frac{2}{\sqrt{\pi}} \int_0^\infty \frac{y^v dy}{1 + \exp(y - x)}. \tag{2.21}$$

Thus, the electron concentration at thermal equilibrium is expressed as

$$n_0 = \frac{1}{4} \left(\frac{2m_n^* k_B T}{\pi \hbar^2} \right)^{3/2} F_{1/2}(\eta_F). \tag{2.22}$$

Equation (2.22) can be further simplified by defining

$$N_c = \frac{1}{4} \left(\frac{2m_n^* k_B T}{\pi \hbar^2} \right)^{3/2}. \tag{2.23}$$

This quantity is called the effective density of states in the conduction band since it is the electron concentration when the Fermi energy is located at the conduction band edge ($E_F = E_c$). Finally, we obtain a compact equation for the electron concentration.

$$n_0 = N_c F_{1/2}(\eta_F) \tag{2.24}$$

To calculate the hole concentration, we can use a similar procedure with a few necessary changes. The Fermi–Dirac distribution function, $f(E)$, needs to be replaced by $[1 - f(E)]$, since the existence of a hole may be considered the absence of an electron. We also have to use $g_v(E)$ instead of $g_c(E)$.

$$p_0 = \int_{-\infty}^{E_v} g_v(E)[1 - f(E)] dE$$
$$= \frac{1}{2\pi^2} \left(\frac{2m_p^*}{\hbar^2} \right)^{3/2} \int_{-\infty}^{E_v} \frac{(E_v - E)^{1/2} dE}{1 + \exp[(E_F - E)/k_B T]} \tag{2.25}$$

If we define normalized energies for holes as

$$\eta' = \frac{E_v - E}{k_B T} \tag{2.26a}$$

$$\eta'_F = \frac{E_v - E_F}{k_B T}, \tag{2.26b}$$

Equation (2.25) becomes

$$p_0 = \frac{1}{4} \left(\frac{2m_p^* k_B T}{\pi \hbar^2} \right)^{3/2} \frac{2}{\sqrt{\pi}} \int_0^\infty \frac{\eta'^{1/2} d\eta'}{1 + \exp(\eta' - \eta'_F)}. \tag{2.27}$$

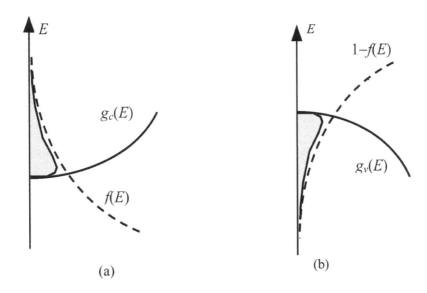

Figure 2.5. Density of states and Fermi–Dirac distribution function for (a) electrons and (b) holes. The product of the two functions is also sketched.

Since this equation has exactly the same form as Eq. (2.22) except for a few parameter changes ($m_n^* \to m_p^*$, $\eta \to \eta'$, and $\eta_F \to \eta_F'$), we can obtain

$$p_0 = N_v F_{1/2}(\eta_F'), \tag{2.28}$$

where the effective density of states in the valence band is defined as

$$N_v = \frac{1}{4}\left(\frac{2m_p^* k_B T}{\pi \hbar^2}\right)^{3/2}. \tag{2.29}$$

In general, the Fermi integral does not have a simple analytical solution and has to be calculated by a numerical method. However, when the Fermi energy is located within the energy gap and is separated from the band edge at least by a few $k_B T$, η_F becomes a relatively large negative number ($-\eta_F \gg 1$) and we can derive a very convenient approximation to the Fermi integral:

$$F_\nu(x) \cong \frac{2}{\sqrt{\pi}}\int_0^\infty y^\nu \exp(x - y)dy = \frac{2}{\sqrt{\pi}}\exp(x)\Gamma(\nu + 1), \tag{2.30}$$

where $\Gamma(\nu)$ is a Gamma function (see Appendix III).

For $\nu = 1/2$ and $x = \eta_F$, we obtain

$$F_{1/2}(\eta_F) = \exp(\eta_F).$$ (2.31)

Using this approximation, the electron and hole concentrations can be expressed as

$$n_0 \cong N_c \exp(\eta_F) = N_c \exp[(E_F - E_c)/k_B T]$$ (2.32a)
$$p_0 \cong N_v \exp(\eta_F') = N_v \exp[(E_v - E_F)/k_B T].$$ (2.32b)

If we take the product of the electron and hole concentrations, it becomes a constant for a given material (energy gap) and temperature.

$$n_0 p_0 = N_c \exp[(E_F - E_c)/k_B T] N_v \exp[(E_v - E_F)/k_B T]$$
$$= N_c N_v \exp(-E_g/k_B T)$$ (2.33)

That is, it is independent of the Fermi level. This phenomenon is known as the **mass action law** in chemistry. The reason behind this law is that the reaction between electrons and holes is a recombination process, and the reverse process is an electron–hole pair generation. Thermal equilibrium is maintained when the recombination rate is the same as the generation rate. The recombination rate is proportional to the product of the electron and hole concentrations, but the generation rate should be independent of the Fermi level as long as the Fermi level is far away from the band edges (non-degenerate semiconductor). Thus, in a non-degenerate semiconductor, the product of the electron and hole concentrations is constant at thermal equilibrium. The non-degeneracy condition is important here. If the Fermi energy is too close to the band edge or is located inside a band (degenerate semiconductor), the generation rate may be affected by the presence of electrons or holes in one of the bands.

2.1.5 *Intrinsic Semiconductors*

If a semiconductor has no impurities or defects, it is called an **intrinsic** semiconductor. In this type of semiconductors, electron–hole pairs are generated thermally and their concentrations should be exactly the same. This particular concentration is called "intrinsic carrier concentration" and denoted by n_i.

$$n_0 = p_0 = n_i$$ (2.34)

We can find out the position of the Fermi level in an intrinsic semi-conductor by using Eqs. (2.32) and (2.34).

$$N_c \exp[-(E_c - E_F)/k_B T] = N_v \exp[-(E_F - E_v)/k_B T] \quad (2.35)$$

After taking the logarithm of Eq. (2.35) and rearranging a few terms, we obtain

$$E_F = \frac{E_c + E_v}{2} + \frac{k_B T}{2} \ln \frac{N_v}{N_c}. \quad (2.36)$$

Thus, the position of the Fermi level in an intrinsic semiconductor is somewhat different from the midgap energy. Since the energy gap is much larger than the thermal energy in most semiconductors, however, the intrinsic Fermi level is close to the midgap. The intrinsic Fermi energy can also be expressed as a function of effective masses.

$$E_F = \frac{E_c + E_v}{2} + \frac{3k_B T}{4} \ln \frac{m_p^*}{m_n^*} \quad (2.37)$$

Using Eq. (2.35), we can express the intrinsic carrier concentration as a function of the energy gap and the effective density of states.

$$n_i^2 = N_c N_v \exp(-E_g/k_B T) \quad (2.38)$$

The intrinsic carrier concentration is

$$n_i = \frac{1}{4} \left(\frac{2\sqrt{m_n^* m_p^*} k_B T}{\pi \hbar^2} \right)^{3/2} \exp\left(-\frac{E_g}{2k_B T}\right). \quad (2.39)$$

Example 2.1: Intrinsic carrier concentration of silicon at room temperature

Calculate the intrinsic carrier concentration of silicon at room temperature (300 K). The density-of-states (DOS) effective masses of electrons and holes are $m_n^* = 1.08m_e$ and $m_p^* = 0.55m_e$, respectively, where m_e is the free electron mass. The energy gap of silicon is $E_g = 1.1$ eV.

Solution:

$$n_i = \frac{1}{4} \left(\frac{2 \times \sqrt{1.08 \times 0.55} \times 9.1 \times 10^{-31} \times 0.026 \times 1.6 \times 10^{-19}}{3.14 \times (1.06 \times 10^{-34})^2} \right)^{3/2}$$

$$\times \exp\left(-\frac{1.1}{2 \times 0.026}\right)$$

$$= 1.06 \times 10^{16} \,(\text{m}^{-3})$$

$$= 1.06 \times 10^{10} \,(\text{cm}^{-3})$$

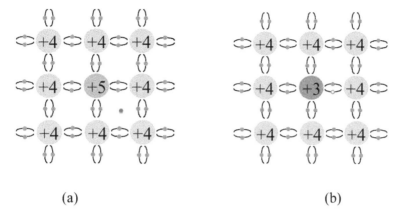

(a)   (b)

Figure 2.6. Dopant impurities in group IV semiconductors: (a) donor (n-type dopant) and (b) acceptor (p-type dopant).

2.1.6 *Carriers in Extrinsic Semiconductors*

If we introduce an impurity atom with one more valence electron than the host atoms into an otherwise perfect semiconductor crystal, we can create an extra electron near the impurity ion after the other valence electrons participate in the covalent bonding with the host atoms. Such a process is called "doping," and the impurity atom is called a "dopant" (Fig. (2.6)). At very low temperatures, the extra electron is bound to the impurity ion since there is an attractive force between the positive charge of the impurity ion and the negative charge of the electron. At room temperature, however, the binding energy can be easily supplied from the environment since the binding energy is just a few tens of meV for many dopants as will be estimated in Example 2.2. Then the extra electron can be freed from the impurity ion. The freed electron stays in the conduction band and contributes to the electrical conduction. Since this electron can be considered as having been donated by the impurity atom, the impurity atom is called a **donor** (Fig. 2.6(a)). When we introduce an impurity atom with one fewer valence electron than the host atoms into a semiconductor crystal, a hole is created because of the deficit in the number of valence electrons (Fig. 2.6(b)). Such an impurity atom is called an **acceptor** because the impurity atom should accept an electron to release the hole originally bound to it. The hole resides in the valence band and contributes to electrical conduction.

Example 2.2: Binding energy of an electron to a donor

Using the hydrogen model, calculate the binding energy, E_d, of an electron to a donor. Assume that the electron effective mass is $m_n^* \cong 0.26m_e$, and that the dielectric constant is $\varepsilon_r = 11.7$ in silicon.

Solution:

We can treat an ionized donor atom as the nucleus of a hydrogen atom. The main difference from the case of the hydrogen atom is that the donor is now imbedded in silicon. We need to consider the effect of silicon atoms that surround the donor. In fact, there are many silicon atoms within the sphere defined by the average radius of the electron wave function. To take into account the interaction between the bound electron and the silicon atoms, we must use the effective mass of electron and the dielectric constant in silicon.

From Eq. (1.55) in Chapter 1, we can obtain

$$E_d = \frac{1}{\varepsilon_r^2} \left(\frac{m_n^*}{m_e} \right) \left[\frac{e^4 m_e}{8\varepsilon_0^2 h^2} \right] = \frac{0.26}{11.7^2} \times 13.6 \cong 0.026 \text{ (eV)}.$$

The measured binding energies for phosphorus (P), arsenic (As), and antimony (Sb) in silicon are 0.044 eV, 0.049 eV, and 0.039 eV, respectively.

As shown in Fig. 2.7(a), the bound state is localized and is usually represented by a short bar within the energy gap. The length of the

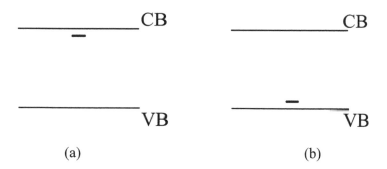

Figure 2.7. Dopant energy levels: (a) donor and (b) acceptor. Since these states are localized, they are represented by a short bar.

bar is short compared with the lines representing the band edges so as to emphasize the fact that it is a localized state. The distance between the dopant energy level and the band edge is the binding energy of the carriers. If an electron is released from a donor state and moves around freely, the energy of the electron must be higher than the conduction band edge. The electron is excited from the bound state to a state in the conduction band, and the energy required for the excitation is much smaller than the energy gap. Thus the donor level is located close to the conduction band in most cases.

In order to release a hole from an acceptor state, the binding energy of the hole should be provided from the environment. The released hole can be thought of as an empty state in the valence band. From the viewpoint of the corresponding electron in the valence band, it is captured by the acceptor state leaving the state of its original occupancy as an empty state. This electron needs to absorb a certain amount of energy from the environment to overcome the repulsive force from the negatively charged acceptor ion. Thus, the hole release process is associated with the valence band and the acceptor level is located close to the valence band, as shown in Fig. 2.7(b).

By doping a semiconductor with dopants having a concentration much higher than the intrinsic carrier concentration, we can upset the balance between electrons and holes in a semiconductor. If we dope the semiconductor with donors, the charge neutrality will force the electron concentration to be more or less the same as the donor concentration, but, due to the recombination of holes with electrons, the hole concentration will decrease dramatically from its intrinsic value. The hole concentration will not decrease indefinitely, however, because the recombination rate will decrease in proportion to the hole concentration. Eventually, the recombination rate will be exactly balanced by the generation rate and a new equilibrium will be established. The mass action law states that the product of electron and hole concentrations is a constant in this case. The calculation method of electron and hole concentrations is illustrated in Example 2.3.

Example 2.3: Electron and hole concentrations as a function of the doping concentration

(a) Using the mass action law and the charge neutrality, calculate the electron and hole concentrations as a function of the doping concentration.
(b) Assuming that the intrinsic carrier concentration in silicon is 1×10^{10} cm^{-3} at room temperature, calculate the electron and hole concentrations when the donor concentration is 1×10^{16} cm^{-3}.

Solution:

(a) Let us assume that a silicon substrate is doped with N_d donors per cubic centimeter and all the donors are ionized. From the charge neutrality, $p_0 - n_0 + N_d = 0$. The mass action law requires $p_0 n_0 = n_i^2$. Solving these two equations, we obtain the carrier concentrations as follows:

$$n_0 = \frac{N_d}{2} \left(1 + \sqrt{1 + \frac{4n_i^2}{N_d^2}} \right)$$

$$p_0 = \frac{N_d}{2} \left(-1 + \sqrt{1 + \frac{4n_i^2}{N_d^2}} \right)$$

For $N_d \gg n_i$, $n_0 \cong N_d$, $p_0 \cong n_i^2/N_d$, while, for $N_d \ll n_i$, $n_0 \cong p_0 \cong n_i$.
(b) $n_0 = 1 \times 10^{16}$ (cm^{-3}), $p_0 = 1 \times 10^4$ (cm^{-3})

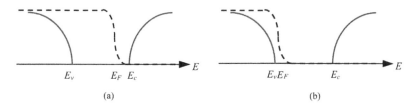

Figure 2.8. Position of the Fermi level in a doped semiconductor: (a) n-type and (b) p-type.

As we have seen in Example 2.3, doping dramatically upsets the balance between electron and hole concentrations. Since the doping concentration is at least a few orders of magnitude higher than the intrinsic carrier concentration, the electron (hole) concentration in an n-type (a p-type) semiconductor is several orders of magnitude higher than the hole (electron) concentration. In a doped semiconductor, there is a huge difference between the higher-concentration carriers and the lower-concentration carriers. The carriers with higher concentration are called **majority carriers**. and those with lower concentration are called **minority carriers**.

2.1.7 *Position of the Fermi Level*

The Fermi level in a doped semiconductor should be different from the intrinsic Fermi level obtained in the previous section. In an n-type semiconductor, the Fermi level moves toward the conduction band edge, so that the electron concentration exceeds the hole concentration (Fig. 2.8(a)). In a p-type semiconductor, it moves toward the valence band edge for the hole concentration to exceed the electron concentration (Fig. 2.8(b)).

If we know the position of the Fermi level, we can easily calculate the carrier concentrations using Eq. (2.32). Defining E_i (intrinsic Fermi level or intrinsic level) as the Fermi level of an intrinsic semiconductor, we can express the electron and hole concentrations as a function of the intrinsic carrier concentration and the Fermi level as follows:

$$
\begin{aligned}
n_0 &= N_c \exp[(E_F - E_c)/k_B T] \\
&= N_c \exp\{[-(E_c - E_i) + (E_F - E_i)]/k_B T\} \\
&= n_i \exp[(E_F - E_i)/k_B T]
\end{aligned}
\tag{2.40a}
$$

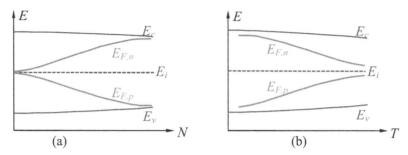

Figure 2.9. Position of the Fermi level as a function of (a) doping concentration and (b) temperature.

$$p_0 = N_v \exp[(E_v - E_F)/k_B T]$$
$$= N_v \exp\{[(E_v - E_i) - (E_F - E_i)]/k_B T\}$$
$$= n_i \exp[-(E_F - E_i)/k_B T] \qquad (2.40b)$$

On the other hand, the Fermi level positions of n-type and p-type semiconductors can also be easily obtained from Eq. (2.41).

$$E_F = E_c - k_B T \ln \frac{N_c}{n_0} = E_i + k_B T \ln \frac{n_0}{n_i} \quad \text{(n-type)} \quad (2.41a)$$

$$E_F = E_v + k_B T \ln \frac{N_v}{p_0} = E_i - k_B T \ln \frac{p_0}{n_i} \quad \text{(p-type)} \quad (2.41b)$$

From Eq. (2.41), we can estimate the dependence of the position of the Fermi level on the doping concentration and the temperature (Fig. 2.9). At a given temperature, the Fermi levels for both n-type and p-type semiconductors approach the intrinsic level as the doping concentration is decreased. The Fermi level increases (decreases) as we increase the doping concentration in an n-type (a p-type) semiconductor. In other words, the Fermi level approaches the band edge as the doping concentration increases. The energy gap, $(E_c - E_v)$, decreases as a function of doping because of dopant band formation and band tailing.

If we keep the doping concentration fixed and change the temperature, the intrinsic carrier concentration becomes a dominant term for the change in Eq. (2.41). At low temperatures, the intrinsic carrier concentration is very small, and the Fermi level is far away from the intrinsic level. As we increase the temperature, the intrinsic carrier concentration increases exponentially, and the Fermi level approaches the intrinsic level. The energy gap narrows as a function of temperature because of the thermal expansion of semiconductors and the local energy gap narrowing due to lattice vibrations.

2.2 Carrier Transport

2.2.1 *Drift of Carriers*

Carriers in a solid are in constant thermal motion with random scattering from lattice vibrations, impurities, and defects. If an electric field $\vec{E}$ is applied, the carriers are accelerated during the free flight between scattering events (collisions). The small deflection of the carrier motion caused by this acceleration results in a net drift cumulatively. As shown in Fig. 2.10, the trajectory of an electron is completely random, and the average position after a sufficiently long time will be more or less the same as the original position if there is no electric field. If there is an electric field, however, the small deflections in the direction of the electric force during the flight between collisions accumulate, and the average position after a time sufficiently longer than the collision time will be proportional to that time.

Before we describe the drift motion of electrons quantitatively, let us examine the concept of collision time, τ_c. The collision time is defined as the expectation value of the time during which an electron would not undergo a collision. Thus, if we pick an electron and trace its motion until it suffers from a collision, the average time that it takes for the electron to experience the collision is the collision time. We can also think of a time-reversed process. If we pick an electron

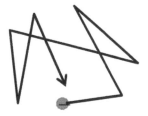 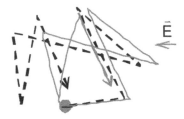

Figure 2.10.   Trajectory of an electron: (a) without electric field and (b) with electric field.

and trace its motion backward to the time when it suffered from a collision, the average will be the same collision time, τ_c.

Based on this definition of collision time, let us evaluate the drift velocity of an electron under the influence of an electric field, E_x, in the x direction. The force on the electron due to the electric field, E_x, is

$$F_x = -eE_x \tag{2.42}$$

According to Newton's law of motion, the acceleration in the x-direction is

$$a_x = F_x/m_n^* = -eE_x/m_n^*. \tag{2.43}$$

Due to collisions, electrons are accelerated for an average time τ_{cn} ($\tau_{cn} \approx 2 \times 10^{-14}$ s at 300K), so that we can obtain the drift velocity of an electron as

$$v_{dn} = a_x \tau_{cn} = -\left(\frac{e\tau_{cn}}{m_n^*}\right) E_x. \tag{2.44}$$

In this equation, the drift velocity is simply proportional to the electric field, and we can simplify the equation by defining the electron **mobility**, μ_n, as

$$\mu_n = \frac{e\tau_{cn}}{m_n^*}. \tag{2.45}$$

Finally, we can obtain the drift current density, $J_{drf,n}$, by multiplying the charge density, $-en$, by the drift velocity.

$$J_{drf,n} = -env_{dn} = en\mu_n E_x \tag{2.46}$$

For the hole drift current, we can follow similar steps. The drift velocity with the same electric field is

$$v_{dp} = a_x \tau_{cp} = \frac{e\tau_{cp}}{m_p^*} E_x. \tag{2.47}$$

The hole mobility is defined as

$$\mu_p = \frac{e\tau_{cp}}{m_p^*}. \tag{2.48}$$

The hole drift current density is given as

$$J_{drf,p} = epv_{dp} = ep\mu_p E_x. \tag{2.49}$$

The total drift current density is the sum of the electron and hole drift current density.

$$J_{drf} = J_{drf,n} + J_{drf,p} = e(n\mu_n + p\mu_p)E_x \qquad (2.50)$$

From this equation, we can obtain the conductivity of the semiconductor as

$$\sigma = \frac{J_{drf}}{E_x} = e(n\mu_n + p\mu_p). \qquad (2.51)$$

By taking the inverse of Eq. (2.51), we can also obtain the resistivity.

$$\rho = \frac{1}{\sigma} = \frac{E_x}{J_{drf}} = \frac{1}{e(n\mu_n + p\mu_p)} \qquad (2.52)$$

2.2.2 Diffusion of Carriers

If there is a concentration gradient of carriers, random motion due to the thermal energy and collisions leads to the diffusion of carriers. Figure 2.11 illustrates the concept of diffusion. Let us consider an imaginary surface that divides the space populated by carriers into two parts at $x = 0$. If we count the number of carriers that pass this surface of unit area per unit time, we can determine the flux of carriers. When there is no concentration difference, there will be no net flux, since, on the average, the number of carriers crossing the surface from the left hand side (LHS) to the right hand side (RHS) will be the same as that crossing the surface in the opposite direction. When there is a concentration difference, however, there will be a net flux. As shown in Fig. 2.11, let us assume that the concentration

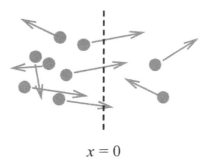

$$x = 0$$

Figure 2.11. Diffusion due to concentration gradient of carriers.

of carriers on the LHS is higher than that on the RHS. Even though each carrier moves in the random direction, the number of carriers crossing the surface from the LHS to the RHS will be larger than that crossing the surface in the opposite direction.

Let us describe the diffusion of electrons quantitatively. If we define l_{nx} as the **mean free path** in the x-direction, it is related to the collision time, τ_{cn} as follows:

$$l_{nx} = v_{th,nx}\tau_{cn}, \tag{2.53}$$

where $v_{th,nx}$ is the electron thermal velocity in the x-direction. The concept of mean free path will be useful in the calculation of the flux of carriers due to a concentration difference.

The flux of electrons can be evaluated by counting the number of electrons crossing the unit area of the surface at $x = 0$ per unit time. The number of electrons crossing the surface of unit area from the LHS to the RHS per unit time is the product of half the electron concentration times the thermal velocity in the x-direction. The factor of one-half(1/2) reflects the fact that, probabilistically, half of the electrons are moving toward the RHS while the other half are moving toward the LHS. One important issue is how to select the representative concentration of electrons. In order for an electron to cross the surface, the electron should not suffer from a collision during its flight to the surface. Since the mean free path is the average distance that an electron can move without a collision, it can be an appropriate distance from $x = 0$ for the selection of the representative concentration. Once this number is calculated, we can obtain the number of electrons crossing the surface of unit area from the RHS to the LHS per unit time by changing the sign of l_{nx}. The net flux of electrons is given by

$$\begin{aligned}\Phi_n &= \tfrac{1}{2}n(-l_{nx})v_{th,nx} - \tfrac{1}{2}n(+l_{nx})v_{th,nx} \\ &= \tfrac{1}{2}v_{th,nx}[n(-l_{nx}) - n(+l_{nx})]\end{aligned} . \tag{2.54}$$

For small l_{nx}, we can approximate the electron concentrations using the Maclaurin series expansion.

$$\Phi_n = \frac{1}{2}v_{th,nx}\left\{\left[n(0) - l_{nx}\frac{dn}{dx}\right] - \left[n(0) + l_{nx}\frac{dn}{dx}\right]\right\} = -v_{th,nx}l_{nx}\frac{dn}{dx} \tag{2.55}$$

If we define the electron **diffusion constant**, D_n, as

$$D_n = v_{th,nx}l_{nx} = v_{th,nx}^2\tau_{cn}, \tag{2.56}$$

we can calculate the electron diffusion current density as follows:

$$J_{diff,n} = -e\Phi_n = eD_n\frac{dn}{dx} \tag{2.57}$$

Following the same procedure, we can calculate the hole diffusion current density in the x-direction.

$$J_{diff,p} = e\Phi_p = -eD_p\frac{dp}{dx} \tag{2.58}$$

In this equation, D_p is the hole diffusion constant. Because of the positive charge of holes, the hole diffusion current has the same direction as the hole flux, while the electron diffusion current flows in the opposite direction to the electron flux. By combining Eqs. (2.57) and (2.58), we obtain the total diffusion current density of carriers.

$$J_{diff} = eD_n\frac{dn}{dx} - eD_p\frac{dp}{dx} \tag{2.59}$$

We can see that the electron current flows in the direction in which the electron concentration increases, while the hole current flows in the direction in which the hole concentration decreases.

2.2.3 Total Current in Semiconductors

Combining Eqs. (2.50) and (2.59), we can obtain the total current density of carriers in the x-direction.

$$J_x = e(n\mu_n + p\mu_p)E_x + eD_n\frac{dn}{dx} - eD_p\frac{dp}{dx} \tag{2.60}$$

We can easily generalize this result into the three-dimensional (3D) case by using a vector electric field and the concentration gradients of carriers.

$$\vec{J} = e(n\mu_n + p\mu_p)\vec{E} + eD_n\nabla n - eD_p\nabla p \tag{2.61}$$

Two observations related to carrier transport in semiconductors deserve our attention. One is that, for a given type of carriers, their mobility and diffusion constant are related. We can glimpse the possibility of such relationship by examining Eq. (2.45) and (2.56). The mobility and the diffusion constant of electrons have the collision time as a common factor, because the random movement of electrons originates from collisions with various scattering sources. If we take the ratio of the diffusion constant and the mobility, it is proportional to the kinetic energy of an electron. Since the kinetic energy is simply the thermal energy, which is proportional to the temperature, we can conclude that the ratio is just a function of temperature. This relationship is called the **Einstein relationship** because Einstein established this relationship in his theory of Brownian motion.

Another observation is that, in most cases, we do not have to consider all the terms in Eq. (2.61). Depending on the charge neutrality or operating conditions, either the drift or diffusion current is dominant. In a doped semiconductor, the concentration of the majority carriers is several orders of magnitude higher than that of the minority carriers. The conductivity defined in Eq. (2.51) will be dominated by the term determined by the majority carriers, so that the drift current will be determined almost completely by them. The diffusion current is not dependent on the carrier concentration itself but rather on the slope of carrier concentration. The huge difference between the majority and minority carrier concentrations will not make majority carriers dominate the diffusion. Due to the tendency of maintaining charge neutrality, however, the diffusions of the two types of carriers are often correlated, so that we can focus on the diffusion of only one type of carriers.

2.2.4 Invariance of Fermi Level in Equilibrium

If two materials are in an intimate contact such that electrons can move between the two (p-n junction, heterojunction, metal–semiconductor junction), their Fermi levels tend to be adjusted to the same value. Starting from the definition of Fermi level, we can easily attain an intuitive understanding of this phenomenon. The Fermi energy is defined as the energy at which the occupation probability is exactly one half (1/2). Above this energy, the occupation probability is less than 1/2, and, below it, more than 1/2. Roughly speaking, a quantum state whose energy is lower than the Fermi energy tends to be filled with an electron, while a state with a higher energy tends to be empty. If there are two isolated materials with different Fermi levels, the electrons in the material with the higher Fermi energy tend to move to the material with the lower Fermi energy as soon as the two materials are brought into contact. This

	material 1	material 2
Density of states	$N_1(E)$	$N_2(E)$
Fermi distribution	$f_1(E)$	$f_2(E)$

Figure 2.12. Two materials in thermal contact. In thermal equilibrium, there is no current and no net charge or energy transfer between the two materials.

movement occurs because the total energy of the system can be lowered by relocating the high-energy electrons in the material with the higher Fermi energy to the empty states in the other material with the lower energy.

Let us analyze this phenomenon quantitatively. In thermal equilibrium, there is zero current and no net charge or energy transfer occurs between the two materials. Such a balance is dynamic, not static. That is, there are still movements of electrons from one material to the other. The net transfer of electrons is zero since the movement of electrons in one direction is exactly balanced by the movement of electrons in the opposite direction. Let us designate the two materials as material 1 and material 2. For each material, we need to consider the density of states and the Fermi distribution function. Figure 2.12 shows these parameters in each material.

For an electron to jump from a state in material 1 to a state in material 2, there should be an electron in material 1 and an empty state in material 2. In addition, energy conservation implies that the electron and the empty state should have the same energy. For a given energy E, the rate of electron transfer from material 1 to material 2 should be proportional to the number of electrons with energy E in material 1. It should also be proportional to the number of empty states with energy E in material 2. Thus, the rate of electron transfer from material 1 to 2 is given as

$$\text{rate from 1 to 2} \ \propto\ N_1(E)f_1(E) \cdot N_2(E)[1 - f_2(E)]. \qquad (2.62)$$

In a similar way, the rate of electron transfer from material 2 to 1 is given as

$$\text{rate from 2 to 1} \ \propto\ N_2(E)f_2(E) \cdot N_1(E)[1 - f_1(E)]. \qquad (2.63)$$

At equilibrium, these two rates should be the same.

$$N_1(E)f_1(E) \cdot N_2(E)[1 - f_2(E)] = N_2(E)f_2(E) \cdot N_1(E)[1 - f_1(E)] \tag{2.64}$$

By rearranging and simplifying Eq. (2.64), we obtain

$$f_1(E) = f_2(E). \tag{2.65}$$

Since $f_1(E)$ and $f_2(E)$ are none other than the Fermi distribution functions, we obtain

$$\frac{1}{1 + \exp[(E - E_{F1})/k_B T]} = \frac{1}{1 + \exp[(E - E_{F2})/k_B T]}. \tag{2.66}$$

In order for this equation to hold for any energy E, the two Fermi energies should be the same.

$$E_{F1} = E_{F2} \tag{2.67}$$

In other words, the Fermi level should not change throughout the entire system in thermal equilibrium.

$$\frac{dE_F}{dx} = 0 \tag{2.68}$$

2.2.5 Non-Uniform Doping

Often, the doping concentration in a semiconductor changes as a function of position. When there is a gradient in doping concentration, diffusion occurs because of the difference in carrier concentration. The diffusion of carriers breaks the charge neutrality between electrons and donors, leading to the formation of a space charge. The space charge induces an electric field, which causes the drift of carriers. Since the direction of drift is opposite to that of diffusion, the net movement of carriers decreases. This process continues until there is an exact balance between diffusion and drift. That is the point where we reach thermal equilibrium. At equilibrium, the Fermi level is constant, as shown in Fig. 2.13.

Whenever the doping concentration is not uniform, a built-in electric field will be formed, as described in the previous paragraph. The electric field is the negative of the electric potential slope, and we can obtain the potential energy of an electron by simply multiplying the electric potential with the electronic charge, $-e$. Since the conduction band edge is the energy where the kinetic energy of an

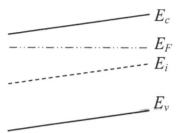

Figure 2.13. Band diagram of a non-uniformly doped semiconductor in thermal equilibrium.

electron becomes zero, the position of conduction band edge should be determined by the potential energy of an electron alone. Hence, we have to change the position of the conduction band edge in accordance with the potential energy of an electron, as shown in Fig. 2.13. In a semiconductor, the valence band edge is always lower than the conduction band edge by the amount of the energy gap, E_g. In addition, Eq. (2.37) tells us that the intrinsic level depends on the band edges and the effective mass of carriers at a given temperature, so that it is away from the conduction band edge by the same distance for a given semiconductor. Thus, the band edges (E_c, E_v) and the intrinsic level (E_i) move together in a band diagram, while keeping their distance in energy. Such a property is shown in Fig. 2.13, and we can easily relate the change in the electric potential to that in any of the three levels (E_c, E_v, and E_i): that is, we can write

$$-ed\psi = dE_c = dE_v = dE_i. \qquad (2.69)$$

Using Eq. (2.69) and the definition of electric field, we can express the electric field as a function of the intrinsic level.

$$E_x = -\frac{d\psi}{dx} = \frac{1}{e}\frac{dE_i}{dx} \qquad (2.70)$$

As long as the gradient in doping concentration is small, the deviation from charge neutrality is not large and we can assume a quasi-neutrality. Then the electron concentration should still be close to the donor concentration.

$$n = n_i \exp\left[\frac{E_F - E_i}{k_B T}\right] \cong N_d(x) \qquad (2.71)$$

After some rearrangements, we can obtain the intrinsic level as a function of donor concentration.

$$E_F - E_i = k_B T \ln \left[\frac{N_d(x)}{n_i} \right] \qquad (2.72)$$

If we differentiate Eq. (2.72) with respect to the position, x,

$$-\frac{dE_i}{dx} = \frac{k_B T}{N_d(x)} \frac{dN_d(x)}{dx}. \qquad (2.73)$$

Using Eq. (2.70), the electric field can be expressed as a function of the donor concentration.

$$E_x = -\frac{k_B T}{e} \frac{1}{N_d(x)} \frac{dN_d(x)}{dx} \qquad (2.74)$$

2.2.6 Einstein Relations

In Section 2.2.3, we noticed that there is a relationship between the mobility and the diffusion constant for a given type of carriers. That relationship is called the Einstein relation. In this section, we will derive this relationship from a completely different point of view. Let us start with a semiconductor in which the doping concentration is changing slowly.

At thermal equilibrium, the electron current should be 0.

$$J_n = e n \mu_n E_x + e D_n \frac{dn}{dx} = 0 \qquad (2.75)$$

In addition, charge neutrality should nearly hold in this case of gradual doping, even though it may not be perfect.

$$n \cong N_d(x) \qquad (2.76)$$

Plugging Eq. (2.76) into Eq. (2.75), we obtain

$$J_n = e \mu_n E_x N_d(x) + e D_n \frac{dN_d(x)}{dx} = 0. \qquad (2.77)$$

Using the built-in field E_x given in Eq. (2.74), Eq. (2.77) can be rewritten as

$$-e \mu_n N_d(x) \frac{k_B T}{e} \frac{1}{N_d(x)} \frac{dN_d(x)}{dx} + e D_n \frac{dN_d(x)}{dx} = 0. \qquad (2.78)$$

In order for Eq. (2.78) to hold for any doping concentration, $N_d(x)$, the following Einstein relation should be valid:

$$\frac{D_n}{\mu_n} = \frac{k_B T}{e} \qquad (2.79)$$

Using a p-type semiconductor with non-uniform doping, we can also establish the Einstein relationship for holes.

$$\frac{D_p}{\mu_p} = \frac{k_B T}{e} \tag{2.80}$$

Thus, for a given temperature, the electron (hole) diffusion constant of a material is always proportional to its electron (hole) mobility.

2.3 Generation, Recombination, and Continuity

2.3.1 *Generation and Recombination Mechanisms*

In a semiconductor, carriers can be generated by various mechanisms. If a piece of semiconductor is illuminated by light, electrons in the valence band can obtain enough energy from a photon to overcome the energy gap and jump into the conduction band, creating electron–hole pairs. This process is called optical generation (Fig. 2.14(a)). The electron–hole pairs can also be generated by thermal energy. Since the expectation value of thermal energy is much smaller than the energy gap, the thermal generation mechanism is often mediated by traps within the energy gap (Fig. 2.14(b)). In this process, an electron in the valence band jumps to a trap and later jump to the conduction band. Another mechanism that can generate electron–hole pairs is **impact ionization**. This process occurs when an electron obtains high energy from a high electric field so that the electron can provide a valence electron with an energy larger

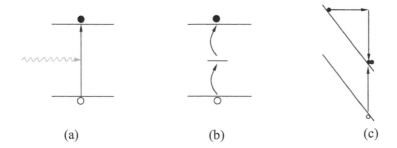

(a)   (b)   (c)

Figure 2.14. Carrier generation mechanisms: (a) optical generation, (b) trap-mediated thermal generation, and (c) generation by impact ionization.

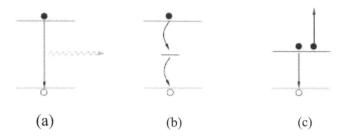

(a) (b) (c)

Figure 2.15. Carrier recombination mechanisms: (a) recombination with optical emission, (b) recombination mediated by a trap, and (c) Auger recombination.

than the energy gap. The valence electron jumps into the conduction band, leaving a hole in the valence band.

Recombination mechanisms are the reverse of generation mechanisms. In a direct-gap material, electrons and holes have similar k values. The electrons and holes have a relatively high chance of possessing a matching momentum so that they can recombine by giving up the extra energy of the electron to the emitted photon (Fig. 2.15(a)). In an indirect-gap material, however, electrons and holes have quite different k values, and direct recombination rarely occurs. Trap-mediated recombination is the dominant mechanism in an indirect-gap material (Fig. 2.15(b)). In this process, an electron in the conduction band is captured by a trap and, later, a hole is captured by the same trap from the valence band. The reverse process of impact ionization is **Auger recombination** (Fig. 2.15(c)). In this case, an electron in the conduction band recombines with a hole in the valence band while giving the extra energy and momentum to another electron in the conduction band. Since this process involves three particles, it occurs only when the carrier concentration is high.

2.3.2 *Excess Carriers and Recombination*

Carriers generated by optical excitation or external injection in excess of the thermal equilibrium values are called **excess carriers**. As defined in Section 2.1.5, the electron and hole concentrations at thermal equilibrium are denoted by n_0 and p_0, respectively. If we add the excess carrier concentration to the carrier concentration at thermal equilibrium, we obtain the carrier concentrations under non-equilibrium condition.

$$n = n_0 + \delta n \tag{2.81a}$$

$$p = p_0 + \delta p \tag{2.81b}$$

As long as the excess carrier concentrations are not zero, the mass action law does not hold.

$$np \neq n_0 p_0 = n_i^2 \tag{2.82}$$

The recombination rate of carriers in the direct recombination process will be proportional to the electron and hole concentrations as given in Eq. (2.83).

$$r = \alpha_r np \tag{2.83}$$

On the other hand, the generation rate of carriers by thermal energy will be constant regardless of the amount of excess carriers, as long as the carrier concentrations are not too high.

$$g = g(T) = \alpha_r n_i^2 \tag{2.84}$$

The net rate of change in electron concentration is the difference between the generation rate and the recombination rate.

$$\frac{dn}{dt} = g - r = \alpha_r (n_i^2 - np) \tag{2.85}$$

Using the mass action law and the fact that n_0 and p_0 are not a function of time and that δn must be the same as δp to maintain charge neutrality, we obtain

$$\frac{d(\delta n)}{dt} = \alpha_r [n_i^2 - (n_0 + \delta n)(p_0 + \delta p)] = -\alpha_r \delta n(n_0 + p_0 + \delta n) \tag{2.86}$$

For a p-type semiconductor under the low-level injection of carriers, the hole concentration, p, should be more or less the same as the hole concentration in thermal equilibrium, p_0, and it should be much larger than the electron concentration $n_0 + \delta n$. Thus, Eq. (2.86) can be simplified to

$$\frac{d(\delta n)}{dt} = -\alpha_r p_0 \delta n. \tag{2.87}$$

If the excess electron concentration at $t = 0$ is Δn, we can obtain the solution to Eq. (2.87) as follows:

$$\delta n = \Delta n \exp(-\alpha_r p_0 t) = \Delta n \exp(-t/\tau_n) \tag{2.88}$$

Here, the lifetime of excess minority carriers (electrons in the case of a p-type semiconductor), τ_n, is defined as

$$\tau_n = 1/(\alpha_r p_0). \tag{2.89}$$

Equation (2.88) states that without sustained optical generation or carrier injection, the excess carriers will disappear by recombination and thermal equilibrium will be restored.

In general, we have to retain both n_0 and p_0 in Eq. (2.86). and the carrier lifetime should be defined as

$$\tau_n = \frac{1}{\alpha_r (n_0 + p_0)}. \tag{2.90}$$

This definition can be used for any type of semiconductor.

2.3.3 Carrier Generation and Quasi-Fermi Levels

Let us consider a steady state maintained by illumination with uniform light. The generation rate should be the sum of the thermal generation rate, $g(T)$, and the optical generation rate, g_{op}. In a steady state, the generation rate should be the same as the recombination rate.

$$g(T) + g_{op} = \alpha_r np = \alpha_r (n_0 + \delta n)(p_0 + \delta p) \tag{2.91}$$

In order to maintain charge neutrality, δn should be the same as δp.

$$g(T) + g_{op} = \alpha_r n_0 p_0 + \alpha_r [(n_0 + p_0)\delta n + \delta n^2] \tag{2.92}$$

As long as the excess carrier concentration is significantly lower than the doping concentration, we can ignore the δn^2 term compared with $(n_0 + p_0)\delta n$. In addition, the thermal generation rate, $g(T)$, will be the same as its equilibrium value (Eq. (2.84)). Then, we can obtain the optical generation rate from Eq. (2.92).

$$g_{op} = \alpha_r (n_0 + p_0)\delta n = \frac{\delta n}{\tau_n} \tag{2.93}$$

The excess carrier concentration is proportional to the optical generation rate. The proportionality constant is the carrier lifetime defined in Eq. (2.90).

$$\delta n = \delta p = g_{op}\tau_n \tag{2.94}$$

Unlike the case of exponential decay described in the previous section, the excess carrier concentration remains constant under uniform illumination. This is because the recombination of excess carriers is completely balanced by the optical generation. The total concentrations of electrons ($n = n_0 + \delta n$) and holes ($p = p_0 + \delta p$) are also constant. Even though thermal equilibrium is destroyed by the light illumination, the carrier concentrations remain constant in this steady state. In terms of carrier concentrations, the steady state is quite similar to thermal equilibrium. Utilizing this property, we can define a **quasi-Fermi level**, which is a steady-state analogue of the Fermi level in thermal equilibrium. That is, the quasi-Fermi level under a steady state plays the role of the Fermi level under thermal equilibrium.

Using the concept of quasi-Fermi level, carrier concentrations in steady state can be expressed in the same form as for thermal equilibrium. One major difference between a quasi-Fermi level and a Fermi level is that the quasi-Fermi level should be defined separately for electrons and holes, while the Fermi level is uniquely defined in a given semiconductor. This difference arises because both the electron and the hole concentrations are increased by the amount of excess carrier concentration, and such an increase requires movements of the quasi-Fermi level away from the Fermi level. If we designate the electron quasi-Fermi level by F_n and the hole quasi-Fermi level by F_p, the electron and hole concentrations can be expressed as follows:

$$n = n_i \exp[(F_n - E_i)/k_B T] \qquad (2.95a)$$
$$p = n_i \exp[(E_i - F_p)/k_B T] \qquad (2.95b)$$

Under illumination, the electron quasi-Fermi level is higher than the Fermi level in thermal equilibrium and the hole quasi-Fermi level is lower than the Fermi level in thermal equilibrium. This fact can be easily deduced from the relationship between the steady-state carrier concentration and the thermal equilibrium carrier concentration ($n > n_0$, $p > p_0$). The difference between the quasi-Fermi levels indicates the deviation from thermal equilibrium.

2.3.4 Continuity Equations

Let us consider a differential volume element shown in Fig. 2.16. In this differential volume element, the rate of increase in hole

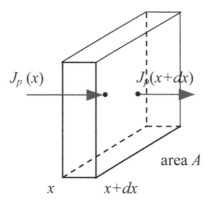

$J_p(x)$ $J_p(x+dx)$

area A

x $x+dx$

Figure 2.16. Differential volume element of cross-sectional area A and thickness dx. $J_p(x)$ and $J_p(x + dx)$ are the incoming and outgoing flux of holes.

concentration is the sum of the flux difference and the net generation rate inside the volume element.

$$\frac{\partial p}{\partial t}Adx = \frac{1}{e}[J_p(x) - J_p(x + dx)]A - \frac{\delta p}{\tau_p}Adx, \qquad (2.96)$$

where $J_p(x)$ and $J_p(x + dx)$ are the incoming and outgoing flux of holes, respectively, A is the cross-sectional area of the volume element, and dx is its thickness. Using the relationship

$$J_p(x) - J_p(x + dx) = (-\partial J_p/\partial x)dx, \qquad (2.97)$$

we can simplify Eq. (2.96):

$$\frac{\partial p}{\partial t} = \frac{\partial \delta p}{\partial t} = -\frac{1}{e}\frac{\partial J_p}{\partial x} - \frac{\delta p}{\tau_p} \qquad (2.98)$$

Thus, we obtain the continuity equation for holes, which is given as

$$\frac{\partial \delta p}{\partial t} = -\frac{1}{e}\frac{\partial J_p}{\partial x} - \frac{\delta p}{\tau_p}. \qquad (2.99)$$

Following similar steps, we can obtain the continuity equation for electrons:

$$\frac{\partial \delta n}{\partial t} = \frac{1}{e}\frac{\partial J_n}{\partial x} - \frac{\delta n}{\tau_n}. \qquad (2.100)$$

So far, we have considered the case where there is thermal generation only. If there are other generation sources such as optical excitation or impact ionization, we have to use the general expression for the net generation rate $(g - r)$ and include the generation rate of the other sources in the generation rate term, g. For this general generation-recombination case, the hole and electron continuity equations should be written as

$$\frac{\partial p}{\partial t} = -\frac{1}{e}\frac{\partial J_p}{\partial x} + g - r \qquad (2.101a)$$

$$\frac{\partial n}{\partial t} = \frac{1}{e}\frac{\partial J_n}{\partial x} + g - r. \qquad (2.101b)$$

2.3.5 Diffusion and Recombination

If there is no electric field in a semiconductor, the current will have a diffusion component only, as in the following equations:

$$J_p = J_{p,diff} = -eD_p\frac{\partial \delta p}{\partial x} \qquad (2.102a)$$

$$J_n = J_{n,diff} = eD_n\frac{\partial \delta n}{\partial x} \qquad (2.102b)$$

Equations (2.99) and (2.100) become

$$\frac{\partial \delta p}{\partial t} = D_p\frac{\partial^2 \delta p}{\partial x^2} - \frac{\delta p}{\tau_p} \qquad (2.103a)$$

$$\frac{\partial \delta n}{\partial t} = D_n\frac{\partial^2 \delta n}{\partial x^2} - \frac{\delta n}{\tau_n}. \qquad (2.103b)$$

In a steady state, there should be no change in carrier concentration. That is, the time derivative of the excess carrier concentration should be zero. We are now left with ordinary differential equations that have spatial derivatives only.

$$\frac{d^2 \delta p}{dx^2} = \frac{\delta p}{D_p\tau_p} = \frac{\delta p}{L_p^2} \qquad (2.104a)$$

$$\frac{d^2 \delta n}{dx^2} = \frac{\delta n}{D_n\tau_n} = \frac{\delta n}{L_n^2}, \qquad (2.104b)$$

where L_p and L_n are the diffusion lengths of holes and electrons, respectively.

$$L_p = \sqrt{D_p \tau_p} \qquad (2.105a)$$

$$L_n = \sqrt{D_n \tau_n} \qquad (2.105b)$$

Let us calculate the steady-state concentration of holes when holes are injected into the right hand side at $x = 0$. The injected holes will experience recombination while diffusing. To solve the differential equation, we need boundary conditions. Often, the carrier concentration at the injection point is given as a boundary condition. Once the hole concentration is given, we can easily obtain the excess hole concentration.

$$\delta p(x = 0) = \Delta p \qquad (2.106)$$

If the length of the semiconductor is infinite, the excess holes should have zero concentration at $x = \infty$, after experiencing enough recombination during their diffusion.

$$\delta p(x = \infty) = 0 \qquad (2.107)$$

With these boundary conditions, we can solve the differential equation (Eq. (2.104a)). The general solution to the differential equation is given as

$$\delta p(x) = C_1 \exp(x/L_p) + C_2 \exp(-x/L_p). \qquad (2.108)$$

Because of the boundary condition at $x = \infty$ (Eq. (2.107)), C_1 should be zero. If C_1 were not zero, the first term would have grown indefinitely as x increases. Once we have the second term left, the boundary condition at $x = 0$ yields $C_2 = \Delta p$ immediately. The solution now takes a very simple form. After satisfying the boundary condition at $x = 0$, it decays exponentially.

$$\delta p(x) = \Delta p \exp(-x/L_p) \qquad (2.109)$$

Figure 2.17 shows the plot of the hole concentration. $p(0)$ is maintained as $p_0 + \Delta p$. As x increases, the hole concentration decreases exponentially and approaches its thermal equilibrium value, p_0. The exponential decrease originates from the recombination of holes during their diffusion.

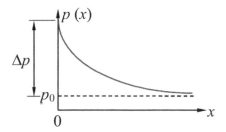

Figure 2.17. Steady-state concentration of holes as a function of distance when holes are injected at $x = 0$.

2.3.6 *Gradients in the Quasi-Fermi Levels*

In Section 2.3.3, we defined quasi-Fermi levels in a steady state in analogy with a Fermi level in thermal equilibrium. Pragmatic readers might have doubted the usefulness of quasi-Fermi levels. If we create a new concept and find out that there is not much use for that concept, it should be dismissed as a waste of effort. In this section, however, we will find that the concept of quasi-Fermi level not only is very useful but also has an important physical meaning.

Let us start with the electron concentration expressed as a function of the quasi-Fermi level (Eq. (2.95a)). Using Eq. (2.95a), the electron current density can be expressed in a simpler form as shown in the following calculation:

$$J_n = en\mu_n \text{E}_x + eD_n \frac{dn}{dx}$$

$$= n\mu_n \left(\frac{dE_i}{dx}\right) + eD_n n \frac{1}{k_B T} \left(\frac{dF_n}{dx} - \frac{dE_i}{dx}\right) \quad (2.110)$$

$$= n\mu_n \frac{dF_n}{dx}$$

For the hole current density, we can obtain a similar expression.

$$J_p = p\mu_p \frac{dF_p}{dx} \quad (2.111)$$

In the above equations, the current density is proportional to the product of the carrier concentration and the slope of quasi-Fermi level. It appears as if the intrinsic energy level in the drift current equation is substituted for by the quasi-Fermi level. If we modify Eqs. (2.110) and (2.111) slightly, the physical meaning of the quasi-Fermi level stands out clearly.

$$J_n = -en\mu_n \frac{d(-F_n/e)}{dx} \qquad (2.112a)$$

$$J_p = -ep\mu_p \frac{d(-F_p/e)}{dx} \qquad (2.112b)$$

Now we can see that the quantities $-F_n/e$ and $-F_p/e$ (quasi-Fermi levels divided by the electronic charge) play exactly the same role as the electrostatic potential, ψ, in the case of drift current. We define these as **quasi-Fermi potentials** in the following equations:

$$\phi_n = -F_n/e \qquad (2.113a)$$

$$\phi_p = -F_p/e \qquad (2.113b)$$

The electron and hole current can be expressed as

$$J_n = -en\mu_n \frac{d\phi_n}{dx} \qquad (2.114a)$$

$$J_p = -ep\mu_p \frac{d\phi_p}{dx}. \qquad (2.114b)$$

The gradient of the quasi-Fermi potential, $-d\phi_n/dx$, plays the role of the electric field in Ohm's law (more accurately, the point form of Ohm's law). That is why these equations (Eq. (2.114a) and Eq. (2.114b)) are called modified Ohm's laws. The quasi-Fermi potential is also called an **electrochemical potential**.

The concept of electrochemical potential is very useful and convenient when we deal with a case where an electric field and a concentration gradient coexist. Instead of considering the drift and diffusion current components separately and adding them together, we can use the electrochemical potential and treat it just like an electric potential in a drift problem. The difference or slope of the electrochemical potential is the single driving force behind currents. The biasing (application of a bias voltage) between two nodes of an electric circuit is indeed the act of enforcing a difference in electrochemical potential between those two nodes. As will be discussed in the following example, a gradient in the electrochemical potential is produced in a uniformly doped semiconductor when we apply a bias voltage across it. When we remove the external bias voltage, thermal equilibrium will be restored, and the electrochemical potential will be constant. Then the quasi-Fermi level will be none other than the Fermi level itself.

Example 2.4: Band diagram of a uniformly doped semiconductor with an external bias voltage

Consider an n-type semiconductor that is uniformly doped. Draw the band diagram and the electron quasi-Fermi level of the semiconductor as a function of the position when a positive voltage is applied to the left-hand side of the sample.

Solution:

The electric field caused by the bias voltage will generate a drift current in the sample. Since the majority carrier concentration is constant, the electron quasi-Fermi potential should have a constant slope. Thus, the band diagram and the electron quasi-Fermi level should have the following form:

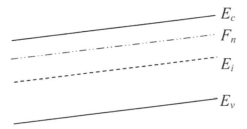

2.4 p-n Junctions

2.4.1 *Junction Formation and Built-In Potential*

When we have an n-type region and a p-type region in contact to each other in a semiconductor, it is called a p-n junction. Such a junction can be formed by introducing opposite type dopants into a pre-doped semiconductor. The concentration of newly introduced dopants should be higher than that of the original dopants, so that the newly introduced dopants compensate for the original dopants and convert the region to the opposite type. The actual doping process during device fabrication is done by solid-state diffusion, or ion implantation followed by annealing.

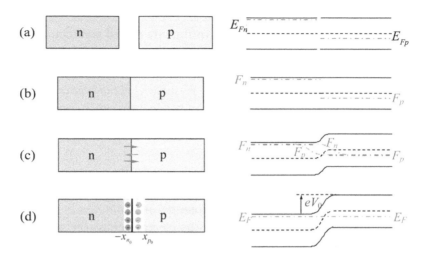

Figure 2.18.   Hypothetical junction formation. If we bring n- and p- doped semiconductors together, there will be a large difference in their quasi-Fermi levels initially. The movement of carriers will generate depletion regions, and the electric field produced by the depletion charge will build up a potential across the depletion region.

Let us consider a hypothetical junction formation method where we form a p-n junction by bringing an n-type semiconductor and a p-type semiconductor together (Fig. 2.18). Of course, this method is not a practical way of forming a p-n junction, but it is just a hypothetical method in a thought experiment to facilitate the understanding of the transient process leading toward thermal equilibrium. We assume that the n-type semiconductor has a higher doping concentration and indicate it as n^+.

When the two semiconductors of opposite types are isolated, Fermi levels can be defined independently. If they are brought together, however, the combined system is no longer at equilibrium, so we cannot define the Fermi level. Only quasi-Fermi levels can be defined. At the moment of contact, the quasi-Fermi levels (of electrons and holes) should maintain the position of the original Fermi levels, since no movement of carriers can occur in zero time. At the metallurgical junction (the boundary between the n^+ and p regions), there is a jump in the quasi-Fermi levels, and their slope becomes infinite. The infinite slope of the quasi-Fermi levels at the boundary will induce strong diffusion currents of electrons and holes. The diffusing electrons (holes) leave the ionized donors (acceptors) in the n-type (p-type) region near the boundary, so that the positive (negative) charge of the ionized donors (acceptors) is exposed. Since the regions of exposed charge (space charge) are formed by the diffusing carriers that have left the region, the carrier concentrations in those regions are significantly lower than that of the neutral region. Thus, the space charge region is often called a **depletion region**. The formation of depletion region is further assisted by the recombination of the diffusing carriers with the majority carriers. Once the depletion regions are formed, the space charges start to build up an electric field and counteract the diffusion of carriers. As the depletion region expands into the n-type and p-type regions near the boundary, the amount of space charges increases, and the electric field also increases. The increase of the electric field will continue until the drift current due to the electric field becomes exactly the same as the diffusion current. At this point, the net current becomes zero, and equilibrium is established.

The process of establishing equilibrium can also be described in terms of the change in the energy band diagram (Fig. 2.18). The existence of an electric field means a change in the electrostatic potential, which should be reflected in the band diagram. The electrostatic potential change can be represented by a potential energy barrier in the band diagram. Because of the potential energy barrier formation, the energy bands in the n-type region are pushed downward, while those in the p-type region are pushed upward. Since the carrier concentration in the neutral region does not change, the quasi-Fermi levels will follow the movement of energy bands, so that the large slope of quasi-Fermi levels near the boundary starts to be reduced. In addition, the difference in the quasi-Fermi levels of the neutral n-type and p-type regions decreases. Eventually, the quasi-Fermi levels will become constant (Fig. 2.18(d)), and a new equilibrium will be established. The potential energy barrier reaches its maximum height (eV_0) at equilibrium, and the corresponding potential (V_0) is called a **built-in potential** since this potential is self-generated within the p-n junction and will stay there as long as equilibrium is maintained.

In the neutral n-type and p-type regions sufficiently far from the depletion region, the values of $(F_n - E_i)$ and $(F_p - E_i)$ are kept the same as those of the isolated case during the transient process toward equilibrium, since the carrier concentrations do not change in these regions. Before the junction is formed (Fig. 2.18(a)), the Fermi levels are different, while the intrinsic level is constant. After the junction is formed and thermal equilibrium is reached, the Fermi level is constant while the intrinsic levels of the neutral n-region and neutral p-region are different. Since the distance between the quasi-Fermi level and the intrinsic level in the neutral region is kept the same during the whole process, the quasi-Fermi level difference in the isolated system should be the same as the intrinsic level difference in the p-n junction under thermal equilibrium. In other words, the Fermi level difference is converted to the built-in potential.

Using the fact that the Fermi level difference in isolated semiconductors is the same as the built-in potential in the p-n junction, we can calculate the built-in potential with given n-type and p-type doping concentrations. From Eq. (2.41), we obtain

$$E_{Fn} = E_i + k_B T \ln \frac{n_0}{n_i} = E_i + k_B T \ln \frac{N_d}{n_i} \qquad (2.115a)$$

$$E_{Fp} = E_i - k_B T \ln \frac{p_0}{n_i} = E_i - k_B T \ln \frac{N_a}{n_i}. \qquad (2.115b)$$

Thus, the built-in potential is

$$V_0 = \frac{E_{Fn} - E_{Fp}}{e} = \frac{k_B T}{e} \ln \frac{N_a N_d}{n_i^2}. \qquad (2.116)$$

Example 2.5: Another method of calculating the built-in potential

Using the fact that no current flows in equilibrium, calculate the built-in potential.

Solution:

Both the electron and hole currents should be zero.

$$J_n = e D_n \frac{dn}{dx} - e n \mu_n \frac{d\psi}{dx} = 0$$

$$J_p = -e D_p \frac{dp}{dx} - e p \mu_p \frac{d\psi}{dx} = 0$$

Combining these two equations and using the Einstein relationship, we obtain

$$\frac{D_n}{\mu_n} \frac{dn}{n} = -\frac{D_p}{\mu_p} \frac{dp}{p} = d\psi.$$

Integrating from the neutral p-region to the neutral n-region, we obtain

$$V_0 = \psi(n) - \psi(p)$$
$$= \frac{k_B T}{e} \ln \frac{n_n}{n_p} = \frac{k_B T}{e} \ln \frac{N_a N_d}{n_i^2}.$$

2.4.2 *Space Charge Region: Depletion Approximation*

In order to calculate the width of the space charge region, we often use the depletion approximation. In this approximation, we assume that there are no carriers in the space charge region and that the transition from the space charge region to the neutral region is abrupt. As shown in Fig. 2.19(a), the charge density has a box shape and can be expressed as

$$\rho(x) = \begin{cases} eN_d & (-x_{n0} < x < 0) \\ -eN_a & (0 < x < x_{p0}) \end{cases}. \tag{2.117}$$

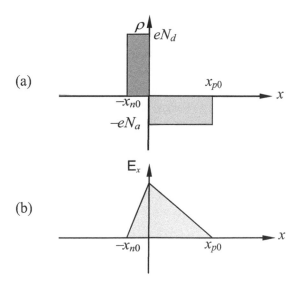

Figure 2.19. (a) Charge density and (b) electric field distribution under the depletion approximation.

The electric field in the x-direction can be calculated from Gauss' law:

$$\frac{dE_x}{dx} = \frac{\rho}{\varepsilon_s} \tag{2.118}$$

By integrating Eq. (2.118) with respect to x, we obtain

$$E_x = \frac{eN_d}{\varepsilon_s}(x + x_{n0}) \quad (-x_{n0} < x < 0) \tag{2.119a}$$

$$E_x = -\frac{eN_a}{\varepsilon_s}(x - x_{p0}) \quad (0 \le x < x_{p0}). \tag{2.119b}$$

As shown in Fig. 2.19(b), the electric field has a triangular shape.

By integrating the electric field once more, we can obtain the electric potential at x.

$$\psi(x) = -\int_{x_{p0}}^{x} E_x dx$$
$$= \begin{cases} \frac{eN_a}{2\varepsilon_s}(x - x_{p0})^2 & (0 \le x < x_{p0}) \\ -\frac{eN_d}{\varepsilon_s}\left(\frac{1}{2}x^2 + x_{n0}x\right) + \frac{eN_a}{2\varepsilon_s}x_{p0}^2 & (-x_{n0} \le x < 0) \end{cases} \tag{2.120}$$

The built-in potential is obtained at $x = -x_{n0}$ in the depletion approximation.

$$V_0 = -\int_{x_{p0}}^{-x_{n0}} E_x dx = \frac{eN_d}{2\varepsilon_s}x_{n0}^2 + \frac{eN_a}{2\varepsilon_s}x_{p0}^2 \tag{2.121}$$

The shape of the electric potential and the value of the built-in potential are shown in Fig. 2.20(a). The spatial change of energy band edges and the intrinsic level can be traced by multiplying the electric potential by the electronic charge, $-e$. The band diagram obtained by such a method is shown in Fig. 2.20(b).

In order to calculate the depletion width, we need to solve two simultaneous equations because there are two unknowns, x_{n0} and x_{p0}. So far, however, we have only one equation (Eq. (2.121)). We can obtain another equation by requiring that the total charge in the space charge region should be zero.

$$eN_d x_{n0} = eN_a x_{p0} \tag{2.122}$$

Then, using Eqs. (2.121) and (2.122), we can obtain

$$x_{n0} = \sqrt{\frac{2\varepsilon_s}{eN_d}\left(\frac{N_a}{N_a + N_d}\right)V_0} \tag{2.123a}$$

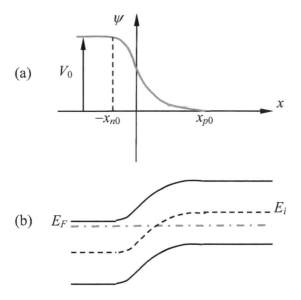

Figure 2.20. (a) Electric potential and (b) band diagram under the depletion approximation

$$x_{p0} = \sqrt{\frac{2\varepsilon_s}{eN_a}\left(\frac{N_d}{N_a + N_d}\right)V_0} \qquad (2.123\text{b})$$

The total depletion width is

$$W = x_{n0} + x_{p0} = \sqrt{\frac{2\varepsilon_s}{e}\left(\frac{1}{N_a} + \frac{1}{N_d}\right)V_0}. \qquad (2.124)$$

For a one-sided junction where $N_d \gg N_a$, the depletion region is confined mostly to the lightly doped region (p-type region in this case).

$$W \cong x_{p0} = \sqrt{\frac{2\varepsilon_s}{eN_a}V_0} \qquad (2.125)$$

When an external bias voltage, V, is applied, the effect of the bias appears almost completely at the depletion region as long as the current is not too large. The voltage drop in the neutral region is small in most cases, since the existence of majority carriers keeps the resistance of the neutral region relatively small. The reduction of carrier concentration in the space charge region creates a large resistance in the depletion region. Since the resistances of the neutral regions and

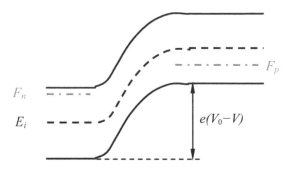

Figure 2.21. Band diagram of a p-n junction under a reverse bias.

the resistance of the depletion region are connected in series, the voltage drop occurs mainly in the depletion region. The bias changes the height of the built-in potential, and the depletion width changes accordingly. The depletion width with an external bias voltage, V, is given as

$$W = \sqrt{\frac{2\varepsilon_s}{e} \left(\frac{1}{N_a} + \frac{1}{N_d} \right) (V_0 - V)} \qquad (2.126)$$

Figure 2.21 shows the band diagram of a p-n junction under a reverse bias (that is, a higher voltage at the end of the n-type region, $V < 0$). The barrier height changes from V_0 to $(V_0 - V)$ and the depletion width increases accordingly. If a forward bias (a higher voltage at the end of the p-type region, $V > 0$) is applied, the barrier height decreases and the depletion width also decreases. One thing we should notice in the case of the forward bias is the reduction in the barrier height when the forward-bias voltage approaches the built-in voltage. If the barrier height is reduced significantly, the current through a p-n junction tends to increase drastically, and the voltage drops in the neutral regions can no longer be ignored. Once the resistance in the neutral regions becomes dominant, the current dependence on the bias voltage becomes linear rather than exponential.

2.4.3 *Junction Rule*

As discussed in the previous subsection, an applied bias voltage appears mostly across the depletion region rather than in the neutral regions as long as the current level is moderate. To calculate the steady-state current as a function of bias voltage, we first need to focus on the behavior of the carriers within and near the depletion region. It is not easy to obtain the carrier concentration directly from the barrier height change caused by the bias voltage, since, as given in Eq. (2.95), the carrier concentration is not only a function of the intrinsic level, E_i, but also a function of the quasi-Fermi level, F_n. The barrier height is the same as the difference in E_i between the depletion edge ($x = -x_{n0}$) of the n-side and the depletion edge ($x = x_{p0}$) of the p-side. The quasi-Fermi level within and near the depletion region cannot be determined as easily as the intrinsic level, however. The only fact we can be sure is that the quasi-Fermi level should change by the amount of the bias voltage from the neutral n-type region to the neutral p-type region, both of which are far away from the depletion region.

Let us consider the quasi-Fermi level of electrons within and near the depletion region. We assume that the p-n junction is forward biased. In the neutral n-type region near the depletion region, the electron quasi-Fermi level should be constant as long as the current level is moderate. Otherwise, the finite dF_n/dx would induce too high a current when it was multiplied by the majority carrier concentration. Now, if the net recombination within the depletion region is negligible, the continuity equations (Eq. (2.101)) imply that the steady-state current of electrons should be constant ($dJ_n/dx = 0$). Thus, $\mu_n n(dF_n/dx)$ at $x = -x_{n0}$ should be equal to $\mu_n n(dF_n/dx)$ at $x = x_{p0}$. Since $n(-x_{n0})$ is much larger than $n(x_{p0})$, and the electron mobilities are comparable at both points, $|dF_n/dx|$ at $x = -x_{n0}$ should be much smaller than $|dF_n/dx|$ at $x = x_{p0}$. Inside the depletion region, $|dF_n/dx|$ should increase exponentially from $-x_{n0}$ to x_{p0}, since the electron concentration decreases exponentially. The exponential increase of $|dF_n/dx|$ makes its value appreciable only when x is close to x_{p0}. In other words, the change of F_n is negligible inside the depletion region except at the depletion edge in the p-region. Once the electrons are injected into the neutral p-region, the electrons will experience recombination with majority carriers while diffusing. Since the electron concentration drops exponentially (Section 2.3.5) while the intrinsic level remains flat in this region, F_n should change linearly, so that $|dF_n/dx|$ would be more or

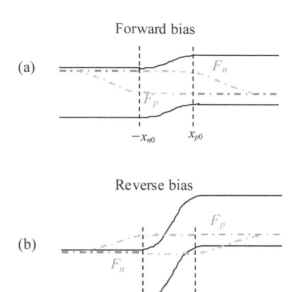

Figure 2.22. Electron and hole quasi-Fermi levels in a p-n junction.

less constant with the same value as $|dF_n/dx|$ at $x = x_{p0}$. When F_n is close to F_p, F_n becomes flat again and stays that way afterward.

Based on the discussion in the previous paragraph, we can conclude that the quasi-Fermi level is almost flat inside the depletion region and changes linearly in the neutral p-region near the depletion region. Such a behavior of the electron quasi-Fermi level is depicted in Fig. 2.22(a). For the hole quasi-Fermi level, F_p, a similar conclusion can be drawn and F_p is also shown in Fig. 2.22(a). If we consider a reverse-biased p-n junction, all the facts we have found about the quasi-Fermi levels in the previous paragraph still hold. Thus, the quasi-Fermi levels are almost flat inside the depletion region and change linearly in the neutral regions when the corresponding carriers become minority carriers. The quasi-Fermi levels under a reverse bias are shown in Fig. 2.22(b).

In Fig. 2.22, it is clear that the quasi-Fermi level of electrons (holes) retains the value it has in the neutral n-type (p-type) region up to the depletion edge of the p-side (n-side). Inside the depletion region, both F_n and F_p are flat and separated by eV. We can express this fact with the following equation.

$$F_n - F_p = eV \quad \text{for} \quad -x_{n0} \leq x \leq x_{p0} \tag{2.127}$$

In Fig. 2.22(a), we can see that $F_n(x_{p0})$ is much closer to the conduction band edge than $F_p(x_{p0})$. This means that under the forward bias, the minority carrier concentration is increased by orders of magnitude compared with the equilibrium concentration. Since the carrier concentration is exponentially dependent on the difference between the quasi-Fermi level and the intrinsic level, the electron concentration at x_{p0} should be increased by a large exponential factor.

Let us derive a quantitative relationship between the electron concentration, $n(x_{p0})$, and its equilibrium value, n_p. Since, in the p-region, the distance between the hole quasi-Fermi level and the intrinsic level is the same as in the neutral region, which in turn is the same as that of the equilibrium case, the electron concentration at equilibrium can be expressed as

$$n_p = n_i \exp[(F_p - E_i)/k_B T]. \tag{2.128}$$

The electron concentration under the bias is

$$n(x_{p0}) = n_i \exp[(F_n - E_i)/k_B T]. \tag{2.129}$$

Using Eqs. (2.127) and (2.128) in Eq. (2.129), we obtain

$$n(x_{p0}) = n_i \exp[(F_n - F_p + F_p - E_i)/k_B T] = n_p \exp(eV/k_B T). \tag{2.130}$$

Thus, the minority carrier concentration at the depletion edge is an exponential function of the applied bias voltage, V. We can derive a similar equation for holes injected into the n-side.

$$p(-x_{n0}) = p_n \exp(eV/k_B T) \tag{2.131}$$

Even though we have derived the junction rule for a forward-biased junction (the case of carrier injection), the junction rules given in Eqs. (2.130) and (2.131) are valid for reverse-biased junctions as well. Under a reverse bias, however, V becomes negative and the minority carrier concentration is much smaller than its equilibrium value.

2.4.4 *Minority Carrier Distribution in Neutral Regions*

The carriers injected into the neutral region become minority carriers as long as the injection level is moderate. Since the drift current of the minority carriers is negligible compared with that of the majority carriers, we need to consider only the diffusion current of the minority carriers. For the majority carriers, we have to consider both the diffusion and the drift current in the neutral region.

Let us calculate the diffusion current of electrons in the neutral p-region. The electrons will recombine with the holes while diffusing, so that Eq. (2.104b) must be used. From the junction rule and the fact that the excess minority carriers will eventually disappear due to recombination, we can set up the boundary conditions as follows.

$$n(x_p = 0) = n_p \exp(eV/k_B T) \qquad (2.132a)$$

$$n(x_p \to \infty) = n_p \qquad (2.132b)$$

where x_p is defined as $(x - x_{p0})$. Using these boundary conditions and modifying Eq. (2.109) appropriately, we can express the excess electron concentration as

$$\delta n(x_p) = n_p[\exp(eV/k_B T) - 1] \exp(-x_p/L_n). \qquad (2.133)$$

For the diffusion of holes in the neutral n-region, we can calculate the hole concentration using Eq. (2.104a). The boundary conditions are

$$p(x_n = 0) = p_n \exp(eV/k_B T). \qquad (2.134a)$$

$$p(x_n \to \infty) = p_n \qquad (2.134b)$$

where x_n is defined as $(-x - x_{n0})$. Using these boundary conditions and Eq. (2.109), the excess hole concentration is obtained as

$$\delta p(x_n) = p_n[\exp(eV/k_B T) - 1] \exp(-x_n/L_p). \qquad (2.135)$$

The minority carrier concentrations in the neutral regions for a forward bias are shown in Fig. 2.23. The excess carrier concentration decreases exponentially as we move deeply into the neutral region. Since the carrier concentration is changing exponentially, its gradient is also changing exponentially. Thus, the minority carrier flux decreases exponentially as we move away from the depletion region. At a position far enough away from the depletion edge, the minority carrier flux will be negligible. What has happened to the flux of

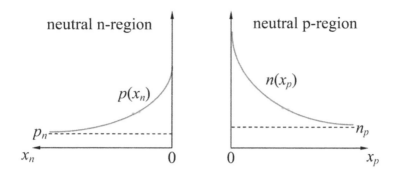

Figure 2.23. Minority carrier concentrations in the neutral regions.

holes? It is due to the recombination of minority carriers with majority carriers during their diffusion. If we calculate the total number of
minority carriers recombining in the neutral region at a given time,
that should be the same as the number of minority carriers injected
into the neutral region at the depletion edge.

Another point to notice is the concentration of majority carriers
in the neutral regions. For charge neutrality to be upheld, the concentration of excess majority carriers should be the same as that of
the minority carriers. As a function of position, the diffusion current
of majority carriers should have the same magnitude as that of minority carriers, while their directions are opposite to each other. The
drift current of majority carriers should be almost constant in the
neutral region, since the excess carrier concentration is negligible
compared with the majority carrier concentration as long as the injection level is moderate.

2.4.5 *Current in Ideal p-n Junctions*

To calculate the current in a p-n junction, let us assume an ideal p-n
junction. In an ideal p-n junction, there is no net recombination in
the depletion region. If there is no net recombination in the depletion region, the electron and hole currents should remain constant
within it. Such preservation of electron and hole current is very useful in simplifying the calculation of current.

Let us calculate the total current. The total current is the sum of
the electron and hole currents at one location. According to the ideal

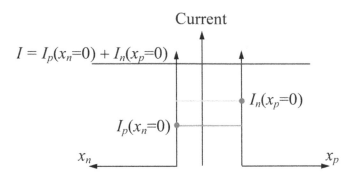

Figure 2.24. Current components in the depletion region of an ideal p-n junction. There is no net recombination in the depletion region, so that the electron and hole currents are constant.

p-n junction assumption, the electron current at the depletion edge of the p-region should be the same as the electron current at the depletion edge of the n-region ($I_n(x_n = 0) = I_n(x_p = 0)$). The same relationship holds for the holes. Now, if we want to calculate the total current at the depletion edge of the n-region, we need simply to add the hole current to the electron current at that location: that is, $I = I_p(x_n = 0) + I_n(x_n = 0) = I_p(x_n = 0) + I_n(x_p = 0)$. In an ideal p-n junction, the total current is simply the sum of the minority carrier currents at the depletion edges. Figure 2.24 shows this relationship clearly. In calculating the current in this way, care must be taken with the current direction. The direction of all currents should be defined consistently and the standard approach is to define the current as positive when it flows from the positively biased electrode to the other electrode.

The minority carrier current at $x_p = 0$ is the electron current given by

$$I_n(x_p = 0) = -e A D_n \frac{dn(x_p)}{dx_p}\Big|_{x_p=0} = -e A D_n \frac{d\delta n(x_p)}{dx_p}\Big|_{x_p=0}. \quad (2.136)$$
$$= \frac{e A D_n n_p}{L_n}\left[\exp(eV/k_B T) - 1\right]$$

The minority carrier current at $x_n = 0$ is the hole current given by

$$I_p(x_n = 0) = \frac{e A D_p p_n}{L_p}\left[\exp(eV/k_B T) - 1\right]. \quad (2.137)$$

Therefore, the total current is

$$I_{ideal} = I_p(x_n = 0) + I_n(x_p = 0)$$

$$= \left(\frac{eAD_p p_n}{L_p} + \frac{eAD_n n_p}{L_n} \right) [\exp(eV/k_BT) - 1] \quad (2.138)$$

$$I_{ideal} = I_0[\exp(eV/k_BT) - 1] \quad (2.139a)$$

$$I_0 = eA \left(\frac{D_p}{L_p} p_n + \frac{D_n}{L_n} n_p \right). \quad (2.139b)$$

Thus, the ideal p-n junction has a simple current vs. voltage relationship and the reverse saturation current, I_0, can be expressed as a function of the equilibrium minority carrier concentrations, p_n and n_p. I_0 is called the reverse saturation current since the current approaches this value asymptotically as the magnitude of the reverse-bias voltage gets large. I_0 depends on the equilibrium minority carrier concentrations, so a low doping concentration is required for a high current level. In a one-sided junction, the doping in the low-concentration region determines the reverse saturation current.

Once the total current in an ideal p-n junction is calculated, we can calculate the current component for each type of carriers. The minority carrier current in the neutral region can be obtained directly from Eqs. (2.133) and (2.135).

$$I_n(x_p) = \frac{eAD_n n_p}{L_n} [\exp(eV/k_BT) - 1] \exp(-x_p/L_n) \quad (2.140a)$$

$$I_p(x_n) = \frac{eAD_p p_n}{L_p} [\exp(eV/k_BT) - 1] \exp(-x_n/L_p) \quad (2.140b)$$

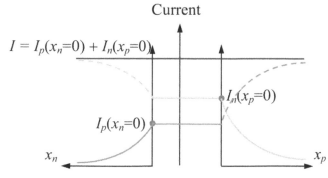

Figure 2.25. Electron and hole current components in an ideal p-n junction.

The majority carrier current can be calculated by subtracting the minority carrier current from the total current. The electron and hole current components are shown in Fig. 2.25. The reduction in the majority current component near the depletion region is caused by the excess majority carrier diffusion. This component is the same as that of the minority carrier diffusion in magnitude, but the direction is opposite because of the opposite charge of the carriers.

2.4.6 Leakage Current and Breakdown in p-n Junctions

In the previous subsection, we considered an ideal p-n junction in which there is no recombination or generation in the depletion region. In real p-n junctions, however, we cannot ignore the recombination-generation in the depletion region, especially when we reverse-bias the p-n junction. In this subsection we will consider the effect of recombination-generation in the depletion region. In order to deal with this phenomenon, we need to go back to the original coordinate system with just one positional variable x instead of x_p and x_n.

In steady state, there is no change in carrier concentrations, so the continuity equation (Eq. (2.101b)) can be rewritten as

$$\frac{1}{e}\frac{\partial J_n}{\partial x} + g - r = 0. \tag{2.141}$$

If we integrate this equation in the depletion region considering the fact that the forward current is defined as the positive current, we obtain

$$I_n(x_{p0}) - I_n(-x_{n0}) = -eA \int_{-x_{n0}}^{x_{p0}} (r - g)dx. \tag{2.142}$$

Now, the total current becomes

$$I = I_p(-x_{n0}) + I_n(-x_{n0}) = I_p(-x_{n0}) + I_n(x_{p0}) + eA \int_{-x_{n0}}^{x_{p0}} (r - g)dx. \tag{2.143}$$

Using Eq. (2.138) for the ideal p-n junction, the total current can be expressed as

$$I = I_{ideal} + I_{r-g}, \tag{2.144}$$

where

$$I_{r-g} = eA \int_{-x_{n0}}^{x_{p0}} (r - g)dx. \tag{2.145}$$

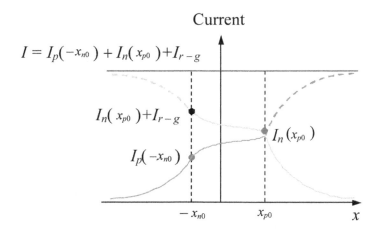

Figure 2.26. Electron and hole current components in a realistic p-n junction.

Thus, in a realistic p-n junction, recombination in the depletion region adds a new current component to the ideal p-n junction current. Figure 2.26 shows the electron and hole current components in a realistic p-n junction.

Let us calculate the recombination-generation current in the depletion region. The generation-recombination in silicon (or any semiconductor material with an indirect bandgap) can be described by Shockley–Read–Hall (SRH) theory [6], [7]. We will not go through the derivation of the net recombination rate here. SRH theory gives us the net recombination rate as

$$r - g = CN_t \frac{np - n_i^2}{n + p + 2n_i \cosh \frac{E_t - E_i}{k_B T}}, \qquad (2.146)$$

where C is a constant, N_t is the trap density, and E_t is the trap energy level.

If a reverse bias is applied to the junction, the electron and hole concentrations, n and p, will be much smaller than their equilibrium values. In the central part of the depletion region, both n and p should be negligible compared with n_i. If we assume that the trap energy level is located near the midgap level ($E_t \approx E_i$), Eq. (2.146) is reduced to

$$r - g \cong -CN_t \frac{n_i}{2}. \qquad (2.147)$$

The negative sign in this equation indicates that there is a net generation. We can obtain the generation current by integrating $eA(r-g)$ with respect to x.

$$I_{r-g} = eA \int (r-g)dx \cong -eACN_t \frac{n_i}{2}W \qquad (2.148)$$

If we define the generation-recombination lifetime, τ_0, as

$$\tau_0 = \frac{1}{CN_t}, \qquad (2.149)$$

we can simplify Eq. (2.148):

$$I_{r-g} = -\frac{eAn_iW}{2\tau_0}. \qquad (2.150)$$

If a forward bias is applied to the junction, there will be excess carriers in the depletion region and recombination current will flow. In the depletion region, the quasi-Fermi levels satisfy the relationship, $F_n - F_p \cong eV$. Using this relationship along with Eqs. (2.95a) and (2.95b), we can express np as a function of the bias voltage.

$$np = n_i^2 \exp[(F_n - F_p)/k_B T] = n_i^2 \exp(eV/k_B T) \text{ (const.)} \quad (2.151)$$

If $np = $ const., $(n + p)$ has its minimum value when $n = p$. For maximum net recombination,

$$n = p = n_i \exp(eV/2k_B T). \qquad (2.152)$$

The recombination current can be obtained by integrating $eA(r - g)$ with respect to x. Even though $(r - g)$ is a function of x, we can estimate the recombination current by approximating it with $(r - g)_{max}$.

$$I_{r-g} \cong eA \int (r-g)_{max}dx = eACN_t \frac{n_i^2[\exp(eV/k_B T) - 1]}{2n_i[\exp(eV/2k_B T) + 1]}W$$

$$= \frac{eAn_iW}{2\tau_0}[\exp(eV/2k_B T) - 1] \qquad (2.153)$$

This equation is reduced to the generation current (Eq. 2.150) for a reverse bias ($V < 0$). Thus, we can use this general form for all bias conditions.

Now, the total current in a real p-n junction is

$$I = I_{ideal} + I_{r-g} = I_s[\exp(eV/k_B T) - 1] + I_{r0}[\exp(eV/2k_B T) - 1], \qquad (2.154)$$

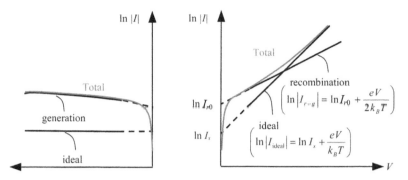

Figure 2.27. Various current components as a function of the applied bias voltage in a realistic p-n junction: (a) under reverse bias and (b) under forward bias

where

$$I_s = eA \left(\frac{L_n n_p}{\tau_n} + \frac{L_p p_n}{\tau_p} \right) \tag{2.155}$$

$$I_{r0} = \frac{eAWn_i}{2\tau_0}. \tag{2.156}$$

In general, $I_{r0} \gg I_s$. Thus, the recombination-generation current is dominant in the low-current region with small forward bias or reverse bias. Since the exponential factor, $\exp(eV/k_BT)$, for the ideal current increases much faster than $\exp(eV/2k_BT)$ for the recombination current, the ideal current catches up with the recombination current at a certain forward bias and becomes dominant. Figure 2.27 illustrates such trends clearly. The $\ln|I|$ vs. V curve in the forward region does not have a constant slope. For a low forward bias, the slope is close to $e/2k_BT$, but it becomes e/k_BT, eventually. We can describe the forward current–voltage characteristic by

$$I = I_{sT}[\exp(eV/nk_BT) - 1], \tag{2.157}$$

where n is an ideality factor $(1 < n < 2)$.

If the reverse-bias voltage increases too much, a p-n junction suffers from breakdown. As shown in Fig. 2.28, there are two breakdown mechanisms in the p-n junction: **avalanche breakdown,** caused by impact ionization, and **Zener breakdown,** caused by tunneling. Avalanche breakdown is initiated and sustained by impact ionization, which is described in Section 2.3.1. When the impact

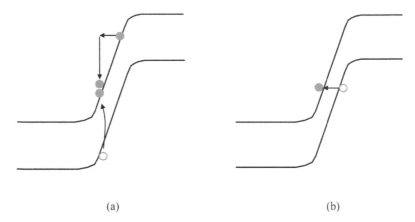

(a)   (b)

Figure 2.28.   Two breakdown mechanisms in a p-n junction: (a) avalanche breakdown (impact ionization) and (b) Zener breakdown (tunneling).

ionization occurs repeatedly, the number of carriers generated increases exponentially, as in the case of avalanche (avalanche multiplication). This type of drastic increase in current is called an avalanche breakdown. A Zener breakdown is caused by tunneling through the triangular energy barrier formed by the energy gap subjected to a strong electric field. The electrons in the valence band tunnels to the conduction band, and the current increases exponentially as the reverse-bias voltage increases.

Both avalanche and Zener breakdown can occur in a reverse-biased p-n junction, but avalanche breakdown is dominant for doping concentration lower than $10^{18}\,\text{cm}^{-3}$, while Zener breakdown is dominant for doping concentration higher than $10^{18}\,\text{cm}^{-3}$. This influence of the doping concentration on which mechanism is dominant stems from the difference in requirements for the depletion width. Avalanche breakdown requires a wide depletion region in order to secure a space for avalanche multiplication, so this mechanism is dominant when the doping concentration is low. Zener breakdown requires a narrow depletion region to maintain a small barrier width for inter-band tunneling.

Since both breakdown mechanisms require a high electric field, we can estimate the breakdown voltage from the maximum electric field in a n^+-p one-sided junction. Figure 2.29 shows the electric field

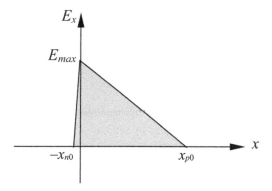

Figure 2.29. Maximum field in n⁺-p one-sided junction.

intensity as a function of the position. In this figure, we can immediately tell that the maximum electric field is

$$E_{max} = \frac{eN_a x_{p0}}{\varepsilon_s}. \qquad (2.158)$$

The depletion width at the point of breakdown is given as

$$x_{p0} \cong \sqrt{\frac{2\varepsilon_s}{eN_a}(V_0 + V_B)}. \qquad (2.159)$$

If the breakdown occurs when $E_{max} = E_{crit}$, we can calculate the breakdown voltage by combining Eqs. (2.158) and (2.159):

$$V_B = \frac{\varepsilon_s E_{crit}^2}{2eN_a} - V_0 \qquad (2.160)$$

Since E_{crit} is a gradually increasing function of doping for both breakdown mechanisms, V_B decreases as the doping concentration increases.

2.4.7 *Junction Tunnel Diodes*

The junction tunnel diode has a relatively long history and is often called the Esaki diode after its inventor [8]. Junction tunnel diodes are p⁺-n⁺ junction diodes in which both sides of the junction are degenerately doped. When they are forward-biased, they can show negative differential resistance. As shown in Fig. 2.30, the energy alignment of carrier source energy in the n⁺ region and the available state energy (hole energy levels) in the p⁺ region results in a

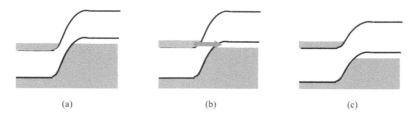

(a)   (b)   (c)

Figure 2.30. Operating principles of a junction tunnel diode: (a) $V = 0$, equilibrium, (b) $V = V_p$, maximum tunneling current due to the full alignment of electron energy levels in the n^+ region and available state levels (hole energy levels) in the p^+ region, and (c) $V = V_v$, no inter-band tunneling due to the alignment of the electron energy levels and the energy gap. Even in case (c), there can be a small trap-assisted tunneling current and a conventional diode current.

peak current at a relatively small forward-bias voltage ($V = V_p$). If the bias voltage increases further, the electron energy levels in the n^+ region are aligned to the energy gap in the p^+ region, so that no inter-band tunneling can occur. Even in this case, there can be a small current caused by trap-assisted tunneling and conventional diode operation (carrier injection over the potential barrier) under a forward bias. With a still higher forward-bias voltage, the current increases exponentially because of the potential barrier lowering.

    Figure 2.31 shows the current vs. voltage characteristics of a junction tunnel diode. The unique characteristic of a junction

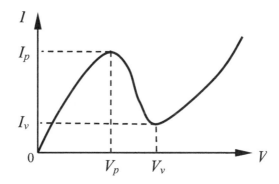

Figure 2.31. Current vs. voltage characteristic of a junction tunnel diode. V_p and V_v are the peak and valley voltage, respectively. I_p and I_v are the peak and valley current, respectively.

tunnel diode is the **negative differential resistance** (NDR) or negative differential conductance (NDC). The NDR is generated when the diode current decreases between the peak and the valley. It is caused by the gradual misalignment of electron energy levels in the n^+ region and available state levels in the p^+ region. The ratio of the peak current to the valley current is called the **peak-to-valley ratio** (PVR). PVR is usually an important figure of merit for an NDR device, since it determines the on-off current ratio in various circuits which include an NDR device.

In the early days of junction tunnel diodes, germanium was used for the substrate. Recently, however, there has been a renewed interest in silicon junction tunnel diodes because of their compatibility with the mainstream CMOS (complementary metal–oxide–semiconductor) technology [9]. One of the issues with silicon junction tunnel diodes is the relatively low peak current caused by the large bandgap of silicon compared with that of germanium. To overcome this issue, Si/SiGe heterojunctions are considered.

2.5 Metal–Semiconductor Contacts and Heterojunctions

2.5.1 *Metal–Semiconductor Contacts*

When we have a metal–semiconductor contact, it forms either a Schottky or an Ohmic junction. A Schottky junction shows a current-rectifying behavior that is similar to a p-n junction. An Ohmic junction shows a resistor-like behavior. These characteristic behaviors are mainly determined by the band alignment between the metal and the semiconductor. Let us consider the Schottky junction first.

Figure 2.32 shows a Schottky junction formed by bringing a metal and a semiconductor together. We assume that the semiconductor is n-type and that the metal has a larger work function than the semiconductor. $e\Phi_m$ is the metal work function, $e\Phi_s$ is the semiconductor work function, $e\chi$ is the electron affinity of the semiconductor, and E_{vac} is the vacuum energy level that is used as a reference energy for all materials. Since the Fermi

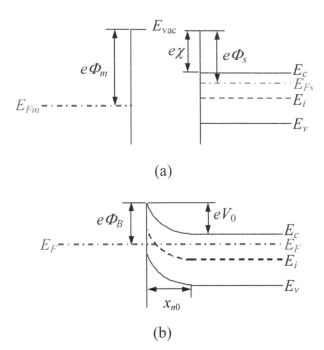

Figure 2.32. Band alignment in a metal–semiconductor system: (a) before contact of materials and (b) after contact of materials.

energy of the semiconductor is higher than that of the metal, the electrons in the silicon are transferred to the metal once the metal is in contact with the n-type semiconductor. Due to the transfer of electrons, a depletion region will be formed in the semiconductor near the metal–semiconductor interface, as shown in Fig. 2.32(b). In the depletion region of the semiconductor, the positive charges of donors are exposed, and the same amount of negative charges will be accumulated in the metal near the metal–semiconductor interface.

As the electron transfer continues, the depletion region in the semiconductor expands and the quasi-Fermi energy of the semiconductor decreases as shown in Fig. 2.32(b). The transfer of electrons stops when the quasi-Fermi energy of the semiconductor becomes the same as the quasi-Fermi energy of the metal. The metal–semiconductor system is now in equilibrium. The built-in potential, V_0, can be expressed as follows:

$$V_0 = |\Phi_m - \Phi_s| \qquad (2.161)$$

Since the width of the accumulation region in the metal is negligible, we can consider the depletion region in the semiconductor only, when we calculate the built-in potential from the charge distribution in the Schottky junction. That is, the Schottky junction can be treated in the same way as a one-sided p-n junction. Using Eq. (2.125), we can obtain the built-in potential after some appropriate change of parameters:

$$V_0 = \frac{eN_d}{2\varepsilon_s} x_{n0}^2 \qquad (2.162)$$

If we apply a bias voltage, V, across the Schottky junction, the potential barrier for the electrons in the semiconductor changes to $(V_0 - V)$ and the depletion width, x_n, is adjusted accordingly. The bias is applied to the metal side of the junction with the bulk semiconductor side as a reference. Thus, the Schottky junction is forward-biased if $V > 0$, and reverse-biased otherwise.

$$V_0 - V = \frac{eN_d}{2\varepsilon_s} x_n^2 \qquad (2.163)$$

The depletion width, W, can be calculated as

$$W = x_n = \sqrt{\frac{2\varepsilon_s}{eN_d}(V_0 - V)}. \qquad (2.164)$$

With the increase of forward bias, the barrier height for the electrons in the n-type semiconductor is reduced, so that more electrons can be injected to the metal. Since the amount of injected electrons will be exponentially dependent on the bias voltage, the current will increase rapidly as the forward bias increases. If $V < 0$, the electrons in the metal has to be injected to the semiconductor, but there is a high energy barrier, $e\Phi_B (= e(\Phi_m - \chi))$, that does not change as a function of the bias voltage. Thus, there will be only a small leakage current with the reverse-bias voltage. This is the origin of rectifying characteristic in the Schottky junction. We can notice that the forward current flows due to the injection of majority carriers from the semiconductor into the metal. The absence of minority carrier injection and the associated delay in the response is a major advantage of Schottky junctions.

Let us calculate the current–voltage characteristics of the Schottky junction [10]. In order to simplify the calculation, we assume that the dominant transport mechanism is **thermionic emission**. That is, all the electrons that have the energy required to overcome the barrier in the semiconductor can be injected into the metal with their velocity, v_x, in the direction of transport. The current density, $J_{s \to m}$, from the semiconductor to the metal can be evaluated by

$$J_{s \to m} = e \int_{E_{min}}^{\infty} v_x dn, \qquad (2.165)$$

where E_{min} is the minimum energy, $e(V_0 - V)$, required for thermionic emission into the metal.

The electron concentration in an incremental volume, $dv_x dv_y dv_z$, in the velocity space is given by

$$dn = g_{\vec{k}}(\vec{v}) f(E) dv_x dv_y dv_z, \qquad (2.166)$$

where $g_{\vec{k}}(\vec{v})$ is the density of states and $f(E)$ is the Fermi–Dirac distribution function. In order to calculate the density of states as a function of the velocity, $\vec{v}$, we need to relate the velocity to the wave vector, $\vec{k}$. For an n-type semiconductor, we can start with the parabolic approximation near the band edge.

$$E - E_c = \frac{\hbar^2 \vec{k}^2}{2m_n^*}. \qquad (2.167)$$

Using the definition of the velocity in Section 1.3.8 (Eq. (1.75)), we can establish the following relationship:

$$\hbar \vec{k} = m_n^* \vec{v} \qquad (2.168)$$

The number of states per unit volume of semiconductor within the volume element, $dk_x dk_y dk_z$, of k-space can be calculated as

$$\frac{2}{L^3} \frac{dk_x dk_y dk_z}{(2\pi/L)^3} = g_{\vec{k}}(\vec{v}) dv_x dv_y dv_z, \qquad (2.169)$$

According to Eq. (2.168), $dk_x = \frac{m_n^*}{\hbar} dv_x$, $dk_y = \frac{m_n^*}{\hbar} dv_y$, and $dk_z = \frac{m_n^*}{\hbar} dv_z$. Thus,

$$g_{\vec{k}}(\vec{v}) = \frac{2}{(2\pi)^3} \left(\frac{m_n^*}{\hbar} \right)^3 = 2 \left(\frac{m_n^*}{h} \right)^3. \qquad (2.170)$$

Using Eq. (2.170) and the Maxwell–Boltzmann approximation of the Fermi–Dirac distribution function, we obtain

$$dn = 2 \left(\frac{m_n^*}{h} \right)^3 \exp \left(-\frac{E - E_F}{k_B T} \right) dv_x dv_y dv_z. \qquad (2.171)$$

Substituting Eq. (2.168) into Eq. (2.167) gives

$$E - E_c = \frac{1}{2} m_n^* (v_x^2 + v_y^2 + v_z^2). \qquad (2.172)$$

Finally, we obtain

$$dn = 2 \left(\frac{m_n^*}{h} \right)^3 \exp \left(-\frac{E_c - E_F}{k_B T} \right)$$
$$\times \exp \left[-\frac{m_n^*}{2k_B T} (v_x^2 + v_y^2 + v_z^2) \right] dv_x dv_y dv_z. \quad (2.173)$$

Eq. (2.165) becomes

$$J_{s \to m} = e \int_{v_{x,min}}^{\infty} \int_{-\infty}^{\infty} \int_{-\infty}^{\infty} 2v_x \left(\frac{m_n^*}{h} \right)^3 \exp \left(-\frac{E_c - E_F}{k_B T} \right)$$
$$\times \exp \left[-\frac{m_n^*}{2k_B T} (v_x^2 + v_y^2 + v_z^2) \right] dv_x dv_y dv_z, \quad (2.174)$$

where $v_{x,min}$ is the minimum speed required for thermal emission in the direction of transport ($\frac{1}{2} m_n^* v_{x,min}^2 = e(V_0 - V)$). By integrating Eq. (2.174), we obtain

$$J_{s \to m} = A^* T^2 \exp(-e\Phi_B / k_B T) \exp(eV / k_B T), \qquad (2.175)$$

where Φ_B is the barrier height ($= (E_c - E_F)/e + V_0$). The total current is the sum of this component and the current by the electron injection from the metal to the semiconductor.

$$J = J_{s \to m} - J_{m \to s} \qquad (2.176)$$
$$J = A^* T^2 \exp(-e\Phi_B / k_B T) [\exp(eV / k_B T) - 1] \quad (2.177)$$

where A^* is the Richardson constant given as

$$A^* = \frac{4\pi e m_n^* k_B^2}{h^3}. \qquad (2.178)$$

In order to make an Ohmic contact to a semiconductor, the work function of a metal should be chosen appropriately. For an n-type semiconductor, the work function of the metal should be smaller than that of the semiconductor, so that before the contact is made,

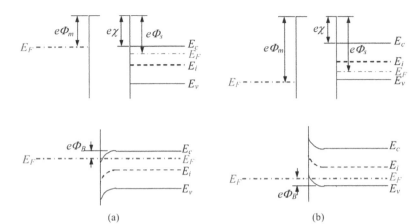

Figure 2.33. Band diagram of a metal–semiconductor system: (a) Ohmic contact with an n-type semiconductor, and (b) Ohmic contact with a p-type semiconductor.

the Fermi level of the metal is higher than that of the semiconductor. As can be seen in Fig. 2.33(a), there is no barrier for the electrons in the semiconductor after the contact is made. The small barrrier for the electrons in the metal does not pose a serious problem, since the potential increases gradually rather than changing abruptly. For a p-type semiconductor, the work function of the metal should be larger than that of the semiconductor, so that before the contact is made, the Fermi level of the metal is lower than that of the semiconductor. After the contact, there is no barrier for the holes in the semiconductor as shown in Fig. 2.33(b).

In reality, however, the formation of Ohmic contacts to a semiconductor is not as simple as the idealized case described in the previous paragraph. For most semiconductors, it is not easy to find a compatible metal with a work function appropriate to the given type. Furthermore, the Fermi level at the interface is often "pinned" by the surface states at the position inside the energy gap. Then, the barrier height is almost independent of the metal work function and is determined mainly by the doping and surface properties of the semiconductor. We can solve the issue of large barrier height by a heavy doping of the region near the semiconductor surface, as shown in Fig. 2.34. The heavy doping reduces the width of the depletion

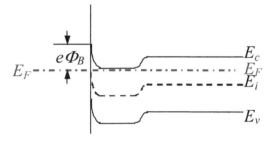

Figure 2.34. Band diagram for an Ohmic junction formed by a heavy doping of the semiconductor surface region.

region and enhances tunneling due to the thin tunneling barrier. The high tunneling current in both direction lowers the resistance of the junction dramatically and facilitates Ohmic contacts.

2.5.2 *Heterojunctions*

A heterojunction is a junction between two different semiconductors. We can take numerous examples of heterojunctions such as Si_xGe_{1-x}–Si, Ge–GaAs, $Al_xGa_{1-x}As$–GaAs, and AlN–GaN. Depending on the bandgap alignment, heterojunctions can be divided into three categories as shown in Fig. 2.35. The type found most frequently is the straddling-gap type (Fig. 2.35(a)). In this type, the conduction band edge of the narrow-gap material is lower than that of the wide-gap material and the valence band edge of the narrow-gap material is higher than that of the wide-gap material. That is, the bandgap of the narrow-gap material is entirely located within the bandgap of the wide-gap material. In staggered-gap heterojunctions

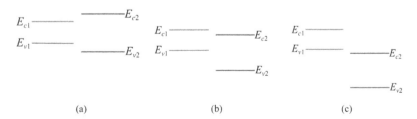

Figure 2.35. Bandgap alignment of heterojunctions: (a) straddling gap, (b) staggered gap, and (c) broken gap.

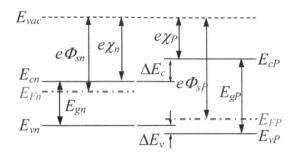

Figure 2.36. Band alignment in a heterojunction system before the contact of materials

(Fig. 2.35(b)), there is an overlap between the bandgaps of the narrow- and wide-gap materials, but a part of the narrow bandgap is outside the wide bandgap. In broken-gap heterojunctions (Fig. 2.35(c)), there is no overlap between the bandgaps of the narrow- and wide-gap materials.

Each material of a heterojunction can have the same (**isotype**) or different (**anisotype**) types of doping. There are two isotype junctions (n-N, p-P), and two anisotype junctions (n-P, N-p). Here, capital letters are used to indicate the doping type of the wide-gap material. Figure 2.36 shows the band alignment of an n-P heterojunction system before the contact of materials.

When the two materials are in contact, the band offsets, ΔE_c and ΔE_v, need to be determined. For clean semiconductor surfaces, it is generally accepted that the rule of electron affinity should be applied to the band offset of heterojunctions. The rule of electron affinity states that the conduction band offset, ΔE_c, is given as the difference in the electron affinity of the two materials. In the case of Fig. 2.36, the band offset is given as

$$\Delta E_c = e(\chi_n - \chi_P). \tag{2.179}$$

Once the conduction band offset is given, the valence band offset is automatically determined, since the bandgaps of the two materials are fixed. The sum of the band offsets is the difference of the bandgaps.

$$\Delta E_c + \Delta E_v = E_{gP} - E_{gn} = \Delta E_g \tag{2.180}$$

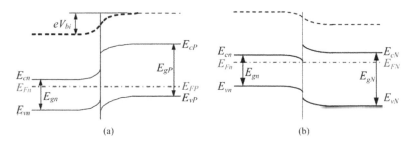

Figure 2.37. Band diagram of heterojunction systems after the contact of materials: (a) n-P heterojunction and (b) n-N heterojunction.

Figure 2.37 shows the band diagrams of n-P and n-N heterojunctions after the contact is made. Due to the charge transfer and formation of depletion (or accumulation) regions, the bands start to bend, while maintaining the band offset at the interface. The charge transfer will stop when an equilibrium is reached. The built-in potential is the same as the difference in the Fermi levels of isolated materials.

Various quantum structures can be formed by utilizing heterojunctions. If a narrow bandgap material is placed between two wide bandgap materials, a quantum well structure can be formed (Fig. 2.38(a)). Both electrons and holes can form bound states in the square potential well with a finite depth. In the quantum well, the carriers can be bound in the direction perpendicular to the interface, but move freely in the in-plane directions. Such a system of

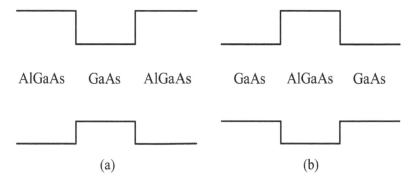

Figure 2.38. Band diagrams of quantum structures based on heterojunctions: (a) quantum well and (b) tunnel barrier.

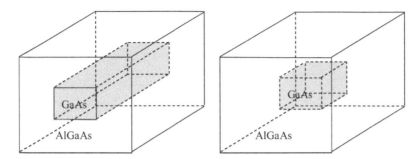

Figure 2.39. Quantum structures based on heterojunctions: (a) quantum wire and (b) quantum dot.

electrons can be called a 2DES. If a wide bandgap material is sandwiched in between two narrow bandgap materials, a tunnel barrier can be formed (Fig. 2.38(b)).

If a narrow bandgap material is surrounded by a wide bandgap material in two perpendicular directions (Fig. 2.39(a)), a quantum wire can be formed. The carriers can be bound in the two directions perpendicular to the wire direction, while moving freely in the wire direction. Such a system of electrons is called 1DES. If a narrow bandgap material is surrounded by a wide bandgap material in all directions (Fig. 2.39(b)), a quantum dot can be formed. The carriers can be bound in the all directions without any degree of freedom. Such a system of electrons is called 0DES.

PROBLEMS

1. **Carrier statistics and dopant statistics:**

   (a) When we calculated the Fermi–Dirac distribution function for carriers, we considered electrons as indistinguishable particles that follow the exclusion principle, but we did not think about their spin carefully. A quantum state with a particular wave vector $\vec{k}$ can accommodate no electron, one electron with its spin up, one electron with its spin down, or two electrons with different spins. Explain that the

Fermi–Dirac distribution is exactly the same, even when we consider all these possibilities.

(b) The quantum state of a dopant (a donor or an acceptor) can accommodate only one electron that has up spin or down spin, since an electron is electrically bound to a dopant ion. Considering this fact, derive the Fermi–Dirac distribution function for donor states.

2. **Hole effective masses in silicon:**
   Silicon has degenerate valence bands near Γ point. They have spherical symmetry near Γ point and effective masses are $m_{lh}^* = 0.16m_e$, $m_{hh}^* = 0.49m_e$. (m_e: free electron mass)

   (a) Calculate the DOS effective mass for holes in Si.
   (b) Calculate the conductivity effective mass for holes in Si.

3. **Electron effective masses in germanium:**
   Germanium has four equivalent ellipsoids at L points with $m_t^* = 0.082m_e$ and $m_l^* = 1.64m_e$.

   (a) Calculate the DOS effective mass for electrons in germanium.
   (b) Calculate the conductivity effective mass for electrons in germanium.

4. **Calculation of Fermi level and the concept of compensation:**
   We doped silicon with 2×10^{15} cm^{-3} phosphorus atoms and then doped it with 1.2×10^{16} cm^{-3} boron atoms.

   (a) At room temperature (300 K), find the position of the Fermi level. The intrinsic carrier concentration is 1×10^{10} cm^{-3} at room temperature and the Fermi level is expressed with respect to the intrinsic level, E_i. Here, we should use the compensation assumption, that is, the net charge of the dopants can be expressed as $e(N_d - N_a)$.
   (b) In this case, what is the concentration of ionized phosphorus at room temperature? What would be the concentration of ionized boron? (We assume that the dopant energy levels are 0.5 eV away from the intrinsic level.)
   (c) Based on the result of (b), verify that the compensation assumption used for the Fermi level calculation is valid.

5. **Alternative derivation of Einstein relationship:**
   Using Eqs. (2.45) and Eq. (2.56), derive Eq. (2.79). (Hint: Use the equipartition law, which states that an equal average energy, $\frac{1}{2} k_B T$, is assigned to each degree of freedom.)

6. **Built-in potential and depletion region in a p-n junction:**

   (a) Even though there is a built-in potential in a p-n junction, that potential cannot be measured by a voltmeter. Explain why.

   (b) In a one-sided junction, most of the depletion region exists in the lightly doped side. Why?

   (c) If we apply a bias voltage across a p-n junction, the applied bias appears only in the depletion region rather than in the neutral regions. Why?

7. **Electron and hole currents in a p-n junction:**

   (a) Assuming that there is no recombination/generation in the depletion region, plot the electron and hole current as a function of position (including the depletion region).

   (b) In the answer to (a), the majority carrier current decreases as we approach the depletion region. Why?

8. **Excess carrier concentrations in the neutral region:**

   (a) Sketch the excess majority and minority carrier concentrations in a neutral region as a function of position when a p-n junction is forward-biased. Explain why this should be the case.

   (b) Based on the result of (a), explain the majority carrier current density in the neutral region.

9. **Minority carrier flux in the neutral region:**

   (a) Calculate the electron current density at $x_p = 0$ and $x_p \rightarrow \infty$ using Eq. (2.140a).

   (b) In the result of (a), why is the minority carrier current density not conserved?

10. **Temperature dependence of the breakdown voltage on the breakdown mechanism:**

    (a) If avalanche multiplication is the cause of breakdown in a p-n junction, would the breakdown voltage increase as temperature increases? Explain the reason.
    (b) If inter-band tunneling is the cause of breakdown in a p-n junction, would the breakdown voltage increase as temperature increases? Explain the reason.

11. **Calculation of the thermionic emission current in a Schottky diode:**
    By carrying out the integration in Eq. (2.174), prove Eq. (2.175).

Bibliography

1. J.S. Blakemore, *Semiconductor Statistics*, Dover Publications, 1987.
2. D.A. Neamen, *Semiconductor Physics and Devices: Basic Principles*; Irwin, 1992.
3. R.S. Muller and T.I. Kamins, *Device Electronics for Integrated Circuits*, Wiley, 1986.
4. B.G. Streetman and S. Banerjee, *Solid State Electronic Devices*, 5th Ed., Prentice-Hall, 2000.
5. Y. Taur and T.H. Ning, *Fundamentals of Modern VLSI Devices*, Cambridge University Press, 1998.
6. R.N. Hall, "Electron–Hole Recombination in Germanium," *Phys. Rev.*, vol. 87, p. 387, 1952.
7. W. Shockley and W.T. Read, "Statistics of the Recombinations of Holes and Electrons," *Phys. Rev.*, vol. 87, p. 835, 1952.
8. L. Esaki, "Discovery of the Tunnel Diode," *IEEE Trans. Elec. Dev.*, vol. ED-23, p. 644, 1976.
9. J. Wang, D. Wheeler, Y. Yan, J. Zhao, S. Howard, and A. Seabaugh, "Silicon Tunnel Diodes Formed by Proximity Rapid Thermal Diffusion, *IEEE Electron. Device. Lett.*, vol. 24, p. 93, 2003.
10. S.M. Sze, *Physics of Semiconductor Devices*, 2nd Ed., John Wiley & Sons, Inc., 1981.

Chapter 3

MOS Structure and CMOS Devices

3.1 MOS Structure

3.1.1 Basic Concepts of MOS Structure

A metal–oxide–semiconductor (MOS) structure is fabricated by growing a silicon dioxide (SiO_2) layer on a silicon surface and then depositing a metal (or poly-silicon) layer on top of that. Together with p-n junctions, MOS structures have a particularly important role in semiconductor devices. A MOS structure is not only a constituent of field effect transistors (FETs) but also a stand-alone device (MOS capacitor) that is an essential tool for characterizing semiconductor and insulator material properties.

Conceptually, a MOS structure is formed by bringing a metal, an oxide and a semiconductor layer together as shown in Fig. 3.1. We assume that the semiconductor is p-type silicon and that the metal has a smaller work function than the silicon. When the three layers are isolated, each one is at equilibrium and their Fermi levels are different. If we bring them together, however, the structure is not at equilibrium and only quasi-Fermi levels are defined. In general, the quasi-Fermi levels are not constant across the structure and the quasi-Fermi level difference induces carrier movement between the metal and the semiconductor. The movement of carriers continues

Nanoelectronic Devices
Byung-Gook Park, Sung Woo Hwang, and Young June Park
Copyright © 2012 Pan Stanford Publishing Pte. Ltd.
www.panstanford.com

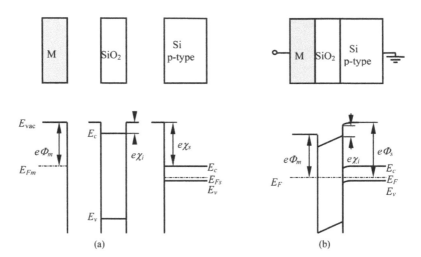

Figure 3.1. Band alignment in a metal–oxide–semiconductor (MOS) system: (a) before contact of materials and (b) after contact of materials.

until a new equilibrium is reached. Once the equilibrium is reached, a Fermi level can be defined as shown in Fig. 3.1(b).

Some readers might wonder about the slowness or perhaps even impossibility of the above process because of the presence of an insulator between the metal and the semiconductor. If there is no path for the carriers between the metal and the semiconductor except for the oxide, the process may be extremely slow indeed. In reality, however, a conduction path can be easily formed between them. Such a path can be formed during the metal deposition process or when the processed wafer is gripped by a pair of tweezers.

Once an equilibrium is established, there will be space charges near the metal–oxide and oxide–semiconductor interfaces. Since the carrier concentration in the metal is extremely high, the width of the space charge region in the metal is negligible. The width of the space charge region in the semiconductor, however, is appreciable in most cases. The finite region of space charge results in band bending near the surface of the semiconductor. If the metal work function, $e\Phi_m$, is smaller than the semiconductor work function, $e\Phi_s$, as in Fig. 3.1(a), the electrons will move from the metal to the semiconductor, leaving positive charges near the metal–oxide interface and negative charges near the oxide–semiconductor interface. The

negative charges in the semiconductor are from the ionized acceptors and they are exposed because of the depletion of holes. The
depletion region is the source of the band bending in the semiconductor. There is also a constant electric field (and a corresponding
potential change) in the oxide because of the positive and negative
charges across it.

If we examine the band diagram of the MOS structure in
Fig. 3.1(b), there exists a large difference between the vacuum level
(E_{vac}) of the metal and that of the neutral semiconductor. Since the
vacuum level corresponds to the potential energy, its change means
a change in the electrostatic potential. In fact, this change in the
electrostatic potential corresponds to the built-in potential in a p-n
junction. The electrostatic potential difference is equal to the Fermi
potential difference between the metal and the semiconductor in the
isolated case (Fig. 3.1(a)). Thus, there is a similarity between a p-n
junction and a MOS structure.

In order to eliminate the band bending and make the semiconductor energy bands flat, we need to apply an appropriate bias voltage across the MOS structure. The required bias voltage is called a
"**flat band**" voltage, and it can be easily obtained from the work functions of the metal and the semiconductor.

$$V_{FB} = \Phi_m - \Phi_s = \Phi_{ms} \tag{3.1}$$

This equation can be interpreted by realizing the fact that the cause
of the band bending is the difference in Fermi levels of the metal and
the semiconductor. If we apply an external bias voltage that can compensate for the difference, we can reverse the carrier movement and
restore the flat bands in the semiconductor and oxide. Throughout
the whole discussion about band bending and flat-band condition,
the metal bands are assumed to be flat since the width of the space
charge region and the amount of band bending within it are both
negligible.

Figure 3.2 shows the band diagram of a MOS structure when the
flat-band voltage is applied to the gate. As expected, all the band
edges and the vacuum levels are flat in this diagram. One point to
notice is the use of the Fermi energy symbols (E_{Fm} and E_{Fs}), even
though the MOS structure is not at equilibrium since we applied the

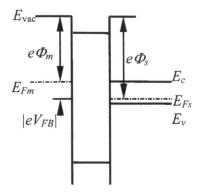

Figure 3.2.   Flat-band condition for a MOS structure.

bias voltage across it. Because of the intervening insulator, however, there is no current flowing through the structure. This makes the metal and the semiconductor behave just like isolated materials in terms of the current flow. Since there is no current, the quasi-Fermi levels should be constant in one material and be the same for electrons and holes. Such a behavior of the quasi-Fermi levels make them identical to the Fermi level, as far as one material is concerned. Thus, we can use the Fermi energy symbol instead of the quasi-Fermi energy symbols (F_n and F_p) without much trouble. In the case of a non-zero current due to a leaky insulator or gated diode structure, however, we must use the quasi-Fermi levels.

Figure 3.3 shows the flat-band voltage (V_{FB}) as a function of doping concentration for various combinations of gate material and semiconductor types. For a given gate material, V_{FB} changes slowly as a function of doping concentration since the Fermi level that determines the work function of the semiconductor is a logarithmic function of doping concentration. In the semi-logarithmic graph of Fig. 3.3, all the plots form straight lines with almost the same absolute value of slope. V_{FB} is affected more strongly by the gate material and semiconductor types. The p^+ poly-silicon (p^+ poly) gate has a V_{FB} that is more than 1.3 V higher than the case for the n^+ polysilicon. At a doping concentration of 10^{16} cm^{-3}, the V_{FB} difference between the n-type silicon (n Si) and p-type silicon (p Si) is about 0.7 V.

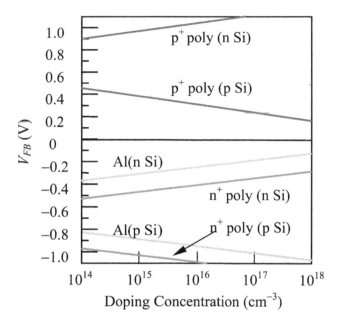

Figure 3.3. Flat-band voltage (V_{FB}) as a function of gate material, semiconductor type, and doping concentration.

In many cases, however, there is some amount of charge within the oxide in a MOS structure. At the interface between the oxide and the semiconductor, interface traps can capture carriers, resulting in the distribution of net charges (Q_{it}) at the interface. There can also be fixed charges (Q_f) in the oxide. Often, the effect of all the oxide charges is represented by an effective oxide charge (Q_{ox}) at the interface per unit area. This effective oxide charge creates a shift (ΔV_{FB}) in the flat-band voltage as shown in Fig. 3.4.

The amount of shift in the flat-band voltage due to the effective oxide charge is given by

$$\Delta V_{FB} = \frac{Q_m}{C_{ox}} = -\frac{Q_{ox}}{C_{ox}}, \tag{3.2}$$

and the flat-band voltage is

$$V_{FB} = \Phi_{ms} + \Delta V_{FB} = \Phi_{ms} - \frac{Q_{ox}}{C_{ox}}. \tag{3.3}$$

Now let us examine the space charge region in a little more detail, but still in a qualitative fashion. We assume that the semiconductor is p-type and that the work function of the gate material is

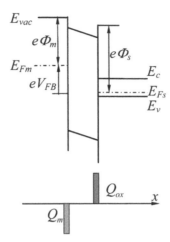

Figure 3.4. Flat-band condition for a MOS structure with an effective oxide charge Q_{ox} at the interface.

smaller than that of the semiconductor. When the gate bias voltage with respect to the neutral bulk region is more negative than the flat-band voltage ($V_G - V_{FB} < 0$), majority carriers (holes) accumulate in the semiconductor near the oxide–semiconductor surface (Fig. 3.5). This phenomenon is called "**accumulation**." The

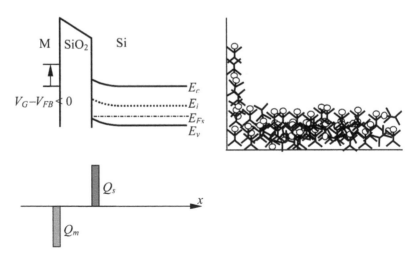

Figure 3.5. Accumulation of carriers. Majority carriers are accumulated at the surface.

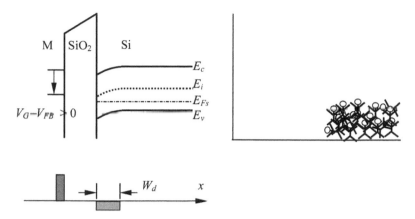

Figure 3.6. Depletion of carriers. Majority carriers are pushed away from the oxide–semiconductor interface.

semiconductor energy bands bend upward as we move from the neutral bulk to the oxide–semiconductor interface. The amount of band bending saturates as the valence band edge (E_v) approaches the Fermi level. To balance the positive charge of accumulated holes in the semiconductor side, electrons are accumulated in the metal near the metal–oxide interface. On the right hand side of Fig. 3.5, we illustrate the concept of accumulation using a drawing of a stick figure crowd. The stick figures with white heads symbolize holes.

When the gate bias voltage is more positive than the flat-band voltage ($V_G - V_{FB} > 0$) but not too high, majority carriers (holes) are pushed away from the oxide–semiconductor interface, leaving a depletion region near the interface (Fig. 3.6). The charge in the semiconductor, Q_s, is now the depletion charge, Q_d, given by

$$Q_s = Q_d = -eN_aW_d, \qquad (3.4)$$

where N_a is the doping concentration and W_d is the depletion width. The surface potential relative to the neutral region (the amount of the band bending in the semiconductor) is given by

$$\psi_s = eN_aW_d^2/2\varepsilon_s \qquad (3.5)$$

where ε_s is the permittivity of the semiconductor. Under the depletion condition, the depletion region contains negative charges from the ionized acceptors and the gate metal has positive charges near

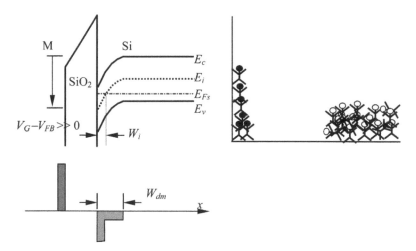

Figure 3.7. Inversion of carriers. Minority carriers become the majority carriers.

the metal–oxide interface. The semiconductor bands bend downward as we approach the oxide–semiconductor interface. In Fig. 3.6, the stick figures show the phenomenon of depletion.

When the gate bias voltage increases further and the intrinsic level, E_i, crosses the Fermi level, E_{Fs}, electrons that used to be minority carriers become majority carriers. Since the originally p-type semiconductor is now inverted to n-type near the oxide–semiconductor interface, we call this phenomenon "**inversion.**" Under the inversion condition, the electron concentration near the interface exceeds the acceptor concentration. The thickness of the inversion layer is W_i as indicated in Fig. 3.7. Since the electron concentration varies exponentially with the distance between E_c and E_{Fs}, the electronic charge in the inversion layer increases rapidly and soon dominates the total charge in the semiconductor. Once the inversion charge becomes the dominant component of the total charge, the depletion width, W_d, starts to saturate. The saturation of the depletion width results from the exponential dependence of the inversion charge on the surface potential. When the gate bias voltage increases, the required charge increase can be provided by increasing the surface potential slightly. According to Eq. (3.5), a slight increase in the surface potential requires only a small change in the depletion width. Thus the change of depletion width becomes negligible and the depletion width reaches its maximum value, W_{dm}. At this point, only the inversion charge responds to the gate bias change, while the depletion width remains more or less the same. In Fig. 3.7, the stick figures illustrate the inversion condition. The stick figures with black heads represent electrons, while those with white heads represent holes. In this cartoon, it is clear that the electrons are spatially separated from the holes in a MOS structure.

The inversion layer plays an important role in a metal–oxide–semiconductor field effect transistor (MOSFET). Since it is used as a current path (channel) between the source and the drain, it influences many critical device parameters of MOSFETs such as the threshold voltage, on-current, and leakage current. Thus, it is worthwhile to investigate its property in more detail here. Before we discuss the inversion phenomenon, we define a quantity ϕ_F (bulk Fermi potential) in the neutral region (bulk semiconductor) as follows:

$$\phi_F = |E_{Fs} - E_i|/e \tag{3.6}$$

Using ϕ_F, we can express the inversion condition for the surface potential. Inversion occurs when $\psi_s > \phi_F$. The inversion can be further divided into two categories, weak and strong inversions. Weak inversion occurs when the surface potential is larger than ϕ_F and smaller than $2\phi_F$ ($\phi_F < \psi_s < 2\phi_F$). With this surface potential, the electron concentration is higher than the hole concentration at the semiconductor surface, so that a surface inversion is achieved. The depletion charge rather than the inversion charge still dominates the space charge because the electron concentration is lower than the acceptor concentration, and the inversion layer thickness is much smaller than the depletion width ($W_i \ll W_d$). As long as electrostatic issues are involved, we can treat the case of weak inversion just like the case of depletion. Figure 3.8 shows the band diagram and the concentrations of dopants and carriers in a MOS structure.

Strong inversion occurs when the surface potential is larger than $2\phi_F$ ($\psi_s > 2\phi_F$) (see Fig. 3.9). At the semiconductor surface, the electron concentration becomes larger than the acceptor concentration. At the beginning of the strong inversion ($\psi_s \approx 2\phi_F$), however, the inversion charge is not the dominant part of the space charge since the inversion layer thickness is much smaller than the depletion width. The inversion charge can dominate the space charge

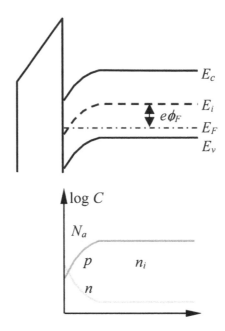

Figure 3.8. Weak inversion. At the surface, the electron concentration is higher than the hole concentration. The depletion charge, however, still dominates the space charge.

only when the electron concentration at the semiconductor surface is much larger than the acceptor concentration. This condition can be obtained by increasing the surface potential significantly higher than $2\phi_F$ (usually several $k_B T$ higher). Once the inversion charge becomes a dominant component of the space charge, the depletion width seldom changes since only a slight change in the depletion width can give rise to a surface potential change that is sufficient for the required change in the inversion charge. The depletion width saturates and reaches its maximum value, W_{dm}. Then, the surface potential reaches its maximum value as well.

Up to now, we have discussed the static aspect of the space charge region: that is, we have dealt with the case of quasi-equilibrium only. When the gate bias voltage changes rapidly, however, we need to consider the dynamic aspects of the space charge region. The accumulation and depletion of majority carriers can occur very rapidly, while the speed of inversion layer formation will depend strongly

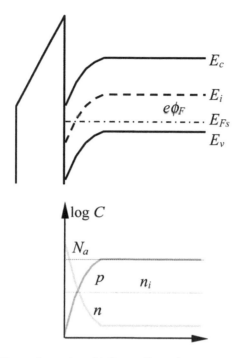

Figure 3.9. Strong inversion. At the surface, electron concentration is so high that the inversion charge is dominating the space charge.

on the availability of the carrier sources. In Example 3.1, we will see that between the majority and minority carrier sources, there is a difference of more than 10 orders of magnitude in the time required to form the same inversion layer.

Example 3.1: Comparison of currents supplied by various mechanisms (see Fig. 3.10)

Let us consider a silicon substrate doped with boron ($N_a = 10^{16}$ cm^{-3}). Assuming that the electric field intensity is 100 V/cm and the generation lifetime (τ_n) is 10 μs, estimate the time required to supply or extract an amount of charge corresponding to the impurity concentration in 10 nm ($Q_s = e \cdot 10^{16} \cdot 10^{-6} = 10^{10} e$) by (a) minority carrier drift current, (b) generation current in the depletion region,

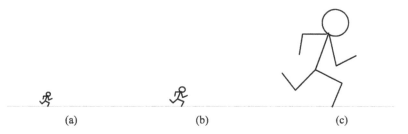

(a)   (b)   (c)

Figure 3.10. Comparison of currents supplied by various mechanisms: (a) minority carrier drift, (b) generation in the depletion region, and (c) majority carrier drift.

and (c) majority carrier drift current (majority carrier concentration $= 10^{16}$ cm^{-3}).

Solution:

For silicon, $n_i \approx 10^{10}$ cm^{-3}. Then, the minority carrier concentration is 10^4 cm^{-3} and the depletion width is about 0.3 μm. From these values, we can estimate the currents as

(a)	(b)	(c)
$J = en\mu_n E$	$J = en_i W_{dm}/\tau_n$	$J = en\mu_n E$
$= e \cdot 10^4 \cdot 10^3 \cdot 10^2$	$= e \cdot 10^{10} \cdot 3 \cdot 10^{-5}/10^{-5}$	$= e \cdot 10^{16} \cdot 10^3 \cdot 10^2$
$= 10^9 e$	$= 3 \cdot 10^{10} e$	$= 10^{21} e$
$\tau = 10$ s	$\tau = 0.3$ s	$\tau = 10^{-11}$ s

3.1.2 *MOS Equations*

So far, we have described the electrical behavior of a MOS structure qualitatively. In order to deal with the behavior of a MOS structure quantitatively, we need to set up a few fundamental equations for electrical variables. If we apply Gauss' law to the two regions shown in Fig. 3.11(a), we obtain

$$-Q_s = \varepsilon_{ox} E_{ox} = \varepsilon_s E_s, \tag{3.7}$$

where Q_s is the total semiconductor charge per unit area, E_{ox} is the oxide electric field, ε_{ox} is the dielectric constant of the oxide, E_s is the electric field at the semiconductor surface, and ε_s is the dielectric constant of semiconductor.

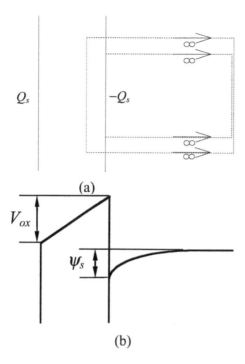

(a)

(b)

Figure 3.11. Application of Gauss' law and the definitions of the oxide voltage and the semiconductor surface potential.

If we consider the voltage distribution in the MOS structure, we obtain

$$V_G - V_{FB} = V_{ox} + \psi_s, \tag{3.8}$$

where V_G is the gate voltage, V_{FB} is the flat-band voltage, V_{ox} is the voltage across the oxide, and ψ_s is the semiconductor surface potential. We can relate V_{ox} to Q_s, using Eq. (3.7), as

$$V_{ox} = t_{ox} \mathsf{E}_{ox} = t_{ox} \frac{\varepsilon_s}{\varepsilon_{ox}} \mathsf{E}_s = -\frac{t_{ox}}{\varepsilon_{ox}} Q_s = -\frac{Q_s}{C_{ox}}, \tag{3.9}$$

where t_{ox} is the oxide thickness and C_{ox} is the oxide capacitance per unit area. Combining Eqs. (3.8) and (3.9), we obtain

$$V_G - V_{FB} = V_{ox} + \psi_s = -\frac{Q_s}{C_{ox}} + \psi_s. \tag{3.10}$$

Now, if we can express Q_s as a function of ψ_s, we will have a complete relationship between the gate bias voltage, V_G and the surface potential, ψ_s. For uniform doping with concentration N_a, Q_s can be exactly calculated as a function of ψ_s. The following subsection will be devoted to such an analysis. If the amount of inversion charge is negligible compared with the depletion charge, we can use the depletion approximation.

$$Q_s \cong Q_d = -\sqrt{2\varepsilon_s e N_a \psi_s} \qquad (3.11)$$

This approximation is valid for the case of depletion and weak inversion, but it can also be used at the beginning of a strong inversion ($\psi_s \approx 2\phi_F$) since the inversion charge is not the dominant part of the space charge yet. Substituting Eq. (3.11) into Eq. (3.10), the relationship between V_G and ψ_s can be established.

$$V_G - V_{FB} = \frac{\sqrt{2\varepsilon_s e N_a}}{C_{ox}}\sqrt{\psi_s} + \psi_s \qquad (3.12)$$

3.1.3 *Analysis of the Space Charge Region*

If the substrate doping of the semiconductor is uniform, we can obtain an exact analytical expression of Q_s as a function of ψ_s. Let us consider a p-type substrate uniformly doped with concentration N_a. The Poisson equation can be written as

$$\frac{d^2\psi}{dx^2} = -\frac{e}{\varepsilon_s}(p - n - N_a) \qquad (3.13)$$

where the hole and electron concentrations are given by

$$p = p_0 \exp(-e\psi/k_B T) \qquad (3.14a)$$

$$n = n_0 \exp(e\psi/k_B T). \qquad (3.14b)$$

Figure 3.12 shows the definitions of the various parameters in a semiconductor band diagram.

Now, let us consider a location far inside the bulk material ($x \rightarrow \infty$). Here ψ is zero, and its derivative with respect to x is also zero since the electric field is zero.

$$\psi(x \rightarrow \infty) = \left.\frac{d\psi}{dx}\right|_{x\rightarrow\infty} = 0 \qquad (3.15)$$

The carrier concentrations retain their thermal equilibrium value for $x \rightarrow \infty$.

$$p_0 = n_i \exp(e\phi_F/k_B T) \qquad (3.16a)$$

$$n_0 = n_i \exp(-e\phi_F/k_B T). \qquad (3.16b)$$

Charge neutrality in this region requires

$$N_a = p_0 - n_0. \qquad (3.17)$$

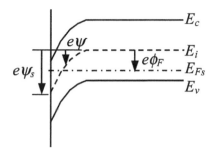

Figure 3.12. Semiconductor band diagram and the definitions of semiconductor potential, surface potential, and Fermi potential.

Using this relationship, we can rewrite Eq. (3.13) as

$$\frac{d^2\psi}{dx^2} = -\frac{e}{\varepsilon_s}\{p_0[\exp(-e\psi/k_BT)-1]-n_0[\exp(e\psi/k_BT)-1]\}.$$

$$(3.18)$$

Since $p_0 \gg n_0$ in a p-type semiconductor, we can approximate p_0 with N_a. Using the mass action law, we obtain

$$\frac{d^2\psi}{dx^2} = -\frac{e}{\varepsilon_s}\left\{N_a[\exp(-e\psi/k_BT)-1]-\frac{n_i^2}{N_a}[\exp(e\psi/k_BT)-1]\right\}.$$

$$(3.19)$$

In order to solve this differential equation, we first multiply both sides of this equation by $\frac{d\psi}{dx}$.

$$\left(\frac{d\psi}{dx}\right)\frac{d^2\psi}{dx^2} = -\frac{e}{\varepsilon_s}\left\{N_a[\exp(-e\psi/k_BT)-1]\right.$$
$$\left.-\frac{n_i^2}{N_a}[\exp(e\psi/k_BT)-1]\right\}\left(\frac{d\psi}{dx}\right) \quad (3.20)$$

Since $\frac{d}{dx}\left(\frac{d\psi}{dx}\right)^2 = 2\left(\frac{d\psi}{dx}\right)\frac{d^2\psi}{dx^2}$,

$$\frac{1}{2}\frac{d}{dx}\left(\frac{d\psi}{dx}\right)^2 = -\frac{e}{\varepsilon_s}\left\{N_a[\exp(-e\psi/k_BT)-1]\right.$$
$$\left.-\frac{n_i^2}{N_a}[\exp(e\psi/k_BT)-1]\right\}\left(\frac{d\psi}{dx}\right). \quad (3.21)$$

Integrating both sides of Eq. (3.21) from x to infinity, we obtain

$$\frac{1}{2}\left(\frac{d\psi}{dx}\right)^2 = \frac{k_BTN_a}{\varepsilon_s}\left\{-[\exp(-e\psi/k_BT)+\frac{e\psi}{k_BT}]\right.$$
$$\left.-\frac{n_i^2}{N_a^2}[\exp(e\psi/k_BT)-\frac{e\psi}{k_BT}]\right\}\Big|_{\psi}^{0}. \quad (3.22)$$

In this integration, the variable of integration on the right hand side changes from dx to $d\psi$. The upper limit changes from ∞ to 0, since $\psi(x \to \infty) = 0$. The lower limit changes from x to ψ. Finally, we obtain the square of the derivative of ψ as a function of ψ.

$$\left(\frac{d\psi}{dx}\right)^2 = \frac{2k_B T N_a}{\varepsilon_s}\left[[\exp(-e\psi/k_B T) + \frac{e\psi}{k_B T} - 1]\right.$$
$$\left. + \frac{n_i^2}{N_a^2}[\exp(e\psi/k_B T) - \frac{e\psi}{k_B T} - 1]\right] \tag{3.23}$$

Since $E_x = -\frac{d\psi}{dx}$,

$$E_x^2 = \frac{2k_B T N_a}{\varepsilon_s}\left\{[\exp(-e\psi/k_B T) + \frac{e\psi}{k_B T} - 1]\right.$$
$$\left. + \frac{n_i^2}{N_a^2}[\exp(e\psi/k_B T) - \frac{e\psi}{k_B T} - 1]\right\} \tag{3.24}$$

If we put $x = 0$ into Eq. (3.24), we can obtain the surface electric field E_s.

$$E_s = \sqrt{\frac{2k_B T N_a}{\varepsilon_s}}\left\{[\exp(-e\psi_s/k_B T) + \frac{e\psi_s}{k_B T} - 1]\right.$$
$$\left. + \frac{n_i^2}{N_a^2}[\exp(e\psi_s/k_B T) - \frac{e\psi_s}{k_B T} - 1]\right\}^{1/2} \tag{3.25}$$

Using Eq. (3.7), we can obtain the semiconductor space charge per unit area as follows:

$$Q_s = -\sqrt{2\varepsilon_s k_B T N_a}\left\{[\exp(-e\psi_s/k_B T) + \frac{e\psi_s}{k_B T} - 1]\right.$$
$$\left. + \frac{n_i^2}{N_a^2}[\exp(e\psi_s/k_B T) - \frac{e\psi_s}{k_B T} - 1]\right\}^{1/2} \tag{3.26}$$

Figure 3.13 shows the semiconductor charge, Q_s. as a function of the surface potential, ψ_s, when $N_a = 1 \times 10^{16}$ cm^{-3}. We can notice that there are three distinctive conditions in this graph: (1) the accumulation condition ($\psi_s < 0$) where the hole charge dominates, (2) the depletion or weak inversion condition ($0 < \psi_s < 2\phi_F$) where the ionized acceptor charge dominates, and (3) the strong inversion condition ($\psi_s > 2\phi_F$) where the electron charge dominates.

Under the accumulation condition ($\psi_s < 0$), the surface hole concentration is larger than the acceptor concentration ($p_s > N_a$).

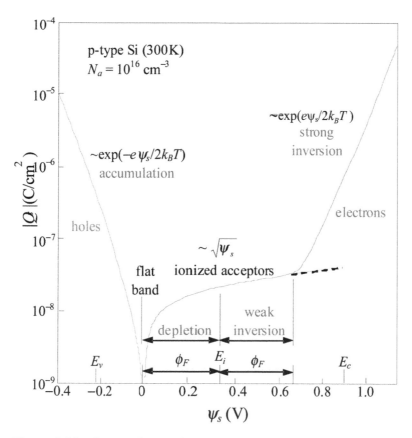

Figure 3.13. Semiconductor charge per unit area as a function of the surface potential. See also Color Insert.

Since the hole concentration is exponentially dependent on the surface potential, the accumulation charge is also exponentially dependent on the surface potential. Thus, Q_s appears as a straight line in the semi-logarithmic plot of Fig. 3.13. Under the depletion or weak inversion condition, the surface concentration of both carriers (electrons and holes) is smaller than the acceptor concentration ($n_s < N_a$ and $p_s < N_a$). The surface carrier concentrations are exponentially dependent on the surface potential, but the ionized acceptor charge dominates the space charge since the carrier concentrations are much smaller than the acceptor concentration. Even though the apparent space charge does not depend on the carrier concentrations, there is a clear distinction between the depletion and the weak inversion condition when we pay attention to the carrier concentrations. Under the depletion condition ($0 < \psi_s < \phi_F$), the electron concentration is smaller than the hole concentration ($n_s < p_s$). Under the weak inversion condition ($\phi_F < \psi_s < 2\phi_F$), the electron concentration is larger than the hole concentration. Because of the exponential dependence, the electron concentration grows rapidly as the surface potential increases. When ψ_s is close to $2\phi_F$, n_s becomes comparable to N_a. Even when $\psi_s = 2\phi_F$, the inversion charge Q_n is still negligible compared with the depletion charge Q_d, since the inversion layer thickness, W_i, is much smaller than the depletion width, W_d. Under the strong inversion condition, the surface electron concentration is larger than the acceptor concentration ($n_s > N_a$). The inversion charge is exponentially dependent on the surface potential, so that Q_s appears as a straight line in the semilogarithmic plot of Fig. 3.13, once Q_n dominates Q_s.

Based on the above observation, we can derive a simple but fairly good approximation to the semiconductor space charge in each region. For the case of accumulation ($\psi_s < 0$), majority carriers form the accumulation charge, and $\exp(-e\psi_s/k_B T)$ is the dominant term in Eq. 3.26. Q_s can be approximated as

$$Q_s = \sqrt{2\varepsilon_s k_B T N_a}\, \exp(-e\psi_s/2k_B T). \qquad (3.27)$$

In the case of depletion and weak inversion ($0 < \psi_s < 2\phi_F$), $\frac{e\psi_s}{k_B T}$ becomes a dominant term, and we can approximate the space charge as

$$Q_s = -\sqrt{2\varepsilon_s e N_a \psi_s}. \qquad (3.28)$$

If we compare this with Eq. (3.11), we find that they are identical. Under this condition, the space charge is almost completely provided by the ionized acceptors in the depletion region. Let us calculate the amount of inversion charge under this condition. We have to retain the term $\frac{n_i^2}{N_a^2}\exp(e\psi_s/k_B T)$ in Eq. (3.26).

$$Q_s = -\sqrt{2\varepsilon_s e N_a \psi_s}\left[1 + \frac{k_B T n_i^2}{e\psi_s N_a^2}\exp(e\psi_s/k_B T)\right]^{1/2} \qquad (3.29)$$

The inversion charge is

$$Q_n = Q_s - Q_d \cong -\sqrt{2\varepsilon_s e N_a \psi_s}\,\frac{k_B T n_i^2}{2e\psi_s N_a^2}\exp(e\psi_s/k_B T). \qquad (3.30)$$

This result will be useful when we calculate the MOSFET current before it is turned on.

In strong inversion, $\frac{n_i^2}{N_a^2}\exp(e\psi_s/k_BT)$ becomes a dominant term. The space charge is given as

$$Q_s = Q_n = -\sqrt{\frac{2\varepsilon_s k_B T n_i^2}{N_a}}\,\exp(e\psi_s/2k_BT). \qquad (3.31)$$

Using Eq. (3.14b) and the relationship $n_0 = \frac{n_i^2}{N_a}$, we can express the surface electron concentration n_s as

$$n_s = \frac{n_i^2}{N_a}\exp(e\psi_s/k_BT). \qquad (3.32)$$

Then, we can relate the inversion charge to the surface electron concentration.

$$|Q_n| = \sqrt{2\varepsilon_s k_B T n_s} \qquad (3.33)$$

3.1.4 *Strong Inversion*

When a MOS structure is deep in strong inversion, the inversion charge Q_n dominates the total semiconductor charge Q_s, as shown in Fig. 3.13. Since Q_n is an exponential function of ψ_s, a small increase in ψ_s brings a large increase in Q_s which is the sum of the inversion charge Q_n and the depletion charge Q_d. Thus, according to Eq. (3.10), a small increase in ψ_s results in a dramatic increase in V_G.

In reality, we cannot directly control ψ_s in a MOS structure, since it is an internal variable defined at the semiconductor surface that we cannot reach from outside. We can control ψ_s only indirectly through the influence of the electric field generated by the gate bias voltage. If we increase the gate bias voltage V_G linearly, ψ_s will increase as a logarithmic function of V_G. As V_G increases, the rate of increase in ψ_s will become smaller, and we can assume that ψ_s is saturated. ψ_s should stay near $2\phi_F$, and W_d would be more or less fixed at W_{dm}.

$$W_{dm} = \sqrt{\frac{4\varepsilon_s \phi_F}{eN_a}} \qquad (3.34)$$

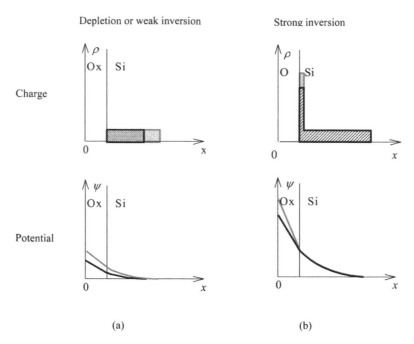

Figure 3.14. Comparison between (a) weak inversion (or depletion) and (b) strong inversion.

Once the depletion width saturates, the amount of depletion charge should also saturate.

$$Q_{dm} = -eN_aW_{dm} = -\sqrt{4\varepsilon_seN_a\phi_F} \qquad (3.35)$$

Because of the saturation of depletion charge, the total charge in the semiconductor, Q_s, changes as a function of the inversion charge under the strong inversion condition.

$$Q_s = Q_n + Q_{dm} = Q_n - \sqrt{4\varepsilon_seN_a\phi_F} \qquad (3.36)$$

In Fig. 3.14, the changes in charge distribution and potential are shown for two cases: (a) depletion or weak inversion, and (b) strong inversion. In depletion or weak inversion, the depletion width changes as a function of the gate bias voltage according to Eq. (3.10). Since the change occurs at the edge of the depletion region, it introduces changes in the electric field throughout the entire depletion region, which, in turn, induces a significant change in ψ_s. For a given change in the gate bias voltage, V_G, the change in ψ_s accounts for a non-negligible portion, as shown in Fig. 3.14 (a). In the case of strong inversion, the depletion width changes little and most of the change occurs in the inversion layer that is located at the semiconductor surface.

3.1.5 Threshold Voltage

The onset of strong inversion has an important function in a MOS structure. Once the strong inversion is started, additional gate bias voltage can be used to increase the inversion charge since the depletion charge is saturated. In a MOS field effect transistor (MOSFET), the inversion layer is used as a channel (current path) between the source and the drain. Thus the starting point of strong inversion can be regarded as the boundary between ON and OFF state of a MOS-FET. The gate bias voltage at which the strong inversion starts is defined as the threshold voltage. We have already established that the surface potential should be $2\phi_F$ at the beginning of the strong inversion. In order to obtain the threshold voltage, we just need to evaluate the gate bias voltage corresponding to the surface potential of $2\phi_F$.

Before we calculate the threshold voltage, we would like to confirm the fact that, even at $\psi_s = 2\phi_F$, the inversion charge is negligible compared with the depletion charge. In Eq. (3.26), the two dominant terms out of the six in the square bracket are $\frac{e\psi_s}{k_B T}$ and $\frac{n_i^2}{N_a^2} \exp(e\psi_s/k_B T)$. At $\psi_s = 2\phi_F$, $\frac{n_i^2}{N_a^2} \exp(e\psi_s/k_B T) = 1$. Since $\frac{e\psi_s}{k_B T} \gg 1$ for doping concentration larger than 10^{15}cm^{-3} at room temperature, the inversion charge should be much smaller than the depletion charge. Thus, we can calculate the semiconductor space charge Q_s as

$$Q_s = Q_{dm}. \tag{3.37}$$

The voltage drop in the oxide is

$$V_{ox} = -\frac{Q_s}{C_{ox}} = -\frac{Q_{dm}}{C_{ox}}. \tag{3.38}$$

Now, we can express the threshold voltage as a function of device parameters.

$$V_T = V_{FB} + \frac{\sqrt{4\varepsilon_s e N_a \phi_F}}{C_{ox}} + 2\phi_F \tag{3.39}$$

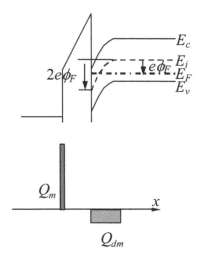

Figure 3.15. Band diagram and charge density at the threshold voltage.

Figure 3.15 shows the band diagram and charge density at the threshold voltage. For the semiconductor space charge, only the depletion charge is shown, ignoring the inversion charge.

Example 3.2: Calculation of the threshold voltage of a MOS structure

Calculate the threshold voltage of a MOS structure with an n^+ poly-silicon gate, 50 nm oxide, and 10^{16} cm^{-3} p-type doped silicon substrate. Assume that there is no oxide charge and use Fig. 3.3 to estimate the flat-band voltage.

Solution:

In Fig. 3.3, the flat-band voltage is about -1.1 V. From Eq. (3.16a), we can calculate the bulk Fermi potential as

$$\phi_F = \frac{k_B T}{e} \ln \frac{p_0}{n_i} = 0.026 \times \ln \frac{10^{16}}{1.5 \times 10^{10}} = 0.348 \text{ (V)}.$$

The oxide capacitance per unit area is calculated to be

$$C_{ox} = \frac{\varepsilon_{ox}}{t_{ox}} = \frac{3.9 \times 8.85 \times 10^{-14}}{5 \times 10^{-6}} = 6.90 \times 10^{-8} \text{ (F / cm}^2\text{)}.$$

Thus, the threshold voltage can be calculated as

$$V_T = -1.1 + \frac{\sqrt{4 \times 11.9 \times 8.85 \times 10^{-14} \times 1.6 \times 10^{-19} \times 10^{16} \times 0.348}}{6.90 \times 10^{-8}}$$
$$+2 \times 0.348 = 0.298 (\text{V}).$$

3.1.6 *Capacitance vs. Voltage (C−V) Characteristics*

A MOS structure is a two-terminal device that is basically a capacitor. Hence, we call it a MOS capacitor. Since the applied voltage is divided into two parts (the oxide voltage drop, V_{ox}, and the semiconductor potential drop, ψ_s) in a MOS capacitor, we can treat the MOS capacitor as an oxide capacitor and a semiconductor capacitor connected in series (Fig. 3.16).

The capacitance of a MOS capacitor can be expressed in terms of the two capacitances, C_{ox} and C_s, as follows:

$$\frac{1}{C} = \frac{1}{C_{ox}} + \frac{1}{C_s} \tag{3.40}$$

When we change the bias voltage, the oxide capacitance, C_{ox}, remains the same, since it is determined by the thickness and permittivity of oxide. If the oxide thickness is t_{ox} and the permittivity is ε_{ox}, the oxide capacitance per unit area (C_{ox}) is given by

$$C_{ox} = \frac{\varepsilon_{ox}}{t_{ox}}. \tag{3.41}$$

The semiconductor capacitance C_s, however, changes as a function of the bias voltage, since the space charge is a nonlinear function of the surface potential. Thus, the capacitance, C, of a MOS capacitor changes as a function of the bias voltage. In other words, MOS capacitors are variable capacitors, and it is important to know their capacitance vs. voltage (C–V) characteristics.

Figure 3.16. Equivalent circuit for a MOS capacitor.

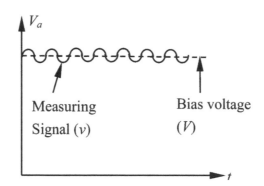

Figure 3.17. Voltage applied to a MOS capacitor during $(C-V)$ measurements.

Figure 3.17 shows the voltage, V_a, applied to a MOS capacitor during $(C-V)$ measurements. It is the superposition of a DC bias voltage, V, and a sufficiently small sinusoidal measuring signal, v.

$$V_a = V + v \qquad (3.42)$$

The bias voltage creates a condition under which the capacitance can be measured. If we apply the bias voltage alone, all the electrical parameters such as band bending and charge distribution take their quasi-equilibrium (static) values as described in the previous subsections. If we add the measuring signal to the bias voltage, however, the quasi-equilibrium is disturbed and the capacitance at the given bias voltage can be extracted from the measured signal current vs. voltage relationship.

Because of the non-negligible delay in carrier responses, there can be two different types of $(C-V)$ curves, depending on the measuring frequency. When the measuring frequency is low (a few Hz or lower), we obtain low-frequency (or quasi-static) $C-V$ curves. At low frequency, even the minority carriers can respond to the signal in time, so that the inversion charge follows the signal properly. Under the strong inversion condition, the capacitance becomes similar to the oxide capacitance, since the change in the semiconductor charge occurs near the oxide–semiconductor interface. However, when the measuring frequency is high (typically 1MHz), we obtain high-frequency $(C-V)$ curves. At high frequency, the minority carriers cannot respond to the signal in time. The inversion charge cannot follow the signal, but the majority carriers at the depletion edge

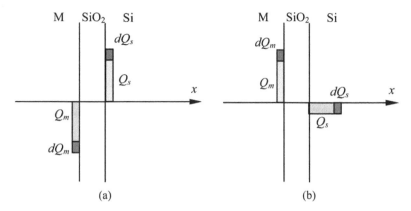

Figure 3.18. Charge distribution in a MOS capacitor during (*C–V*) measurements: (a) under the accumulation condition and (b) under the depletion or weak inversion condition.

can. Under the strong inversion condition, the capacitance remains at its minimum value, since the change in the semiconductor charge occurs at the depletion edge.

Let us examine the MOS (*C–V*) curves under three different conditions: (1) accumulation, (2) depletion or weak inversion, and (3) strong inversion. In accumulation, Q_s is an exponential function of ψ_s. C_s is much larger than C_{ox}, so that the total capacitance per unit area is almost the same as C_{ox}. Figure 3.18(a) shows the distribution of charge in a MOS structure with a p-type semiconductor under the accumulation condition. Holes are accumulated at the semiconductor surface and electrons are accumulated at the metal surface. In addition to the static charge induced by the bias voltage, the figure also shows the incremental change of semiconductor charge (dQ_s) due to the measuring signal. Since the incremental change of semiconductor charge occurs at the surface of the semiconductor, the capacitance is almost the same as the oxide capacitance. In depletion or weak inversion, the incremental change of semiconductor charge occurs at the depletion edge. The semiconductor capacitance, C_s, becomes the depletion capacitance, C_d. As the bias voltage increases, the depletion width increases and C_s decreases. Thus, the total capacitance, C, decreases according to Eq. (3.40). Looking at Fig. 3.18(b), we can see why the total capacitance

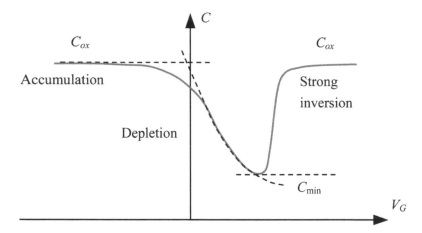

Figure 3.19. Low-frequency $(C-V)$ characteristic for a MOS capacitor with a p-type substrate.

decreases as the bias voltage increases. As the bias voltage increases, the distance increases between the locations where the incremental changes of charge occur. It is as if the distance between two electrodes in a parallel-plate capacitor has been increased. The capacitance is bound to decrease in such a case.

Under the inversion condition, the $(C-V)$ characteristics show a strong dependence on the measuring frequency. At low frequencies, the capacitance is restored to C_{ox} after strong inversion as shown in Fig. 3.19. Since the inversion charge can respond to the signal, Q_s is an exponential function of ψ_s. C_s is much larger than C_{ox}, so the total capacitance per unit area is almost the same as C_{ox}.

Figure 3.20 shows the distribution of charge in the MOS structure under the inversion condition. The incremental change in the semiconductor charge (dQ_s) due to the measuring signal occurs at the inversion layer. It is clear that the total capacitance should be close to the oxide capacitance.

The low-frequency $(C-V)$ characteristic is often called a quasi-static C-V characteristic, since it is obtained at the zero-frequency limit. That is, the low-frequency $(C-V)$ characteristic is an ideal $(C-V)$ characteristic that can be realized when the incremental change in the semiconductor charge obeys the static relationship of Q_s to ψ_s in Eq. (3.26). We can derive an analytical expression for the

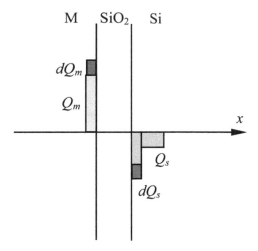

Figure 3.20. Charge distribution in a MOS capacitor during low-frequency (*C–V*) measurement under the inversion condition.

low-frequency semiconductor capacitance as a function of ψ_s by differentiating Eq. (3.26) with respect to ψ_s.

$$C_s = -\frac{dQ_s}{d\psi_s} = \sqrt{\frac{\varepsilon_s e^2 N_a}{2k_B T}}$$

$$\times \frac{[1 - \exp(-e\psi_s/k_B T)] + \frac{n_i^2}{N_a^2}[\exp(e\psi_s/k_B T) - 1]}{\left\{[\exp(-e\psi_s/k_B T) + \frac{e\psi_s}{k_B T} - 1] + \frac{n_i^2}{N_a^2}[\exp(e\psi_s/k_B T) - \frac{e\psi_s}{k_B T} - 1)\right\}^{1/2}}$$

$$(3.43)$$

This equation is valid for all ψ_s except 0. At $\psi_s = 0$ (flat-band condition), the denominator becomes zero, so that C_s cannot be determined by Eq. (3.43). We can make C_s continuous at $\psi_s = 0$ by defining $C_s(0)$ as

$$C_s(0) = \lim_{\psi_s \to 0} C_s = \sqrt{\frac{\varepsilon_s e^2}{k_B T}\left(N_a + \frac{n_i^2}{N_a}\right)}$$

$$= \sqrt{\frac{\varepsilon_s e^2}{k_B T}(N_a + n_0)} = \sqrt{\frac{\varepsilon_s e^2 p_0}{k_B T}} \qquad (3.44)$$

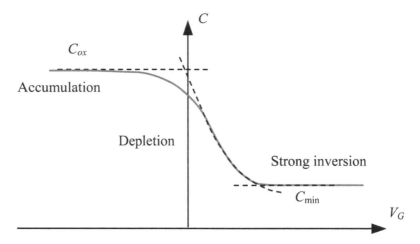

Figure 3.21. High frequency (*C–V*) characteristic for a MOS capacitor with a p-type substrate.

Since $\psi_s = 0$ is the flat-band condition, we can define the flat-band capacitance C_{FB} using Eq. (3.40).

$$C_{FB} = \frac{C_{ox}C_s(0)}{C_{ox} + C_s(0)} \qquad (3.45)$$

At high frequencies, the capacitance stays at C_{min} after the onset of strong inversion, as shown in Fig. 3.21. In a MOS capacitor, minority carriers are supplied mainly by thermal generation in the depletion region. For high-frequency signals, the minority carrier supply from generation-recombination cannot follow the signal. The variation (modulation) must be accommodated by the majority carriers, and it occurs at the depletion edge. Since the depletion width is more or less fixed at its maximum value, W_{dm}, under the strong inversion condition, the depletion capacitance remains at its minimum value, C_{dm}. Thus, the total capacitance reaches its minimum value, C_{min}, which is given as

$$C_{min} = \frac{C_{ox}C_{dm}}{C_{ox} + C_{dm}}, \qquad (3.46)$$

where

$$C_{dm} = \varepsilon_s/W_{dm} = \sqrt{\varepsilon_s e N_a/(4\phi_F)}. \qquad (3.47)$$

Figure 3.22 shows the distribution of charge in the MOS structure under the inversion condition. The incremental change in the

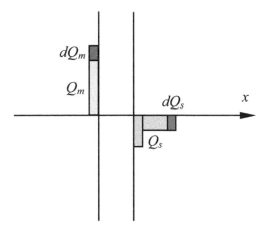

Figure 3.22. Charge distribution in a MOS capacitor during high-frequency (C–V) measurements under the inversion condition.

semiconductor charge (dQ_s) due to the measuring signal occurs at the depletion edge. Note that the inversion layer is clearly formed by the bias voltage, keeping the depletion width almost the same. Incremental change cannot occur in the inversion layer, however, since the required amount of inversion charge cannot be provided by the thermal generation or minority carrier drift. Noticing that dQ_s maintains its maximum distance from dQ_m, we can conclude that the total capacitance should have its minimum value (C_{min}) under the inversion condition.

3.1.7 Quantum Effect on the MOS (C–V) Characteristics

Up to now, we have been ignoring the quantum effects in a MOS structure. As the electric field at the oxide–semiconductor interface increases, however, carriers are strongly confined in a potential well and the quantum effect emerges more conspicuously in the inversion (or accumulation) layer. The confinement of carriers increases the spacing between the energy levels in the direction of confinement (the direction perpendicular to the oxide–semiconductor interface). Combined with the degrees of freedom in the other two dimensions, subbands are formed in the inversion layer. At the same time, the wave functions reveal the effect of the stronger

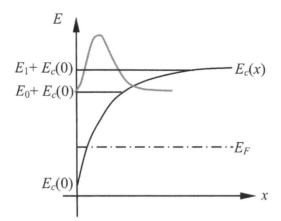

Figure 3.23. Quantized energy levels and the ground-state wave function in the inversion layer of a MOS capacitor.

confinement. As shown in Fig. 3.23, the ground state wave function has a peak below the oxide–silicon interface and becomes nearly zero at the interface. It also decays exponentially towards the bulk semiconductor, once the conduction band minimum $(E_c(x))$ becomes larger than the ground state energy $(E_0 + E_c(0))$. The wave functions with a higher energy level show similar characteristics.

The subband formation and the confinement of wave functions introduce two important changes in the MOS $(C-V)$ characteristics. One is a shift of the threshold voltage. Since E_0 is positive, for strong inversion the bands should bend more in the quantum mechanical case than in the classical case. That is, the surface potential, ψ_s, should be larger when we consider the subband formation. The larger ψ_s will require a larger depletion width, which, in turn, increases the gate bias voltage required for the same amount of inversion charge. Thus, the threshold voltage increases when the quantum effect is included.

Another change in MOS $(C-V)$ characteristics is the reduction of the capacitance in the cases of accumulation and strong inversion. The capacitance is reduced because the average position of electrons in each subband is significantly farther from the oxide–semiconductor interface than that of the classical case. Figure 3.24 shows the normalized electron concentration as a function of the

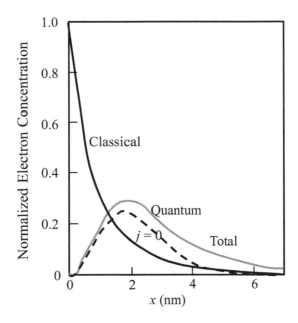

Figure 3.24. Normalized electron concentration as a function of the position in a MOS capacitor.

position, x, starting from the oxide–semiconductor interface. In silicon, there is $1 \sim 1.5$ nm difference between the average electron positions in the classical and quantum mechanical calculations. If we convert it into the equivalent oxide thickness, it becomes $0.3 \sim 0.5$ nm. Such a difference is non-negligible compared with the oxide thickness ($1 \sim 1.5$ nm) of current MOSFET devices.

Figure 3.25 shows the calculated low-frequency MOS (C–V) curves with and without quantum mechanical considerations. The semiconductor is silicon, and the oxide thickness is assumed to be 3 nm. The substrate is p-type, and the doping concentration is 3×10^{17} cm^{-3}. As expected, we can clearly see that the threshold voltage is increased and the accumulation and inversion capacitances are decreased.

In order to calculate the low-frequency capacitance in the case of strong inversion, we need to know the semiconductor charge. For subbands in the inversion layer, we can use the two-dimensional density of states (DOS) for each subband and the Fermi–Dirac distribution. As derived in Section 1.4.2, the two-dimensional DOS for

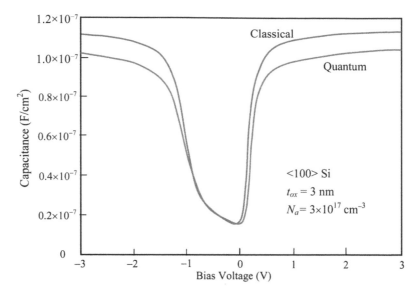

Figure 3.25. Quantum effect on the current–voltage (*C–V*) characteristics of a MOS capacitor. In this simulation, the oxide thickness is 3 nm, and the substrate doping is 3×10^{17} cm^{-3}.

the j-th subband is given by

$$g_j(E) = \frac{d_j m_j}{\pi \hbar^2} \Theta(E - E_j - E_c(0)), \qquad (3.48)$$

where d_j is the degeneracy and m_j is the effective mass of the subband, and $\Theta(E)$ is a step function defined as

$$\Theta(E) = \begin{cases} 1, & \text{for} \quad E \geq 0 \\ 0, & \text{for} \quad E < 0 \end{cases}. \qquad (3.49)$$

The total density of states is the sum of all subband DOS values.

$$g(E) = \sum_{j=0}^{\infty} \frac{d_j m_j}{\pi \hbar^2} \Theta(E - E_j - E_c(0)) \qquad (3.50)$$

After multiplying the Fermi–Dirac distribution function and integrating with respect to energy, we can calculate the surface density of electrons, n_s.

$$n_s = \int_{E_c(0)}^{\infty} g(E)f(E)dE$$

$$= \sum_{j=0}^{\infty} \frac{d_j m_j}{\pi \hbar^2} \int_{E_j+E_c(0)}^{\infty} \frac{1}{1 + \exp[(E - E_F)/k_B T]} dE \quad (3.51)$$

$$= \frac{k_B T}{\pi \hbar^2} \sum_{j=0}^{\infty} d_j m_j \ln\{1 + \exp[(E_F - E_j - E_c(0))/k_B T]\}$$

Since the inversion charge is $-en_s$, we obtain finally the quantum mechanical value of the inversion charge.

$$Q_n^{QM} = -\frac{ek_B T}{\pi \hbar^2} \sum_j d_j m_j \ln\{1 + \exp[(E_F - E_j - E_c(0))/k_B T]\}$$

$$(3.52)$$

Now we would like to express this equation as a function of ψ_s. $E_c(0)$ is related to the bulk value $E_c(\infty)$ as follows:

$$E_c(0) = -e\psi_s + E_c(\infty) \quad (3.53)$$

In the bulk semiconductor, the Fermi level and the conduction band have the following relationship.

$$E_c(\infty) = E_g/2 + \phi_F + E_F \quad (3.54)$$

Combining Eqs. (3.52), (3.53), and (3.54), we obtain

$$Q_n^{QM} = -\frac{ek_B T}{\pi \hbar^2} \sum_j d_j m_j \ln\{1 + \exp[(e\psi_s - E_g/2 - \phi_F - E_j)/k_B T]\}$$

$$(3.55)$$

By differentiating this equation with respect to ψ_s and taking the absolute value, we can obtain the capacitance.

$$C_s = \frac{e^2}{\pi \hbar^2} \sum_j \frac{d_j m_j}{1 + \exp[-(e\psi_s - E_g/2 - \phi_F - E_j)/k_B T]} \quad (3.56)$$

3.2 MOSFET and Its Operation

3.2.1 Concept of MOSFET

Using a MOS structure, we can build a three-terminal device called a metal–oxide–semiconductor field effect transistor (MOSFET). The

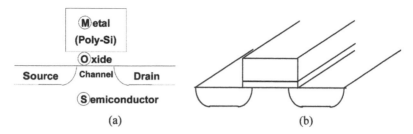

Figure 3.26. MOSFET structure: (a) cross-section and (b) three-dimensional structure.

structure of a MOSFET is depicted in Fig. 3.26. We can easily recognize the MOS structure in Fig. 3.26(a), especially if we scan our focus from the top to the bottom of the figure. In addition to the MOS structure, source and drain regions are introduced into the semiconductor. The source and drain are heavily doped with dopants of the type opposite to those in the body, so that they act as a source and sink of the inversion charge. The inversion layer that connects the source and drain is called a channel, since it forms the current path between the source and drain.

The operation of a MOSFET is based on the operation of a MOS structure. The gate bias voltage controls the surface potential of the semiconductor, so that we can accumulate or deplete the majority carriers and invert the surface by bringing minority carriers toward it. The existence of a source in the MOSFET makes a dramatic difference in the speed of inversion layer formation. Since the source is the reservoir of carriers required for inversion layer formation, it can provide the carriers immediately. The drain acts as a sink for the carriers in the inversion layer. There is a reverse bias between the channel region and the drain, so that the carriers in the inversion layer fall down to the drain whenever they come close to the drain end of the channel.

In order to visualize the operation of a MOSFET, NANOCAD [5] simulation is carried out for a few gate bias conditions while the drain bias is kept constant. Since the MOSFET has an inherently two-dimensional device structure (MOS in one direction and source-channel-drain in the other), we need to use a contour plot or a three-dimensional plot. Figure 3.27 shows the results of the device

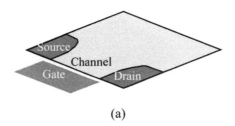

(a)

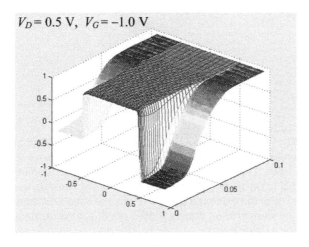

(b)

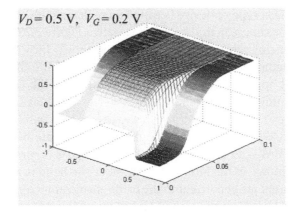

(c)

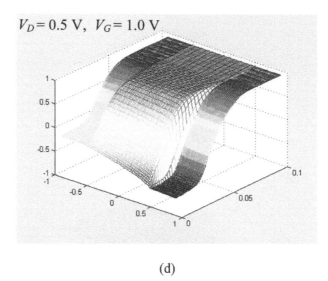

$V_D = 0.5$ V, $V_G = 1.0$ V

(d)

Figure 3.27. Simulated MOSFET band diagrams for various bias conditions: (a) cross-section of the MOSFET shown as it appears in a three-
dimensional plot, (b) conduction band minimum as a function of location
at $V_G = -1$ V, (c) at $V_G = 0.2$ V, and (d) at $V_G = 1$ V. The gate length of the
MOSFET is 1 μm, the oxide thickness is 2 nm, the substrate doping is 10^{18}
cm^{-3}, and the source/drain doping is 10^{20} cm^{-3}. The substrate material is
silicon. See also Color Insert.

simulation for a MOSFET with n$^+$ source/drain and p-type silicon
substrate. This type of device is called an n-channel MOSFET, since
the inversion charge in the channel is negative. Figure 3.27(a) shows
the locations of gate, source, drain, and channel. Such a diagram
can be obtained by rotating the cross section of the MOSFET in
Fig. 3.26(a) appropriately. Figures 3.27(b–d) show the conduction
band minimum as a function of location. The gate bias is changed
from -1 V to 1 V, while the drain bias voltage is kept at 0.5 V.

Figure 3.27(b) shows the band diagram for a gate bias voltage
(V_G) of -1 V. We can see that the gate bias voltage is very close to the
flat-band voltage of the MOS structure, since the band is almost flat
in the direction perpendicular to the oxide–semiconductor interface.
The energy barrier at the source end of the channel is so high that
almost no current is flowing through the channel. The energy barrier
at the drain end of the channel is 0.5 V higher than its counterpart

at the source end, so that any carrier injection from the drain to the channel is even less feasible. Obviously, the MOSFET is in its cutoff mode.

When the gate of the MOSFET is biased at 0.2 V, which is close to but still somewhat lower than the threshold voltage, the barrier at the source end of the channel is lowered significantly, and a non-negligible number of carriers can be injected from the source into the channel (Fig. 3.27(c)). The injected carriers can diffuse through the channel and reach the drain end of the channel. At the drain end, the carriers are easily drifted down to the drain. The current level, however, is still low compared with that of the fully turned-on device. The MOSFET is still in its **cutoff** mode and only the **subthreshold** current is flowing.

When the MOSFET is fully turned on (strong inversion) as shown in Fig. 3.27(d), there is almost no barrier at the source end of the channel. The carriers can be injected into the channel without much hindrance. The current is not determined by the injection, but, instead, mainly by the transport within the channel. Since strong inversion is now achieved throughout the channel, the drain affects the surface potential of the channel significantly and builds up a large lateral electric field along the channel. The lateral field, in turn, enhances the drift current in the channel, and the drain current increases rapidly as a function of the drain bias voltage.

The drain current does not increase indefinitely as we increase the drain bias, however. The reason can be found in Fig. 3.28. Here we show the band diagrams of the same MOSFET with different gate and drain biases. The gate bias is fixed at 0.8 V and only the drain bias is changed from 0.4 V to 1 V. The MOSFET is fully turned on, since the threshold voltage of the MOS structure is about 0.3 V and the gate bias is 0.5 V higher than the threshold voltage. When we compare the band diagrams in Fig. 3.28(a) and Fig 3.28(b), we can see that the surface potential profiles of the two figures almost coincide throughout the channel. The only difference occurs at the boundary between the channel and the drain. When we change the drain bias voltage by 0.6 V, there is negligible potential change throughout the entire channel and the additional voltage drop occurs only at the very end of the channel. Due to this negligible potential change in

$V_G = 0.8$ V, $V_D = 0.4$ V

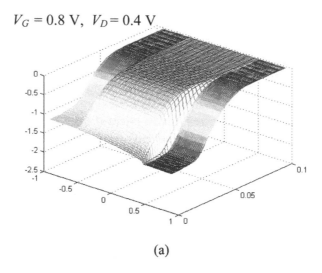

(a)

$V_G = 0.8$ V, $V_D = 1.0$ V

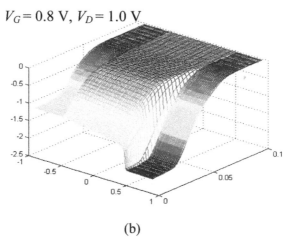

(b)

Figure 3.28. Simulated MOSFET band diagrams for various bias conditions: (a) conduction band minimum as a function of location at $V_G = 0.8$ V, $V_D = 0.4$ V and (b) at $V_G = 0.8$ V, $V_D = 1$ V. All the other device parameters are the same as those for Fig. 3.27. See also Color Insert.

most of the channel region, we expect that there will be a negligible change in current. The channel current saturates and the MOSFET is now operating in the saturation mode.

Some readers might be still curious about the reason behind the apparent loss of the drain control over the channel potential in Fig. 3.28(b). We can understand this phenomenon by considering the condition of strong inversion. The strong inversion can be maintained only when the effective gate bias (the quasi-Fermi energy difference between the channel and the gate) is larger than the threshold voltage. If the drain bias is small, the amount of quasi-Fermi energy drop at the drain end of the channel is not large, and the effective gate bias is still larger than the threshold voltage. If, however, the drain bias is increased to the point where the effective gate bias at the drain end of the channel is the same as the threshold voltage, the strong inversion condition can no longer be maintained. The strong inversion layer is pinched off and will begin to disappear. Once this happens, the drain cannot effectively control the channel quasi-Fermi level and the channel potential since the channel is now "detached" from the drain.

3.2.2 *Understanding MOSFET Operation*

In the previous subsection, we found that both the gate and drain bias voltages play an important role in controlling the inversion charge density and the potential profile, which results in the modulation of drain current. In order to obtain a quantitative relationship between the bias voltages and the inversion charge density, we will use the gated diode structure shown in Fig. 3.29.

A gated diode is an MOS structure combined with a p-n junction. In Fig. 3.29, the purpose of the n^+ region is to control the channel

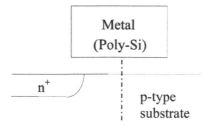

Figure 3.29. Gated diode structure. The dot-dash line is the line along which the band diagram will be drawn in Fig. 3.30.

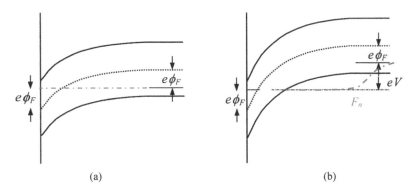

(a)   (b)

Figure 3.30. Band diagram of a gated diode: (a) with zero channel bias and (b) with positive channel bias.

bias (quasi-Fermi level of the inversion layer). We assume that the gate bias voltage is high enough to maintain the strong inversion condition. All the bias voltages are measured relative to the bulk of the substrate. That is, the rightmost end of the substrate is biased at the ground.

Let us first consider the case where the n^+ region has zero bias voltage. The quasi-Fermi level of the n^+ region should be the same as for the p-type substrate. Since the channel is in strong inversion, the quasi-Fermi level of the channel should be the same as for the n^+ region. Otherwise, the carriers in the channel will move around until the quasi-Fermi level becomes flat. Thus, the quasi-Fermi level should be flat throughout the semiconductor as is the case with a MOS structure. Fig. 3.30(a) shows the band diagram and the quasi-Fermi level along the cut-line shown in Fig. 3.29. In this zero channel bias case, $\psi_s = 2\phi_F$ and $Q_{dm} = -\sqrt{4\varepsilon_s e N_a \phi_F}$ as in a MOS structure.

When the n^+ region is biased at V, the quasi-Fermi level of the n^+ region should be at $-eV$. The quasi-Fermi level of the channel should be the same as that of the n^+ region, as long as the channel is in strong inversion. If there is any gradient in the channel quasi-Fermi level, the carriers will move until the gradient disappears. It is as if the n^+ region has a grip on the quasi-Fermi level of the channel and controls its value. The channel is now biased at V because of the control of n^+ region. Thus, not only the n^+-p junction but also the channel (induced n^+ region)-p junction is reverse-biased with

V, when V is positive. We expect that both the surface potential and the magnitude of depletion charge will increase due to this reverse bias across the junction.

$$\psi_s = 2\phi_F + V \tag{3.57}$$

$$Q_{dm} = -\sqrt{2\varepsilon_s e N_a \left(2\phi_F + V\right)} \tag{3.58}$$

Using the MOS equation Eqs. (3.10) and (3.57), we can express the semiconductor charge as a function of the gate bias (V_G) and the channel bias (V).

$$Q_s = -C_{ox}(V_G - V_{FB} - \psi_s) = -C_{ox}(V_G - V_{FB} - 2\phi_F - V) \tag{3.59}$$

By subtracting the depletion charge (Q_{dm}) from the total semiconductor charge (Q_s), we can obtain the inversion charge (Q_n).

$$Q_n = -C_{ox}(V_G - V - V_T'), \tag{3.60}$$

where

$$V_T' = V_{FB} + 2\phi_F + \frac{\sqrt{2\varepsilon_s e N_a(2\phi_F + V)}}{C_{ox}}. \tag{3.61}$$

In Eq. (3.59), we can see that the magnitude of the total semiconductor charge is decreased because of the increase of the surface potential. Compared with the case of zero channel bias, the effective gate bias is reduced by the channel bias, V. If we consider the inversion charge, it is further reduced because of the increase in the magnitude of the depletion charge (Q_{dm}). This increase of $|Q_{dm}|$ is caused by the increase of the depletion width, which, again, is a result of the channel bias, V. This secondary effect of channel bias on the inversion charge is reflected in the modified threshold voltage, V_T'. If V is positive, V_T' is definitely larger than the threshold voltage, V_T, with zero channel bias.

$$V_T = V_{FB} + 2\phi_F + \frac{\sqrt{4\varepsilon_s e N_a \phi_F}}{C_{ox}} \tag{3.62}$$

The double reduction of $|Q_n|$ caused by the channel bias is illustrated in Fig. 3.31. When a positive bias is applied to the channel, the effective bias between the gate and the channel decreases, which results in the reduction of $|Q_s|$. This phenomenon can be clearly seen in the reduction of Q_m, since $Q_m = |Q_s|$. This is the primary cause of the reduction in $|Q_n|$, since $|Q_n|$ is a part of $|Q_s|$. At the same time,

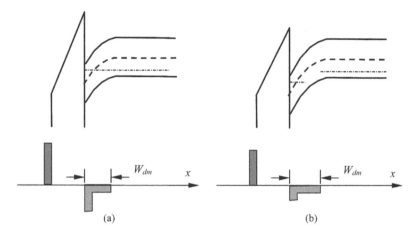

Figure 3.31. Band diagram and charge concentration in a gated diode structure: (a) with zero channel bias and (b) with positive channel bias.

the magnitude of the depletion charge, $|Q_{dm}|$ increases because of the increase of the depletion width. It is the secondary cause of the $|Q_n|$ reduction, since $|Q_n| = |Q_s| - |Q_{dm}|$. One interesting and important question is what would happen if we kept increasing the channel bias. The reduction of $|Q_s|$ and the increase of $|Q_{dm}|$ would soon make them the same and $|Q_n|$ would be zero. In this case, the assumption of strong inversion would not hold any more. The channel is pinched off because there is too much channel bias, and the n^+ region will not be able to control the surface potential of the inversion layer, since the relationship, $-e\psi_s = F_n - e\phi_F$, is maintained only in strong inversion. The channel region is now dominated by the depletion charge and is freed from the control of the n^+ region.

In an n-channel MOSFET, there are two n^+ regions called a source and a drain at the end of the shared channel. Under the strong inversion condition, the channel bias between the source and the drain should depend on both the source and drain bias, while the channel bias at the source and drain end should be the same as the source and drain bias, respectively. Usually, the source is biased at the same voltage as the substrate (or body), which is regarded as the reference for all the other bias voltages. That is, the source and substrate bias voltages are zero (or grounded), as shown in Fig. 3.32. The drain is biased at V_D.

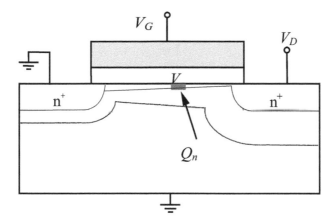

Figure 3.32. Bias voltages and the channel inversion charge in a MOSFET structure. The inversion layer is delineated by a solid line and the depletion region by the dotted line. The channel bias at the marked position is V, and the inversion charge is Q_n.

The channel bias, V, should change gradually from 0 to V_D as we move from the source end to the drain end of the channel. The value of the channel bias can be determined from the requirement for current continuity in the channel, as will be shown in the following subsection. Once the channel bias is determined, we can calculate the inversion charge using Eqs. (3.60) and (3.61). Since the source and body biases are zero, the inversion charge is linearly dependent on the gate bias. If we ignore the dependence of the threshold voltage on the channel bias, the inversion charge is also linearly dependent on the channel bias. This is the reason why the corresponding portion of the current vs. voltage characteristic is called a linear region. $|Q_n|$ will decrease gradually and the depletion width will increase gradually from the source end to the drain end of the channel. Such a change is depicted in Fig. 3.32. The apparent reduction of the inversion layer thickness from the source to the drain is to emphasize the reduction of the inversion charge in that direction. The actual thickness of the inversion layer may not necessarily decrease from the source to the drain.

When the drain bias is increased to $(V_G - V_T')$, the inversion charge at the drain end will be zero according to Eq. (3.60). The inversion charge "disappears" or is "pinched off" at the drain end of

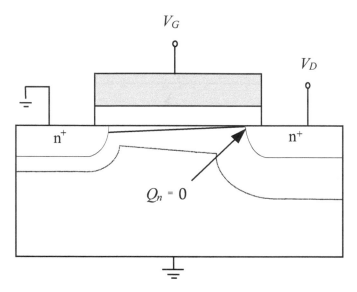

Figure 3.33. Channel inversion charge in a MOSFET under the pinch-off condition. The inversion charge becomes zero at the drain end of the channel.

the channel (Fig. 3.33). In order for the channel current to be continuous, the lateral electric field should become infinite at the site of the pinch-off. (If we calculate the electric field as a function of the lateral position along the semiconductor surface, it is indeed infinite at the drain end of the channel when $V_D = V_G - V_T'$.) Certainly, the pinch-off condition is a limiting case, which is not physically feasible. It is just a mathematically idealized condition, under which an infinitesimally small number of carriers move with an infinite velocity. In reality, there will be a finite number of carriers moving with a finite velocity much smaller than the speed of light. For a long-channel device, however, we often ignore the physical non-feasibility and accept the mathematical idealization since it makes the analysis much simpler and easier. It is when we consider short-channel devices that we must consider the finite field intensity and the finite velocity of carriers.

If the drain bias is increased beyond the pinch-off condition, the additional bias voltage appears only at the drain end of the channel, while most of the channel maintains bias and potential profiles that are nearly the same as those at the onset of pinch-off (Fig. 3.33). The additional potential energy drop at the drain end cannot contribute significantly to the current, since the current is already determined in the non-pinch-off region. It is as if there is a waterfall at the end of a river and the waterfall cannot affect the flow rate of water since the flow rate is already determined in the river. Thus, the MOSFET current saturates when the drain bias increases beyond the pinch-off condition. The corresponding portion of the current vs. voltage characteristic is called a **saturation region**. The current in the saturation region is called the drain saturation current (I_{Dsat}), and the drain voltage at the beginning of the saturation is called the drain saturation voltage (V_{Dsat}).

When the drain bias is increased beyond V_{Dsat}, the length of the pinch-off region will be non-zero. Since the actual electric field cannot be infinite, there should be a pinch-off region with a finite length to accommodate the additional voltage drop ($V_D - V_{Dsat}$). In a long-channel device, this finite length of the pinch-off region is negligible compared with the channel length. In a short-channel device, however, the pinch-off region cannot be ignored and affects the current vs. voltage characteristics. As the length of the pinch-off region increases, the effective channel length decreases, so that V_{Dsat} is applied to a shorter distance. Hence, the average lateral electric field increases, and the current increases slowly as a function of V_D. This phenomenon is called **channel length modulation**.

Up to now, we have been dealing with the case where the gate voltage is larger than the threshold voltage ($V_G > V_T$). When $V_G < V_T$, the strong inversion condition is not maintained, but Q_n does not drop to 0 immediately. As we have seen in Section 3.1.3, there is still a weak inversion layer in the channel. The inversion charge, Q_n, is an exponential function of V_G near the source, since both the source and the body are at ground. At the drain end of the channel, however, the drain bias does not have much control over the channel potential since the strong inversion condition is not maintained. The conduction band minimum or the surface potential remains more or less constant along the channel. With such a potential profile, the carriers move via diffusion through the channel. In this subthreshold region, the drain current has an exponential dependence on V_G. Figure 3.34 shows the linear relationship between

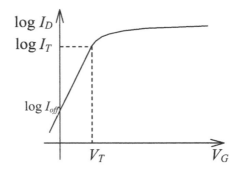

Figure 3.34. Subthreshold characteristic of MOSFET.

In I_D and V_G for $V_G < V_T$. We can see that even at $V_G = 0$, I_D is not zero. It has a finite value, I_{off}. I_{off} is called an "off" current, and it plays the important role of leakage current in the off-state of MOSFET.

3.2.3 *Current–Voltage Characteristics*

Now let us calculate the drain current in an n-channel MOSFET when the strong inversion layer is formed throughout the channel as shown in Fig. 3.35. In preparation for the calculation we need to define two axes, the $x-$ and $y-$axes, as shown in Fig. 3.35; x represents the vertical position and increases as we move from the surface to the bulk of semiconductor, while y represents the horizontal position and increases as we move from the source to the drain. According to this convention, the surface potential, ψ, and the inversion charge, Q_n, depend on y only, as indicated in the figure.

One important step in the calculation of the drain current is called the **gradual channel approximation**. In this approximation, we assume that the variation of the electric field in the y-direction is much slower than the variation in the $x-$direction ($\partial E_x/\partial x \gg \partial E_y/\partial y$). With this approximation, the two-dimensional Poisson equation can be reduced to the one-dimensional Poisson equation, so that we can use the result of the MOS analysis in the x-direction. In other words, we can consider only the vertical field (E_x) for the inversion layer formation and only the lateral field (E_y) for the drain current calculation.

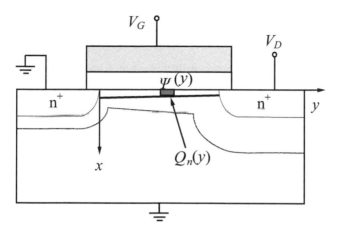

Figure 3.35. n-channel MOSFET with a strong inversion layer formed throughout the channel. Two axes, x and y, are defined. The surface potential ψ and the inversion charge Q_n depend on y only.

Now the current density at a position (x, y) can be expressed as

$$J_n(x, y) = -e\mu_n n(x, y) \frac{dV(y)}{dy}, \tag{3.63}$$

where μ_n is the electron mobility, $n(x, y)$ is the electron concentration, and $V(y)$ is the quasi-Fermi potential. We should note that μ_n is the channel mobility, which is usually smaller than the bulk mobility because of additional scattering mechanisms in the channel. By using the quasi-Fermi (electrochemical) potential instead of the electrostatic potential (ψ_s), we can obtain the sum of the drift and diffusion current, as discussed in Section 2.3.6. Here we are using the symbol $V(y)$ instead of $\phi_n(y)$, following the definition used in Section 3.2.2.

By integrating the current density with respect to x and z (the position in the direction perpendicular to the cross-section shown in Fig. 3.35), we can obtain the drain current. Since none of the parameters depends on z, integration with respect to z reduces to multiplication by the channel width W. In the x-direction, we need to integrate the current density from the semiconductor surface to the inversion layer thickness, W_i.

$$I_D(y) = eW \int_0^{W_i} \mu_n n(x, y) \frac{dV}{dy} dx \tag{3.64}$$

Since V is a function of y only, we can rewrite Eq. (3.64) as

$$I_D(y) = eW \frac{dV}{dy} \int_0^{W_i} \mu_n n(x, y) dx \qquad (3.65)$$

Since the channel electron mobility is a function of x, we cannot arbitrarily assume that it is a constant. Instead, we use the concept of effective mobility, which is the weighted average defined as

$$\mu_{eff} = \frac{\int_0^{W_i} \mu_n n(x, y) dx}{\int_0^{W_i} n(x, y) dx} \qquad (3.66)$$

Using Eq. (3.66) and the definition of the inversion charge,

$$Q_n(y) = -e \int_0^{W_i} n(x, y) dx, \qquad (3.67)$$

we find the following expression for the drain current:

$$\therefore I_D(y) = -\mu_{eff} W \frac{dV}{dy} Q_n(y) = -\mu_{eff} W \frac{dV}{dy} Q_n(V). \qquad (3.68)$$

Multiplying both sides by dy and integrating with respect to y, we get

$$\int_0^L I_D dy = \mu_{eff} W \int_0^{V_D} [-Q_n(V)] dV. \qquad (3.69)$$

Using the fact that I_D should be constant, we have finally:

$$I_D = \mu_{eff} \frac{W}{L} \int_0^{V_D} [-Q_n(V)] dV. \qquad (3.70)$$

Substituting Eq. (3.60) into Eq. (3.70) gives

$$I_D = \mu_{eff} \frac{W}{L} \int_0^{V_D} C_{ox}(V_G - V - V_T') dV. \qquad (3.71)$$

Carrying out the integration, we obtain

$$I_D = \mu_{eff} C_{ox} \frac{W}{L} \left\{ (V_G - V_{FB} - 2\phi_F - \frac{V_D}{2})V_D \right.$$
$$\left. - \frac{2\sqrt{2\varepsilon_s e N_a}}{3C_{ox}} [(2\phi_F + V_D)^{3/2} - (2\phi_F)^{3/2}] \right\} \qquad (3.72)$$

Equation (3.72) has a compact analytical form, but it is still somewhat complicated and looks quite different from the simpler equation in other textbooks. Using the Maclaurin series for $(2\phi_F + V_D)^{3/2}$ and retaining up to the second-order term, we obtain

$$I_D \cong \mu_{eff} C_{ox} \frac{W}{L} \left\{ (V_G - V_{FB} - 2\phi_F - \frac{V_D}{2})V_D \right.$$
$$\left. - \frac{2\sqrt{2\varepsilon_s e N_a}}{3C_{ox}} \left[\frac{3}{2}\sqrt{2\phi_F} V_D + \frac{3}{8} \frac{V_D^2}{\sqrt{2\phi_F}} \right] \right\}. \qquad (3.73)$$

After rearrangement of terms, we recognize V_T, so that we can express the drain current as

$$I_D = \mu_{eff} C_{ox} \frac{W}{L} [(V_G - V_T)V_D - \frac{\gamma}{2}V_D^2], \qquad (3.74)$$

where

$$\gamma = 1 + \frac{\sqrt{\frac{\varepsilon_s e N_a}{4\phi_F}}}{C_{ox}} = 1 + \frac{C_{dm}}{C_{ox}}. \qquad (3.75)$$

Here, γ is called a **body effect** coefficient, since it reflects the effect of the bias between the channel and the body. The body effect coefficient can also be used to express the threshold voltage in a different form.

$$V_T = V_{FB} + 2\phi_F + \frac{\sqrt{4\varepsilon_s e N_a \phi_F}}{C_{ox}} = V_{FB} + (2\gamma - 1)(2\phi_F) \qquad (3.76)$$

Starting from Eqs. (3.60) and (3.61), we can approximate the inversion charge with the Maclaurin series expansion of $\sqrt{2\varepsilon_s e N_a (2\phi_F + V)}$.

$$Q_n = -C_{ox} \left[V_G - V_{FB} - 2\phi_F - V - \frac{\sqrt{2\varepsilon_s e N_a (2\phi_F + V)}}{C_{ox}} \right]$$

$$\cong -C_{ox} \left[V_G - V_{FB} - 2\phi_F - V \right.$$

$$\left. - \frac{1}{C_{ox}} \left(\sqrt{4\varepsilon_s e N_a \phi_F} + \sqrt{\varepsilon_s e N_a / 4\phi_F} V \right) \right] \qquad (3.77)$$

After rearranging terms, we obtain

$$Q_n(V) = -C_{ox}(V_G - V_T - \gamma V). \qquad (3.78)$$

From this equation, we can calculate the saturation voltage (V_{Dsat}) and current (I_{Dsat}) using the pinch-off condition at the drain end of the channel.

$$Q_n(V_{Dsat}) = -C_{ox}(V_G - V_T - \gamma V_{Dsat}) = 0 \qquad (3.79)$$

Thus the saturation voltage is obtained as

$$V_{Dsat} = \frac{V_G - V_T}{\gamma}. \qquad (3.80)$$

By plugging this value into Eq. (3.74), we can obtain the saturation current.

$$I_{Dsat} = \mu_{eff} C_{ox} \frac{W}{L} \frac{(V_G - V_T)^2}{2\gamma} \qquad (3.81)$$

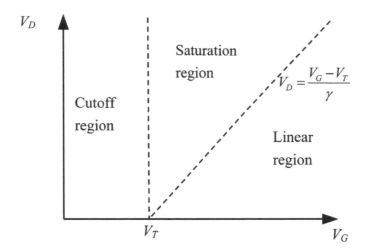

Figure 3.36. Three operating regions of a MOSFET: cutoff, saturation, and linear regions.

Thus, I_{Dsat} is proportional to the square of the gate overdrive, $(V_G - V_T)$.

Figure 3.36 shows the three operating regions of a MOSFET on a $V_D - V_G$ plane. When $V_G < V_T$, the MOSFET is in the cutoff (subthreshold) region regardless of the drain bias voltage. The drain current does not become 0 immediately but rather decreases as an exponential function of V_G. When $V_G > V_T$, the MOSFET is turned on and can be in either the saturation region or the linear region depending on V_D. The boundary between the saturation region and the linear region is a straight line given as

$$V_D = \frac{V_G - V_T}{\gamma}. \tag{3.82}$$

If we fix V_D and sweep V_G starting from somewhere in the cutoff region, we will come across the saturation region first and the linear region later. This type of bias condition occurs when we measure the subthreshold or transfer characteristic shown in Fig. 3.34. If V_D is low, the MOSFET will pass the saturation region quickly and stay longer in the linear region during the sweep. If V_D is high enough, the MOSFET may stay in the saturation region through the entire sweep.

If we fix $V_G (> V_T)$ and sweep V_D from zero to a certain positive value (usually the power supply voltage), we will pass the linear region first and then move through the saturation region. By measuring the drain current under this condition, we obtain the I_D vs. V_D characteristics of the MOSFET. Such characteristics are often called **output characteristics** since the drain is used as an output node in most circuits. Figure 3.37 shows the typical output characteristics of an n-channel MOSFET. As can be expected from Eq. (3.74), each characteristic takes the form of a parabola in the linear region. When we reach the vertex of the parabola, the saturation condition is met. Once we move into the saturation region, the drain current is fixed at I_{Dsat}. All the curves with a different gate bias voltage share the same property. In fact, another parabola is generated when we connect all the starting points of saturation. The origin of this parabola is the relationship between I_{Dsat} and V_{Dsat}.

$$I_{Dsat} = \frac{\mu_{eff} C_{ox} W \gamma}{2L} V_{Dsat}^2 \qquad (3.83)$$

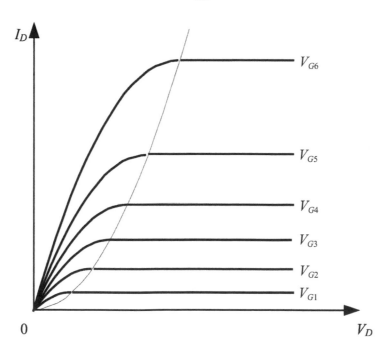

Figure 3.37. Output characteristics of a MOSFET. I_D vs. V_D characteristics are drawn for six gate bias voltages. The dotted line shows the boundary between the linear region and the saturation region.

This is clearly an equation for a parabola with its vertex at $I_{Dsat} = V_{Dsat} = 0$. Thus, in the first quadrant of the $I_D - V_D$ plane, the region above the parabola is the linear region and the rest is the saturation region.

3.2.4 Secondary Effects in MOSFETs

The mobility of carriers in a MOSFET channel is significantly lower than its bulk value. This is due to the additional scattering mechanisms in the MOSFET channel, such as surface roughness scattering, surface phonon scattering, and Coulomb scattering. These scattering mechanisms depend on the intensity of the normal field, since the scattering is mainly an interaction between carriers and the oxide–semiconductor interface. A higher normal field will enhance such an interaction. The field intensity, however, changes as a function of position, x, and so we need a criterion for where to define the effective field. Conceptually, the effective field should be defined at the average position of carriers in the direction perpendicular to the interface (x-direction).

It has been found empirically that the channel mobility of carriers can be expressed as a universal curve when it is plotted against an appropriately defined effective field [6]. The effective field should be defined differently for electrons and holes in silicon, however, in order to make the mobility curve universal. Equation (3.84) gives the best result for electrons, and so does Eq. (3.85) for holes.

$$E_{eff} = \frac{1}{\varepsilon_s}\left(|Q_d| + \frac{1}{2}|Q_n|\right) \qquad (3.84)$$

$$E_{eff} = \frac{1}{\varepsilon_s}\left(|Q_d| + \frac{1}{3}|Q_p|\right) \qquad (3.85)$$

Another important characteristic of a MOSFET is the subthreshold current. We can calculate the subthreshold current using Eq. (3.70). In order to obtain the inversion charge, we need to modify Eq. (3.30) for the case where the channel bias, V, is non-zero. Since the electron concentration is determined by the position of the conduction band minimum relative to the quasi-Fermi level, Eq. (3.30) should be modified to

$$Q_n = -\sqrt{2\varepsilon_s e N_a \psi_s}\, \frac{k_B T n_i^2}{2 e \psi_s N_a^2}\, \exp[e(\psi_s - V)/k_B T], \qquad (3.86)$$

when there is a positive channel bias V. Note that we modified only the argument of the exponential function. This approximation works, because the exponential function plays a dominant role in determining the value of inversion charge. Substituting Q_n into Eq. (3.70) and carrying out the integration, we obtain the subthreshold current.

$$I_D = \mu_{eff}\frac{W}{L}\sqrt{\frac{\varepsilon_s e N_a}{2\psi_s}\frac{(k_B T)^2 n_i^2}{e^2 N_a^2}}\,\exp(e\psi_s/k_B T)[1 - \exp(-eV_D/k_B T)]$$
$$(3.87)$$

In order to express ψ_s as a function of V_G, we first modify Eq. (3.12) to get V_G

$$V_G = V_{FB} + \frac{\sqrt{2\varepsilon_s e N_a}}{C_{ox}}\sqrt{\psi_s} + \psi_s \qquad (3.88)$$

To make the calculation simpler, we focus on the gate voltage close to the threshold voltage. Then, ψ_s should be close in value to $2\phi_F$, and we can use the Taylor series expansion near $2\phi_F$. If we retain the first-order term only, we obtain

$$V_G = V_{FB} + 2\phi_F + \frac{\sqrt{4\varepsilon_s e N_a \phi_F}}{C_{ox}} + \left(1 + \frac{\sqrt{\varepsilon_s e N_a/4\phi_F}}{C_{ox}}\right)(\psi_s - 2\phi_F).$$
$$(3.89)$$

ψ_s can be written as

$$\psi_s = (V_G - V_T)/\gamma + 2\phi_F \qquad (3.90)$$

With this value of ψ_s, Eq. (3.87) becomes

$$I_D = \mu_{eff}\frac{W}{L}\sqrt{\frac{\varepsilon_s e N_a}{2\psi_s}\frac{(k_B T)^2}{e^2}}\,\exp[e(V_G - V_T)/\gamma k_B T]$$
$$[1 - \exp(-eV_D/k_B T)]. \qquad (3.91)$$

The subthreshold current is an exponential function of the gate bias voltage, while it is almost independent of the drain bias voltage larger than a few thermal voltage ($k_B T/e$). As the gate bias voltage decreases below the threshold voltage, the drain current decreases as an exponential function of the gate bias voltage. The change of gate voltage required to change the drain current by an order of magnitude is called a **subthreshold swing**. It can be easily evaluated using Eq. (3.91).

$$S = \left[\frac{d(\log_{10} I_D)}{dV_G}\right]^{-1} = 2.3\frac{\gamma k_B T}{e} = 2.3\frac{k_B T}{e}\left(1 + \frac{C_{dm}}{C_{ox}}\right) \quad (3.92)$$

Since the typical C_{dm} value is $20 \sim 50\%$ of C_{ox}, S is $70 \sim 90$ mV/decade usually. A small S is desirable for easier turn-on and turn-off.

3.3 CMOS Circuits

3.3.1 *Circuit Symbol of MOSFETs*

If we connect active devices (transistors and diodes) with passive devices (resistors, capacitors, and inductors) and power supplies (voltage and current sources), we can form electronic circuits. In a circuit diagram, MOSFETs are often depicted by the circuit symbols illustrated in Fig. 3.38. Circuit symbols resemble the cross section of MOSFETs, albeit in a radically simplified form. The arrow near the source terminal indicates the direction of current flow, since it starts from the p-region and ends at the n-region.

Figure 3.39 shows the advantage of using the circuit symbol instead of the cross-sectional diagram. The circuit is further simplified by substituting the voltage sources with a terminal accompanied with the name of the voltage source (or the value of the bias voltage

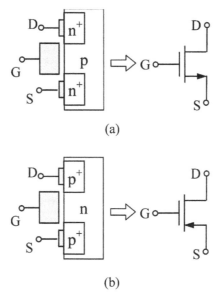

(a)

(b)

Figure 3.38. MOSFET circuit symbols: (a) n-channel MOSFET and (b) p-channel MOSFET.

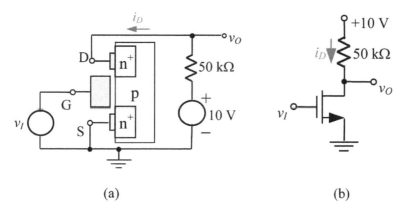

Figure 3.39. A MOSFET circuit: (a) with a cross-sectional MOSFET image and full connection diagram and (b) with a MOSFET circuit symbol and abbreviated connection diagram.

in the case of a DC voltage source). The ground symbol is used for the reference point of voltage.

3.3.2 Complementary MOSFET (CMOS) Circuits

CMOS circuits are composed of n-channel MOSFETs and p-channel MOSFETs. Figure 3.40(a) shows the cross section of a CMOS inverter circuit. The source of the n-channel MOSFET is grounded, while the source of the p-channel MOSFET is connected to the power supply (V_{DD}). The gates of n-channel and p-channel MOSFETs are connected to one node, which forms an input terminal. The drains of two MOSFETs are connected to another node, which functions as an output terminal.

Figure 3.40(b) shows the circuit diagram of a CMOS inverter. If the input voltage (v_I) is low (0 V), the n-channel MOSFET is turned off, while the p-channel MOSFET is turned on, so that the output voltage (v_O) becomes high (V_{DD}). On the other hand, if the input voltage is high ($v_I = V_{DD}$), the p-channel MOSFET is turned off, while the n-channel MOSFET is turned on, so that the output voltage becomes low ($v_O = 0$). Thus, in a CMOS inverter, one of the two devices is OFF except a short period of transition. This is the reason why the power consumption is minimized in CMOS circuits.

Using device isolation and interconnection, integrated circuits (ICs) can be fabricated. Integrated circuits can be divided into analog

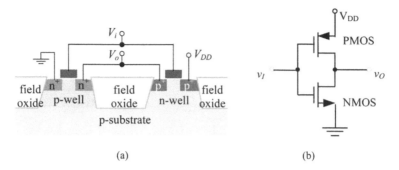

(a)   (b)

Figure 3.40. CMOS inverter: (a) cross section and (b) circuit diagram. See also Color Insert.

and digital ICs. Digital ICs are further divided into logic and memory ICs. Here, we will discuss digital ICs only, since they are the dominant ICs currently.

3.3.3 *CMOS Logic Gates*

By combining n-channel and p-channel MOSFETs appropriately, it is possible to build not only an inverter logic gate but also any other logic gates. In a NAND gate, if either of the two input voltages is low, the corresponding n-channel device is off, so that the output voltage becomes high. If both of the input voltages are high, both of the n-channel devices are off, so that the output voltage is low. Thus, the NAND operation is realized (Fig. 3.41).

3.3.4 *Memory Circuits*

Memory ICs can be divided into volatile and nonvolatile memory ICs. Volatile memories lose their data when the power is turned off, while non-volatile memories retain their data. Volatile memories are further divided into **dynamic random access memory (DRAM)** (Fig. 3.42) and **static random access memory (SRAM)**.

Since DRAM contains only one transistor per cell, it can achieve a high integration density. Because of the leakage current, however, the charge stored in the capacitor decreases and the memory is lost within a few hundred milliseconds (retention time). We have to restore (read and rewrite) the memory state periodically and the

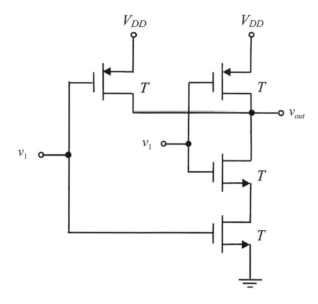

Figure 3.41. CMOS NAND circuit.

period for that process should be shorter than the retention time. This process is called **"refresh"**. Because of the need to refresh, DRAMs consume relatively high power. In addition, the capacitor should maintain a large enough capacitance in order to have sufficient retention time. But a large capacitance means a longer delay time in read and write operations and a slower speed of the DRAM.

SRAM depends on the "latch" mechanism between two circularly connected CMOS inverters, as shown in Fig. 3.43. Including two pass

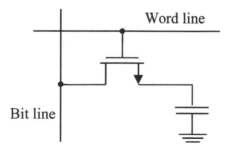

Figure 3.42. DRAM cell that consists of an n-channel MOSFET (pass transistor) and a capacitor.

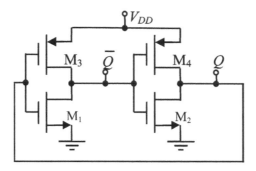

Figure 3.43. Two circularly connected CMOS inverters that can be used as a core of an SRAM cell. If we add two pass transistors, an SRAM cell is constructed.

transistors, SRAM requires total six transistors per cell and this high count of devices per cell is a major roadblock for high-density integration. Since SRAM does not require refresh, however, we do not have to worry about retention time and no large capacitance is involved. Hence, SRAM can achieve both high speed and low power.

If we assemble the memory cells in a regular array of columns and rows, we can make a storage cell array as shown in Fig. 3.44. A memory cell shown in Fig. 3.42 or Fig. 3.43 is located at each crosspoint of the row and the column lines. A row line is often called a **word line**, while a column line is called a **bit line**. We can access each cell by assigning a unique address to it. For example, we can access 2^{M+N} memory cells by using an address with $(M + N)$ bits. The first M bits are used for row addressing and the rest (N bits) are used for column addressing. The row decoder selects the word line (K) that is designated by the row address (the first M bits), while the column decoder selects the bit line (L) that is designated by the column address (the following N bits). Thus, we can randomly access the cell designated by the given address. This is the reason why these types of memories are called random access memory.

Once the designated cell is selected, we can write or read data. The sense amplifiers/drivers perform the action of writing and reading data. In the case of a DRAM, the sense amplifier automatically writes the data back to the cell after reading them, accomplishing the refresh process. Thus, in a DRAM, we just need to read the data

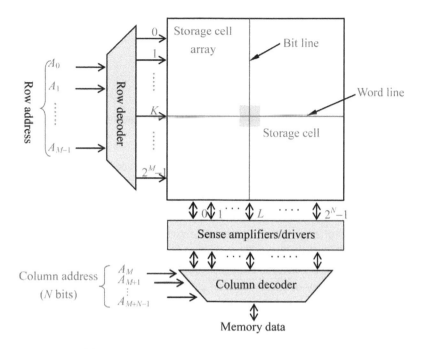

Figure 3.44. Structure of a memory cell array. See also Color Insert.

in a regular period of time that is shorter than the retention time. An SRAM does not require a refresh process, since the latch maintains the state as long as the power is supplied to the cell.

As an example of non-volatile memory, we can take flash memories. Flash memories are based on a MOSFET structure with a floating memory node sandwiched between the gate and the channel. Figure 3.45 shows a typical flash cell transistor structure and its circuit symbol. The circuit symbol contains a bar that represents the floating memory node.

The **tunnel oxide** is used as a tunnel barrier through which **Fowler–Nordheim (F–N) tunneling** occurs. The F–N tunneling mechanism is depicted in Fig. 3.46. The F–N tunneling of carriers occurs when a triangular tunnel barrier is formed due to a strong electric field. As the electric field increases, the effective tunnel barrier thickness decreases, so that the tunneling current increases exponentially.

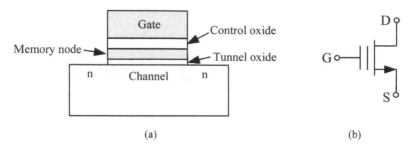

(a)   (b)

Figure 3.45. Flash memory cell transistor: (a) cross section and (b) circuit symbol.

The floating **memory node** is typically made of poly-silicon or other types of conducting material, providing the storage space of carriers. Recently, silicon nitride is also used as a memory node material, since it contains charge traps **(charge trap flash)**. The insulating property of silicon nitride provides a few advantages in reliability, retention, and three-dimensional stackability. The **control**

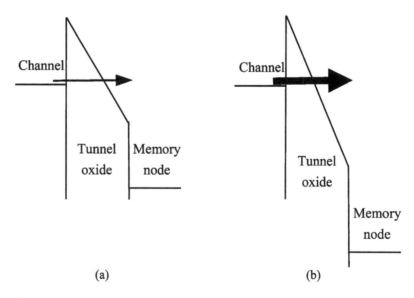

(a)   (b)

Figure 3.46. Fowler–Nordheim tunneling mechanism: (a) A triangular tunnel barrier is formed due to a strong electric field and (b) the tunneling current increases exponentially as the effective barrier thickness decreases due to the increase in the electric field.

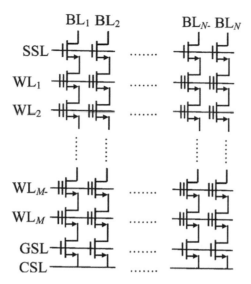

Figure 3.47. NAND flash memory array. The array consists of N SSL, N GSL, and $(M \times N)$ cell transistors.

oxide is used for the blocking of carrier tunneling and is often called a **blocking oxide**. The control oxide should be thicker than the tunnel oxide to block carrier tunneling efficiently. Since the decrease of capacitance in the control oxide due to the increase of its thickness brings about a weaker gate-memory node coupling, its capacitance is boosted by increasing the effective gate area (gate surrounding the memory node) or increasing the dielectric constant of the blocking oxide (high-κ dielectric).

Figure 3.47 shows the circuit diagram of a NAND flash memory array. There are $(M \times N)$ cell transistors with M word lines and N bit lines. All the cell transistors are n-channel MOSFETs. The cell transistors in a bit line are connected in series, just like the case of the CMOS NAND circuit shown in Fig. 3.41. This is the reason why the array structure in Fig. 3.45 is called a NAND flash.

Different from DRAM or SRAM operation that consists of two types (write and read), there are three types of operations (program, erase, and read) in flash memories. For program operation, the string-select line (SSL) is biased high and the selected bit line is set to 0 V. The ground-select line (GSL) is biased low to isolate the cell transistors from the common source line (CSL) and inhibit the program of unselected bit lines. When the program and pass biases are applied to the gate of the selected word line and pass transistors, respectively, the channel potential of unselected bit lines transistors is boosted by capacitive coupling between the control gate and channel. Thus, only the selected bit line transistor(s) is(are) programmed. For erase operation, the gate-to-channel bias becomes a large negative value for the whole block, so that the electrons in the memory node tunnel back to the channel (or the holes in the channel are injected to the memory node).

For read operation, all SSL and GSL transistors are turned on and the read and pass biases are applied to the gate of selected word line and pass transistors, respectively. Then, the current can flow through a bit line if the cell transistor on the selected word line is not programmed. By sensing the current for each bit line, we can read the data from all the memory cells on the selected word line.

PROBLEMS

1. **Formation of MOS structure:**
   In Fig. 3.1, we can notice that the space charge region is formed near the oxide–semiconductor interface, even though the most likely path of the charge movement is not through the oxide layer. Why is the space charge region formed only near the interface? For example, what happens if a space charge region is first formed in the bulk of the semiconductor? Can it maintain its position for a while? Show that such a space charge region will quickly disappear and the space charge will exist only near the interface when an equilibrium is reached.

2. **MOS structure:**
   (a) In a MOS structure, we use the condition $\psi_s = 2\phi_F$ as the starting point of the strong inversion. Under this condition, however, the inversion layer charge does not dominate the charge in the semiconductor. Why?
   (b) Utilizing the conclusion in (a), we can calculate the threshold voltage of the MOS structure. How can this be done? Explain briefly.

(c) Even if the inversion is really strong ($\psi_s \gg 2\phi_F$), the value of ψ_s is not completely fixed but depends weakly on the gate voltage. Express ψ_s as a function of the gate voltage.

3. **Flat-band capacitance of MOS structure:**

   (a) Using the Maclaurin series expansion, prove Eq. (3.44).
   (b) The Debye length is often used as a measure of screening by mobile carriers, and is defined as $L_{Dp} = \sqrt{(\varepsilon_s k_B T)/(e^2 p_0)}$ for holes. Express the flat-band capacitance as a function of the Debye length of holes.
   (c) Using the result of (b), explain why $C_s(0)$ is not infinite but finite.

4. **Flat-band voltage of MOS structure:**

   (a) If the gate is phosphorus doped poly-silicon, the substrate is doped with 10^{16}cm^{-3} boron, and the oxide thickness is 5 nm, what is the flat-band voltage of this MOS structure? (We assume that the work function of poly-silicon is the same as the electron affinity of silicon.)
   (b) In the MOS structure in (a), there exists a sheet of charge with 1.6×10^{-8}C/cm^2 surface density that lies 1 nm away from the oxide–silicon interface. Sketch the band diagram at the flat-band voltage, including the oxide layer. Based on this diagram calculate the flat-band voltage.

5. **Low-frequency MOS C-V curve with a ramp voltage:**
   Often we measure a low-frequency MOS (C–V) curve by applying a ramp voltage ($V_g = \alpha t$). Assuming that the inversion charge is supplied only by generation in the depletion region, derive a formula for maximum α to maintain the low-frequency condition. Oxide capacitance per unit area $= C_i$, depletion width $= W_{dm}$, generation lifetime τ_{gen}, intrinsic carrier concentration $= n_i$, electronic charge e.

6. **Application of MOS C-V curve:**
   Regarding the MOS C-V curve shown below, answer the following questions:

   (a) Is this curve measured at high frequency or low frequency? Explain your answer.
   (b) What is the doping type of the substrate? Give reason.

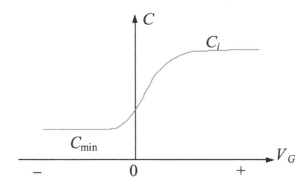

(c) Explain the method of finding the oxide thickness and the doping concentration from this curve.

(d) Explain the method of finding the flat-band voltage and the threshold voltage from this curve.

(e) If we want to obtain the density of interface traps (D_{it}), what sort of additional measurement do we have to perform? Explain the method of finding D_{it} from the given curve and the result of the additional measurement.

7. **Lateral electric field of an n-channel MOSFET:**

   (a) By changing the upper limit of the integral into y in Eq. (3.69) and using Eq. (3.78), calculate the channel bias voltage, V, as a function of y.

   (b) If the strong inversion condition is maintained, we can approximate the electric field, E_y, as $-\frac{dV(y)}{dy}$. Using this approximation, calculate E_y as a function of y.

   (c) Calculate E_y at the drain end of the channel $(y = L)$ when $V_D = (V_G - V_T)/\gamma$.

8. **Current–voltage characteristics of an n-channel MOSFET:**

   (a) When the drain bias voltage is very small, the drain current is proportional to the drain voltage. What is the reason for this behavior?

   (b) In (a), calculate the proportionality constant approximately. The gate voltage is V_G, the threshold voltage is V_T, the oxide capacitance per unit area is C_{ox}, the electron mobility in the channel is μ_n, and the channel width is W.

9. **Drift and diffusion component in the MOSFET current:**

   (a) When the gate bias voltage, V_G, is larger than the threshold voltage, V_T, the dominant component of the drain current is the drift current. Why is this the case?

   (b) In the subthreshold region ($V_G < V_T$), the dominant component of the drain current is the diffusion current. Explain why is this the case.

10. **Body effect:**

    An n-channel MOSFET satisfies the condition, $-V_{FB} = 2\phi_F = 0.8$ V. What should be the body bias with the source as a reference to double the threshold voltage of this MOSFET?

11. **CMOS NOR circuit:**

    In a NOR circuit, the output voltage is high only when all input voltages are low. For all other input combinations, the output voltage should be low. Design a two-input CMOS NOR circuit with two n-channel and two p-channel MOSFETs.

12. **DRAM retention time:**

    In a DRAM cell, we have a pass transistor with 0.1 pA leakage current and a capacitor with 30 fF capacitance. If we define the retention time as the time required to change the voltage across the capacitor by 0.5 V, what is the retention time for this cell?

Bibliography

1. D.A. Neamen, *Semiconductor Physics and Devices: Basic Principles*; Irwin, 1992.

2. R.S. Muller and T.I. Kamins, *Device Electronics for Integrated Circuits*, Wiley, 1986.

3. B.G. Streetman and S. Banerjee, *Solid State Electronic Devices*, 5th Ed., Prentice-Hall, 2000.

4. Y. Taur and T.H. Ning, *Fundamentals of Modern VLSI Devices*, Cambridge University Press, 1998.

5. Seonghoon Jin, NANOCAD Multi-Dimensional Device Simulator Version 1.0 User Manual, Physical Electronics Laboratory, 2004.

6. A.G. Sabnis and J.T. Clemens, "Characterization of electron mobility in the inverted < 100 > Si surface," *IEDM Tech. Dig.*, pp. 613–616, 1979.

7. S. Takagi, M. Iwase, and A. Toriumi, "On universality of inversion-layer mobility in n- and p-channel MOSFETs," *IEDM Tech. Dig.*, pp. 398–401, 1988.

Chapter 4

Quantum Well Devices

4.1 Issues in CMOS Device Scaling

4.1.1 *MOSFET Scaling Trend*

MOSFET scaling has been going on for almost 50 years now. Every three years the average design rule has shrunk to the 70% of the previous generation. Such a rate of shrinkage brings roughly an order of magnitude reduction every 20 years. Starting with a few tens of μm design rules in the 1960s, we already see devices with a physical gate length of a few tens of nanometers in production. In 50 years, some devices have been shrunk down by three orders of magnitude in size (Fig. 4.1).

In early 1990s, people in the United States began their work on a semiconductor technology roadmap (National Technology Roadmap for Semiconductors, Semiconductor Industry Association, 1992) to guide the semiconductor industry and set up milestones for technology development. In 1999, the roadmap was internationalized with a new name, the International Technology Roadmap for Semiconductors (ITRS). According to the recent ITRS (2007 version), we may be able to see microprocessors made of CMOS devices with a 4.5 nm physical gate length in 2022.

Nanoelectronic Devices
Byung-Gook Park, Sung Woo Hwang, and Young June Park
Copyright © 2012 Pan Stanford Publishing Pte. Ltd.
www.panstanford.com

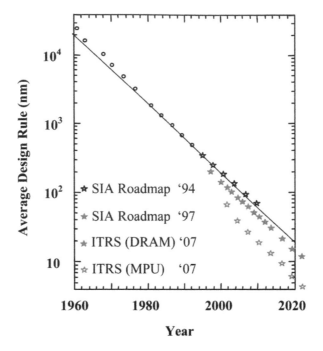

Figure 4.1. Average design rule vs. year. The predictions of the SIA roadmap and the ITRS are also plotted. See also Color Insert.

What are the motivations behind this rapid and incessant scaling down of CMOS devices? The most obvious motivation is the increase of integration density. By fabricating more devices in a given area, we can reduce the cost of production per device. We can enhance the functionality or capacity of a chip by having more devices in the same area. Recent integrated circuits such as the system on a chip (SoC), multi-core microprocessors, and multi-gigabit memories, have all benefited from the exponential increase in integration density.

Another important motivation is the enhancement of circuit performance (speed). Let us estimate the delay in the simple inverter circuit shown in Fig. 4.2. We assume that a step voltage with a step height of V_{DD} is applied to the input of the inverter. Once the power supply voltage, V_{DD}, is applied to the input node, the n-channel MOSFET will be fully turned on, while the p-channel MOSFET will

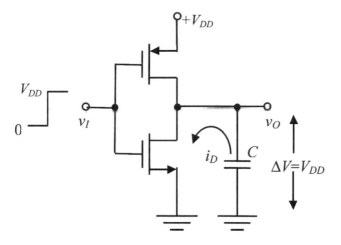

Figure 4.2. MOSFET inverter circuit for the calculation of delay.

be completely turned off. Since the output of the inverter must have been at V_{DD} before the voltage step (when $v_I = 0$), the current i_D through the n-channel MOSFET will discharge the capacitance C. Because of the change in the output voltage, i_D is not constant during the discharge, but for simplicity, let us assume i_D is fixed at its initial value, I_D. Then, during the discharge, I_D has the following relationship with the voltage change ΔV and the discharge time Δt.

$$I_D = \frac{C \, \Delta V}{\Delta t} = \frac{C \, V_{DD}}{\Delta t} \tag{4.1}$$

The time delay τ_d of the inverter is determined by the discharge time Δt, and we can assume that they are roughly the same.

$$\tau_d \approx \Delta t = \frac{C \, V_{DD}}{I_D} \tag{4.2}$$

For the parasitic capacitance, we ignore the interconnection capacitance and assume that the main contribution comes from the next stage which is identical to the inverter shown in Fig. 4.2. Then we can estimate the capacitance as

$$C = C_{ox} W L, \tag{4.3}$$

where C_{ox} is the oxide thickness, W is the channel width, and L is the channel length. If we assume that the threshold voltage is sufficiently small, the gate overdrive ($V_G - V_T$) can be assumed to be close to V_{DD}, so that the MOSFET saturation current becomes

$$I_D = \frac{\mu_n W C_{ox} V_{DD}^2}{2L},$$

(4.4)

where μ_n is the electron mobility.

Substituting Eqs. (4.3) and (4.4) into Eq. (4.2), we obtain the delay.

$$\tau_d \approx \frac{2L^2}{\mu_n V_{DD}}$$

(4.5)

Thus, the shrinkage of channel length is essential to the reduction of delay. If we can reduce the delay, a higher clock frequency can be used, and the integrated circuit performance can be enhanced. In fact, the dramatic reduction in the physical gate length of microprocessors around the end of the 20th century was ignited by the fierce competition among microprocessor makers who wanted to achieve 1 GHz operation earlier than other companies. Before this period, the average design rules for microprocessors were larger than those of DRAMs. Since this drastic reduction in gate length, however, microprocessors have been maintaining their leading role in physical gate length scaling.

The third motivation is the reduction of power consumption per device. We can calculate the power consumption of the inverter circuit shown in Fig. 4.2 as follows. We first evaluate the energy consumed during one complete clock cycle. Multiplying it by the clock frequency, we can obtain the power consumption. Let us first consider a high-low transition (discharging process) at the output node. The instantaneous power, p, that the n-channel MOSFET consumes is

$$p = v_0 i_D = -v_0 C \frac{dv_0}{dt}.$$

(4.6)

The energy lost during this high-low transition can be obtained by integrating p.

$$W_{HL} = \int_0^{\tau_d} p\, dt = -\int_0^{\tau_d} v_0 C \frac{dv_0}{dt} dt$$

$$= \int_0^{V_{DD}} C v_0\, dv_0 = \frac{1}{2} C V_{DD}^2$$

(4.7)

The instantaneous power that the p-channel MOSFET consumes during the low-high transition (charging process) is given as

$$p = (V_{DD} - v_O)i_D = (V_{DD} - v_O)C\frac{dv_O}{dt}. \tag{4.8}$$

The energy loss during the low-high transition is

$$W_{LH} = \frac{1}{2}CV_{DD}^2. \tag{4.9}$$

One clock cycle includes both the high-low and low-high transition, so that we need to add up Eqs. (4.7) and (4.9) in order to obtain the total energy loss per clock. Now, if we multiply it by the clock frequency f, we obtain the power dissipation.

$$P = fCV_{DD}^2 \tag{4.10}$$

In this equation, we can see that the power supply voltage (V_{DD}) plays an important role in power dissipation. For low-power operation, V_{DD} reduction is essential. But in Eq. (4.5), if we reduce V_{DD}, the delay increases so that the speed would be sacrificed. The only way out of this dilemma is to reduce the channel length L along with the reduction in V_{DD}. By scaling the channel length, we can reduce V_{DD} without sacrificing performance.

Device scaling brings us enormous benefits, but at the same time it challenges us with many difficult issues. There are several traditional issues that have been troubling device engineers. Short-channel effects (threshold voltage roll-off and drain-induced barrier lowering), punch-through, and velocity saturation are in this category. To cope with these issues, MOSFET scaling theories have been developed and successfully applied down to 100 nm channel devices. When we moved into the nanoelectronic era ($L < 100$ nm) at the dawn of the 21st century, however, a few new issues started to challenge us. One of them is dopant number fluctuation due to the granularity of the microscopic world, and the other is tunneling due to the wave nature of electrons.

4.1.2 *Short-Channel Effects*

If we change the channel length L while keeping all the other process parameters the same during device fabrication, we observe that the

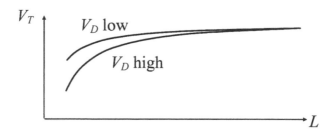

Figure 4.3. Short-channel effects. As the channel length of an n-MOSFET decreases, the measured threshold voltage (V_T) also decreases. Such a decrease becomes more severe when the channel length is shorter or the drain voltage is higher.

measured threshold voltage, V_T, decreases as the channel length decreases (Fig. 4.3). Such a decrease becomes more severe when the channel length is shorter or when the drain voltage is higher. From this observation, we can infer that there are actually two effects: One is due to the reduction in channel length, and the other is due to the influence of the drain bias. Both effects are called short-channel effects. The short-channel effects are serious issues in MOSFET scaling since we cannot control the channel length perfectly. The variation of the channel length induces a variation in the threshold voltage, which can limit the proper operation of CMOS circuits.

Let us first consider the channel length dependence of the measured threshold voltage when the drain voltage is sufficiently low. If we examine the equation of the theoretical threshold voltage (Eq. (3.39), Chapter 3), the channel length does not appear in the equation at all. The threshold voltage is supposed to be independent of the channel length. One may argue that it is independent of the channel length since we have used the MOS structure for the calculation of the threshold voltage. But the definition of the threshold voltage remains the same in the drain current calculation of a MOSFET. Obviously, we ignored something when we calculated the threshold voltage of a MOSFET.

Figure 4.4 shows what we ignored in Chapter 3. We ignored the fact that not only the gate but also the source and the drain can generate depletion regions. Out of the depletion region delineated by the dotted line, the gray region can be generated either by the gate or by

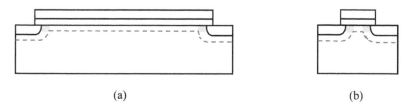

Figure 4.4. Sharing of depletion charge between the gate and the source/drain: (a) long-channel MOSFET and (b) short-channel MOSFET.

the source/drain. Since these parts of depletion region are shared by the gate and the source/drain, this phenomenon is called charge sharing. We can divide each part into halves and associate each half with the gate or the source/drain depending on their proximity. This is a frequently used method when we carry out a quantitative analysis.

For a quantitative analysis, we have drawn a detailed diagram (Fig. 4.5) that shows all the necessary parameters and the geometrical relationship. The source/drain junction depth is x_j and the lateral junction boundary of the source/drain is assumed to be a quarter-circle centered at A (A'). The radius of the quarter-circle is

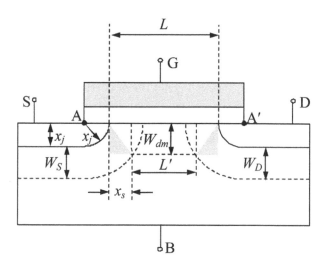

Figure 4.5. Diagram for the quantitative analysis of the short channel effect. The two gray triangular regions are assumed to be the depletion region generated by the source and the drain.

also x_j. We assume all the depletion widths are the same.

$$W_S = W_D = W_{dm} \tag{4.11}$$

The depletion region generated by the quarter-circle junction is assumed to have a boundary at another concentric quarter-circle with a radius $(x_j + W_{dm})$.

Now, the trapezoidal area with an upper side L and a lower side L' is the depletion region generated by the gate. The two gray areas in the shape of right triangle can be assumed to be generated by the source/drain. If it were not for the source and the drain, the gate would have generated a rectangular depletion region with sides L and W_{dm}. Let Q_{dm} be the depletion charge per unit area without charge sharing, and let Q'_{dm} be the depletion charge per unit area with charge sharing. Then, the ratio of Q'_{dm} to Q_{dm} should be the same as the ratio of the trapezoidal area to the rectangular area. From this relationship, we obtain

$$Q'_{dm}L = Q_{dm}\frac{L+L'}{2}. \tag{4.12}$$

Once Q'_{dm} is known, we can calculate the threshold voltage as follows:

$$V_T = V_{FB} + 2\phi_F - \frac{Q'_{dm}}{C_{ox}} \tag{4.13}$$

In order to evaluate Q'_{dm}, we have to know L'. L' can be obtained by calculating x_s first. Applying the Pythagoras' theorem to an appropriate right triangle, which is not the gray triangle, we obtain:

$$(x_j + W_{dm})^2 = (x_j + x_s)^2 + W_{dm}^2 \tag{4.14}$$

Solving this quadratic equation, we obtain

$$x_s = x_j\left(\sqrt{1 + 2W_{dm}/x_j} - 1\right). \tag{4.15}$$

Now, Q'_{dm} is given as

$$Q'_{dm} = Q_{dm}\left[1 - (x_j/L)\left(\sqrt{1 + 2W_{dm}/x_j} - 1\right)\right]. \tag{4.16}$$

Finally, the threshold voltage for an n-channel MOSFET becomes

$$V_T = V_{FB} + 2\phi_F - \frac{Q_{dm}}{C_{ox}}\left[1 - (x_j/L)\left(\sqrt{1 + 2W_{dm}/x_j} - 1\right)\right]. \tag{4.17}$$

Since Q'_{dm} is negative for an n-channel MOSFET, V_T decreases monotonically as L decreases. In this equation, there are three key parameters that affect the rate of change in V_T as a function of L. They are C_{ox}, x_j, and W_{dm}. We would like to keep the rate of change in V_T as small as possible. In order to do this, we need to have large C_{ox}, small x_j and small W_{dm}. These requirements have important implications in MOSFET scaling. To increase C_{ox}, we need to have a thinner oxide. The reduction of x_j means a shallower junction. For small W_{dm} we need higher channel doping. When we discuss the MOSFET scaling theory in Section 4.1.5, we will find that all these need to be done for the proper scaling of MOSFETs.

Now, let us consider the additional reduction of the measured threshold voltage when the drain voltage is increased. The reduction of the measured threshold voltage is larger when the drain voltage increases or the channel length decreases. As shown in Fig. 4.6, this phenomenon occurs because the drain voltage affects the potential barrier between the source and the channel and, consequently, reduces the barrier. Thus, we call this effect drain-induced barrier lowering (DIBL). DIBL occurs much earlier than the punch-through,

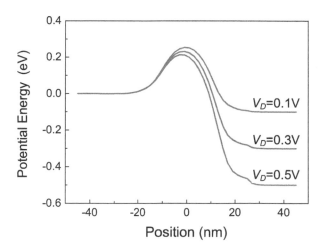

Figure 4.6. Drain-induced barrier lowering (DIBL). The drain voltage affects the potential barrier between the source and the channel and reduces the barrier. The channel length of MOSFET is 50 nm and the channel doping is 3×10^{18} cm^{-3}.

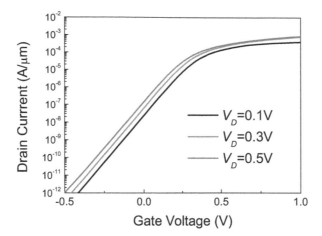

Figure 4.7. Subthreshold characteristics of a MOSFET showing DIBL. The channel length and the doping are the same as those of Figure 4.6. See also Color Insert.

which will be described in the following subsection, even though there are many similarities.

The results of drain-induced barrier lowering show up clearly in the subthreshold characteristics (log I_D vs. V_G curve) of a MOSFET. As shown in Fig. 4.7, the threshold voltage decreases as the drain voltage is increased. At the same time, the subthreshold current at a given gate voltage increases. The subthreshold swing, however, remains more or less the same. This is because the DIBL occurs at the surface of the semiconductor, not in the subsurface or the bulk. Once we know the value of DIBL in the channel, we can translate it to the threshold voltage shift using the following relationship:

$$\Delta V_T = \gamma \Delta \psi_{s,DIBL} \tag{4.18}$$

where γ is the body effect coefficient defined in Eq. (3.75), Section 3.2.3, Chapter 3. This relationship can be confirmed by differentiating V_T with respect to $2\phi_F$ in Eq. (3.62), Section 3.2.2, Chapter 3.

4.1.3 *Punch-Through*

Punch-through occurs when the edge of the depletion region generated by the drain reaches the edge of the source-generated depletion region. If this happens, the potential barrier between the source and the body is reduced and a bulk conducting path (subsurface channel) is formed. Since this bulk conducting path is far from the gate, it is quite difficult for the gate to control the punch-through current. Thus, the subthreshold current and the subthreshold swing increase drastically.

Figure 4.8 shows the progress of punch-through as we increase the drain voltage. The punch-through phenomenon is the most severe if the subsurface doping concentration is low. In order to prevent punch-through, we often increase the subsurface doping by performing a punch-through stop implantation during the MOSFET fabrication.

The subthreshold characteristics clearly show the impact of punch-through on the drain current (Fig. 4.9). Unlike the situation with DIBL, we can see that the subthreshold swing increases significantly when the punch-through occurs. In fact, when the drain voltage is small, we can observe only DIBL. The threshold voltage shifts toward the negative direction, but the subthreshold swing remains the same. When the drain voltage is increased beyond a certain value (1.0 V in Fig. 4.9), the subthreshold swing increases rapidly.

4.1.4 *Velocity Saturation*

According to the drift theory of carriers, the drift velocity is proportional to the applied electric field. The main assumption behind this theory, however, is that the electric field is small enough that the drift velocity remains significantly smaller than the thermal velocity. When the electric field is high, the average carrier energy increases and carriers lose their energy by emitting optical phonons (energy quanta of a certain lattice vibration mode) as soon as they gain it from the electric field. In this case, the carrier velocity cannot increase any further and has to saturate. This phenomenon is called velocity saturation.

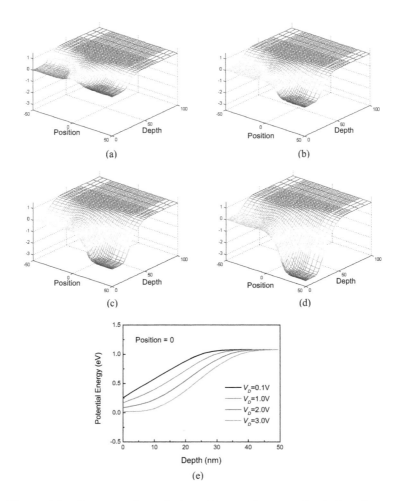

Figure 4.8. Progress of punch-through as we increase the drain voltage: (a) $V_D = 0.1$ V, (b) $V_D = 1$ V, (c) $V_D = 2$ V, (d) $V_D = 3$ V, and (e) potential barrier height as a function of depth from the oxide-silicon interface (position $= 0$). The channel length and the doping are the same as those of Figure 4.6. See also Color Insert.

Figure 4.10 shows the carrier velocity in bulk silicon and a MOSFET channel as a function of the electric field. The saturation velocity is similar regardless of the carrier type or the effective normal field. The critical field, E_c, which can be regarded as an onset point of velocity saturation, depends on the carrier type and the effective

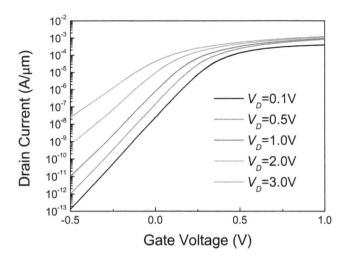

Figure 4.9. Subthreshold characteristics showing the impact of punch-through on the drain current. The channel length and the doping are the same as those of Figure 4.6. See also Color Insert.

normal field. A higher low-field mobility leads to a lower critical field.

The electron velocity, v_{dn}, in a MOSFET channel is often modeled as

$$v_{dn} = -\frac{\mu_{eff}}{1 + \left|E_y\right|/E_c} E_y, \tag{4.19}$$

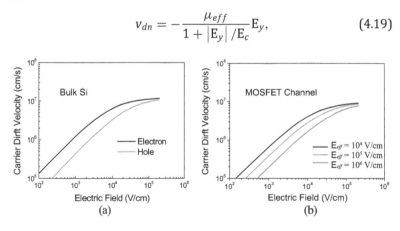

Figure 4.10. Carrier velocity vs. electric field: (a) electron and hole velocity in bulk silicon and (b) electron velocity in a silicon MOSFET channel. See also Color Insert.

where μ_{eff} is the effective mobility in the channel and E_y is the electric field in the y-direction (along the channel). Since $\lim_{E_y \to -\infty} v_{dn} = \mu_{eff} E_c = v_{sat}$ we can relate the critical field to the saturation velocity as follows:

$$E_c = \frac{v_{sat}}{\mu_{eff}} \qquad (4.20)$$

Before we calculate the drain current, we would like to establish the relationship between the quasi-Fermi potential V and the surface potential ψ_s in the channel under the strong inversion condition. In strong inversion, V and ψ_s have the following relationship:

$$\psi_s = V + \phi_F \qquad (4.21)$$

Then we can relate the lateral electric field E_y with V in the channel.

$$E_y = -\frac{d\psi_s}{dy} = -\frac{dV}{dy} \qquad (4.22)$$

Equation (4.19) can be rewritten as

$$v_{dn} = \frac{\mu_{eff}}{1 + (dV/dy)/E_c}(dV/dy). \qquad (4.23)$$

Now, we can calculate the drain current using Eqs. (4.23) and (3.78) from Section 3.2.3, Chapter 3.

$$I_D = -WQ_n v_{dn}$$
$$= WC_{ox}(V_G - V_T - \gamma V)\frac{\mu_{eff}}{1 + (dV/dy)/E_c}(dV/dy) \qquad (4.24)$$

Multiplying both sides by $[1+(dV/dy)/E_c]$ and integrating, we obtain

$$I_D\left[L + (V_D/E_c)\right] = \mu_{eff} WC_{ox}\left[(V_G - V_T)V_D - \frac{\gamma V_D^2}{2}\right]. \qquad (4.25)$$

Finally, the drain current is given as

$$I_D = \frac{\mu_{eff} WC_{ox}}{L\left[1 + (V_D/E_c L)\right]}\left[(V_G - V_T)V_D - \frac{\gamma V_D^2}{2}\right]. \qquad (4.26)$$

If $E_c L \gg V_D$ (long channel or no velocity saturation), this equation becomes identical to Eq. (3.74) in Section 3.2.3, Chapter 3.

Let us now determine the saturation voltage, V_{Dsat}, and the saturation current, I_{Dsat}. V_{Dsat} is the voltage at which the drain current reaches its maximum.

$$\left.\frac{\partial I_D}{\partial V_D}\right|_{V_D=V_{Dsat}} = 0 \qquad (4.27)$$

From this equation, we obtain

$$V_{Dsat} = \frac{2(V_G - V_T)/\gamma}{1 + \sqrt{1 + 2(V_G - V_T)/(\gamma E_c L)}}. \qquad (4.28)$$

Substituting this equation into Eq. (4.26), we can find I_{Dsat}.

$$I_{Dsat} = v_{sat} W C_{ox}(V_G - V_T) \frac{\sqrt{1 + 2(V_G - V_T)/(\gamma E_c L)} - 1}{\sqrt{1 + 2(V_G - V_T)/(\gamma E_c L)} + 1} \qquad (4.29)$$

Both V_{Dsat} and I_{Dsat} are smaller than the corresponding values without velocity saturation. If the channel length, L, approaches zero, V_{Dsat} becomes infinitesimally small and I_{Dsat} becomes

$$I_{Dsat} = v_{sat} W C_{ox}(V_G - V_T). \qquad (4.30)$$

In contrast to the long-channel case where the saturation current is proportional to the square of the gate overdrive, $(V_G - V_T)$, here the saturation current is linearly dependent on the gate overdrive. The drain current will increase much more slowly as a function of the gate bias than is the case for the long-channel MOSFETs.

4.1.5 MOSFET Scaling Theory

In order to cope with the short-channel effects, device engineers have developed a systematic way of shrinking MOSFET devices called **scaling theory**. In such a theory, a maximum of two (λ, α) **scaling factors** are used to vary the device parameters systematically. In 1974, Dennard et al. [1] proposed constant electric field scaling, in which one scaling factor is used to scale all the device dimensions and bias voltages. Ten years later, Baccarani et al. [2] came up with generalized scaling, where two scaling factors are used. One of them is for the device dimensions and the other is for the electric field. Since generalized scaling includes constant electric field scaling as a special case, we would like to describe generalized scaling first and discuss constant electric field scaling and constant voltage scaling as special cases.

In generalized scaling, we scale all of the device dimensions with a multiplication factor, $1/\lambda$, and we allow the electric field to scale by a factor of α. Since the potential is obtained by integrating the electric field with respect to the position, the potential or bias voltage

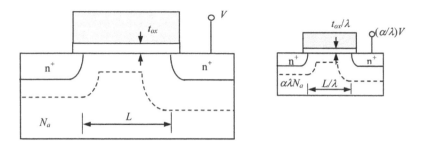

Figure 4.11. Method of generalized scaling.

will scale by a factor of α/λ. After scaling, the device dimensions are expressed as follows:

$$x' = x/\lambda, \quad y' = y/\lambda, \quad z' = z/\lambda. \qquad (4.31)$$

In addition, the potentials and bias voltages become

$$\psi' = \frac{\alpha}{\lambda}\psi, \quad V' = \frac{\alpha}{\lambda}V. \qquad (4.32)$$

Figure 4.11 illustrates the method of generalized scaling.

The main idea behind the generalized scaling is to keep the Poisson equation invariant. If we keep the Poisson equation invariant, the electric field pattern is preserved after scaling, so that the device does not suffer from short-channel effects. The Poisson equation is written as

$$\frac{\partial^2 \psi}{\partial x^2} + \frac{\partial^2 \psi}{\partial y^2} + \frac{\partial^2 \psi}{\partial z^2} = -\frac{e}{\varepsilon_s}(N_d - N_a + p - n). \qquad (4.33)$$

In the neutral region of the device, the Poisson equation is automatically satisfied regardless of the doping concentration. Hence, we will focus here on the depletion region. If the substrate is p-type, we can retain only the acceptor concentration term inside the parenthesis of the right-hand-side of Eq. (4.33). Substituting Eqs. (4.31) and (4.32) into Eq. (4.33), we obtain

$$\frac{\partial^2 \left(\frac{\alpha}{\lambda}\psi\right)}{\partial \left(\frac{x}{\lambda}\right)^2} + \frac{\partial^2 \left(\frac{\alpha}{\lambda}\psi\right)}{\partial \left(\frac{y}{\lambda}\right)^2} + \frac{\partial^2 \left(\frac{\alpha}{\lambda}\psi\right)}{\partial \left(\frac{z}{\lambda}\right)^2} = \frac{eN'_a}{\varepsilon_s}. \qquad (4.34)$$

From this equation, we can extract an appropriate scaling factor for the doping concentration.

$$N'_a = \alpha\lambda N_a \qquad (4.35)$$

The only potential problem concerns the inversion region of a MOSFET. Since the inversion charge is dominant in this region, the scaling of doping concentration does not guarantee the appropriate scaling of the inversion charge. Indeed, there can be some non-scaling parameters caused by this inversion charge non-scaling.

If $\alpha = 1$, generalized scaling becomes constant electric field scaling. In this case, the electric field is constant and all the bias voltages scale by the same factor as the dimensional factor, $1/\lambda$. The fact that the electric field is constant brings many benefits to the device operation. Since the electric field does not increase, the effect of velocity saturation will be negligible and the reliability issue associated with high electric field will disappear. The greatest advantage of constant electric field scaling is probably in its power density (power consumption per unit area). Both the voltage and the current scale by a factor of $1/\lambda$, so that the power consumed by one device scales by a factor of $1/\lambda^2$. Since the circuit density scales by λ^2, the power density remains constant throughout scaling. As long as we follow constant electric field scaling, we do not have to worry about the power density.

There are, however, non-scaling device parameters in constant electric field scaling. The flat-band voltage, V_{FB}, and the bulk Fermi potential, ϕ_F, do not scale with the factor, $1/\lambda$. As a consequence, the threshold voltage does not scale appropriately. The non-scaling of threshold voltage poses a serious problem in voltage scaling, since we cannot lower the gate voltage below the threshold voltage. The subthreshold operation of a MOSFET is not impossible but will reduce the circuit speed drastically.

Due to the existence of non-scaling parameters and circuit compatibility issues, constant electric field scaling was not used for quite a long time. Instead, constant voltage scaling was used until the early 1990s. In constant voltage scaling, α is equal to λ, so the voltage remains constant throughout scaling. Because of the constant voltage, the non-scaling of V_{FB}, ϕ_F, and V_T does not pose a problem here. Since the current scales by a factor of λ before the onset of velocity saturation, the circuit delay was reduced rapidly and the clock frequency increased exponentially during this period.

One of the major concerns related to constant voltage scaling is the reliability issue caused by a high electric field. High electric fields can bring about many detrimental effects such as punch-through, hot carrier effect, and breakdown. Another issue is the high power density. In constant voltage scaling, the power density scales by a factor of λ^3 (before velocity saturation) or λ^2 (after velocity saturation). The lack of cost-effective cooling solution and the sharp increase in the demand of low-power mobile applications brought an end to the era of constant voltage scaling.

Once the barrier of constant voltage (5 V) was broken, the bias voltage began to be reduced rapidly, following more or less the prescriptions of constant electric field scaling. Many reliability issues have been remedied, and the power density has been retained at a manageable level. The reduction of the bias voltage, however, cannot continue endlessly. The main reason is the non-scaling subthreshold swing. In order to reduce the gate voltage, the threshold voltage should be reduced. The threshold voltage may be reduced by adjusting the doping profile in the channel or the device structure as will be discussed in Section 4.2. Due to the non-scaling property of the subthreshold swing, however, the off current (I_{off}) of MOSFET increases exponentially if we reduce the threshold voltage. Since the increase in the off current is the source of standby power increase, we cannot scale the threshold voltage arbitrarily. At present, we are observing a saturation of the bias voltage near 1 V. For further voltage scaling, we need to find out a way to reduce the subthreshold swing significantly.

4.1.6 *Dopant Number Fluctuation*

As the device dimension decreases, the number of dopant atoms in the depletion region decreases. Because of the discrete nature of dopant atoms, there are naturally statistical fluctuations in the number of dopants within a given volume. The standard deviation of the number of dopants around its average value N is given by

$$\sigma_N = \sqrt{N}. \tag{4.36}$$

When N is large, the standard deviation is just a small portion of N. When N is small, however, σ_N becomes a significant portion of N, as will be shown in the following example.

Example 4.1: Dopant number fluctuation

Calculate the standard deviation of the number of dopants in the depletion region for the given geometrical and doping parameters of a MOSFET. In addition, express the standard deviation as the percentage of the average number of dopants.

(a) The channel length (L) and width (W) are 500 nm, and the doping concentration (N_a) is 10^{17} cm^{-3}
(b) $L = W = 50$ nm, and $N_a = 10^{18}$ cm^{-3}

Solution:

(a) For $N_a = 10^{17}$ cm^{-3}, $W_{dm} = 100$ nm. Then, $N = N_a LWW_{dm} = 2500$. $\sigma_N = N^{1/2} = 50$. It is 2% of the average number of dopants in the depletion region.
(b) For $N_a = 10^{18}$ cm^{-3}, $W_{dm} = 35$ nm. Then, $N = N_a LWW_{dm} = 87.5$. $\sigma_N = N^{1/2} = 9.35$. It is 10.7% of the average number of dopants in the depletion region.

Let us calculate the variation in threshold voltage due to dopant number fluctuation. We will derive a simple first-order model of the dopant number fluctuation. We consider an infinitesimal volume element, $dxdydz$, at (x, y, z) in Fig. 4.12(a). The average dopant number is $N_a dxdydz$, and the standard deviation is $\sigma_{dN} = (N_a dxdydz)^{1/2}$. For simplicity, the effect of this doping fluctuation is assumed to be equivalent to that of a uniform delta-function implant of dose ΔD and depth x as shown in Fig. 4.12(b).

The standard deviation of the dose is

$$\Delta D = \frac{\sigma_{dN}}{WL} = \frac{(N_a dxdydz)^{1/2}}{WL}. \tag{4.37}$$

Because of this sheet charge, the depletion width shrinks to W'_{dm}.

$$2\phi_F = \frac{eN_a W'^2_{dm}}{2\varepsilon_s} + \frac{xe\Delta D}{\varepsilon_s} = \frac{eN_a W^2_{dm}}{2\varepsilon_s} \tag{4.38}$$

Solving this equation, we obtain

$$W'_{dm} = \sqrt{W^2_{dm} - \frac{2x\Delta D}{N_a}} \cong W_{dm} - \frac{x\Delta D}{N_a W_{dm}}. \tag{4.39}$$

(a)   (b)

Figure 4.12. Position and shape of charge fluctuation: (a) the infinitesimal volume element where the charge fluctuation occurs, and (b) simplifying assumption of the charge fluctuation. In (b), the effect of the dopant fluctuation in (a) is assumed to be equivalent to that of a uniform delta function implant of dose ΔD and depth x. The top plate represents the semiconductor surface. See also Color Insert.

The threshold voltage for this charge fluctuation is

$$V_T' = V_{FB} + 2\phi_F + \frac{(eN_a W_{dm}' + e\Delta D)}{C_{ox}}$$

$$= V_{FB} + 2\phi_F + \frac{1}{C_{ox}}\left[eN_a W_{dm} + e\Delta D\left(1 - \frac{x}{W_{dm}}\right)\right]. \quad (4.40)$$

Thus, the change in the threshold voltage caused by the charge fluctuation at (x, y, z) is

$$\Delta V_T = \frac{e\Delta D}{C_{ox}}\left(1 - \frac{x}{W_{dm}}\right). \quad (4.41)$$

In this equation, we can see that the change in the threshold voltage is largest when $x = 0$ and it decreases as x approaches W_{dm}. This phenomenon can be understood by examining Fig. 4.13 The blue curve shows the electric field profile without a charge

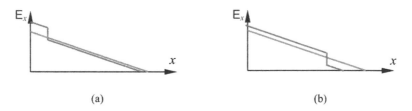

(a)   (b)

Figure 4.13. Dependence of the electric field profile on the position of charge fluctuation: (a) electric field profile when x is small and (b) electric field profile when x is close to W_{dm}. See also Color Insert.

fluctuation, while the red profile shows the electric field with a charge fluctuation. The step at the red curve occurs at the position where the charge fluctuation occurs. The amount of the charge fluctuation is reflected in the height of this step. Due to the condition given in Eq. (4.40), the area under the electric field profile should be the same ($2\phi_F$) regardless of the position or amount of charge fluctuation. If x is small, the surface electric field, $E_x(0)$ is large, so the ΔV_T is large. If x increases, $E_x(0)$ and ΔV_T decrease.

Now the variance of the threshold caused by the dopant number fluctuation at (x, y, z) is

$$\Delta V_T^2 = \frac{e^2 N_a}{C_{ox}^2 L^2 W^2} \left(1 - \frac{x}{W_{dm}}\right)^2 dx dy dz. \qquad (4.42)$$

If we add up all these variances coming from the dopant number fluctuation at each point of the depletion region, we obtain the total variance in the threshold voltage fluctuation.

$$\sigma_{V_T}^2 = \frac{e^2 N_a}{C_{ox}^2 L^2 W^2} \int_0^W \int_0^L \int_0^{W_{dm}} \left(1 - \frac{x}{W_{dm}}\right)^2 dx dy dz \qquad (4.43)$$

Integrating Eq. (4.44), we obtain

$$\sigma_{V_T} = \frac{e}{C_{ox}} \sqrt{\frac{N_a W_{dm}}{3 L W}}. \qquad (4.44)$$

Thus, the variation of the threshold voltage increases as the doping concentration increases, while it decreases as the channel area increases.

4.1.7 p-n Junction and Oxide Tunneling

MOSFET scaling theory prescribes higher channel doping and thinner oxide for a shorter channel device. The continuous reduction of channel length requires extremely high channel doping and ultra-thin oxide for the current generation of devices. Both high channel doping and ultra-thin oxide bring an unwanted quantum phenomenon: tunneling. If we try to follow the conventional scaling theory mindlessly, we will face a serious leakage problem because of tunneling.

According to the scaling theory, the doping concentration required for a 45 nm channel MOSFET is about 5×10^{18} cm^{-3}. With this high concentration, we must take into account the band-to-band tunneling current at the $p^+ - n^+$ junction. The tunneling current is given by Fair and Wivell (1976) [3].

$$J_{B-B} = \frac{\sqrt{2m^*}e^3 E V_R}{4\pi^3\hbar^2 E_g^{1/2}} \exp\left(-\frac{4\sqrt{2m^*}E_g^{3/2}}{3eE\hbar}\right), \qquad (4.45)$$

where m^* is the effective mass of electron, E_g is the energy gap of silicon, V_R is the reverse bias applied to the junction, and E is the peak electric field given by

$$E = \sqrt{\frac{2eN_a(V_0 + V_R)}{\varepsilon_{si}}}. \qquad (4.46)$$

As will be seen in Example 4.2, a doping of 5×10^{18} cm^{-3} results in a very large current density. In order to keep the leakage current low, such a high doping concentration should be avoided.

Example 4.2: Band-to-band tunneling current

We applied 1 V reverse-bias voltage at a p^+-n^+ junction with $N_d = 5 \times 10^{20}$ cm^{-3} and $N_a = 5 \times 10^{18}$ cm^{-3}. Calculate the tunneling current density, J_{B-B}.

Solution:

The built-in potential can be assumed to be almost the same as the energy gap of silicon in this p^+-n^+ junction. Thus, $V_0 \approx 1.1$ eV. The peak electric field is about 1.8×10^6 V/cm. $J_{B-B} \approx 1$ A/cm^2.

A high-performance MOSFET with a channel that is a few decananometers in length requires an ultra-thin oxide of about 1 nm. Such a thin oxide, however, is bound to lose its excellent insulating ability because of tunneling. As shown in Fig. 4.14, there are two types of tunneling mechanisms in oxide. One is Fowler–Nordheim (F–N) tunneling explained in Section 3.3.4, Chapter 3, and the other is direct tunneling. F–N tunneling is a dominant mechanism in thick oxides, since it relies on a reduction in the effective thickness caused by a high electric field. The F–N tunneling current appears at a

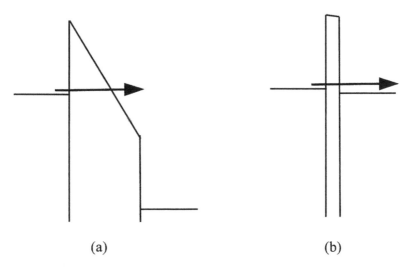

(a)   (b)

Figure 4.14.   Two tunneling mechanisms in oxide: (a) Fowler–Nordheim and (b) direct tunneling.

relatively high gate voltage and shows a strong dependence on the gate voltage. On the other hand, direct tunneling is the dominant mechanism in thin oxides and the dependence of the direct tunneling current on the gate voltage is relatively weak since the tunnel barrier thickness remains the same.

Figure 4.15 shows the oxide tunneling current density as a function of the gate voltage. From Fig. 4.15(a), we can see that the tunneling current in relatively thick oxides ($t_{ox} > 3$ nm) has a strong dependence on the gate voltage [4]. At low gate voltages, the tunneling current is completely negligible. The tunneling current in relatively thin oxides ($t_{ox} < 3$ nm) deviates significantly from the F–N tunneling model and is less strongly dependent on the gate voltage. This phenomenon is caused by the onset of direct tunneling. Figure 4.15(b) depicts the characteristics of thin oxides [5]. As expected, the tunneling current has a relatively weak dependence on the gate voltage but is strongly dependent on the oxide thickness. In this figure, we can also see that the oxide thickness should be thicker than 1.6 nm in order to maintain the leakage current below $1A/cm^2$ at a gate voltage of 1 V.

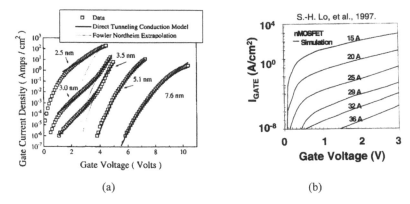

(a)   (b)

Figure 4.15. Tunneling current density as a function of the gate voltage: (a) Thicker oxides. Reproduced with permission from K.F. Schuegraf, D. Park, and C. Hu, "Reliability of Thin SiO$_2$ at Direct Tunneling Voltage," *IEDM Tech. Dig.*, p. 609, 1994. © 1994 IEEE. (b) Thinner oxides. Reproduced with permission from S.-H. Lo, D.A. Buchanan, Y. Taur, and W. Wang, "Quantum Mechanical Modeling of Electron Tunneling Current from the Inversion Layer of Ultrathin Oxide nMOSFETs," *IEEE Electron. Device. Lett.*, vol. 18, no. 5, pp. 209–211, 1997. © 1997 IEEE.

4.2 Approaches to Overcoming Scaling Issues in Nanoscale MOSFETs

4.2.1 *Device Structure Engineering*

As described in the previous section, MOSFET scaling for high integration density, high performance, and low power has created many issues. Conventional scaling theory based on the planar MOSFET structure has been successfully used down to 100 nm channel devices, but it is now confronted with a few new challenges in addition to the traditional ones. To cope with the scaling issues, various new structures and materials have been proposed and implemented.

In Fig. 4.16, three device structures are shown for comparison. As we move from planar to double-gate structure, we can see that the control of the gate over the channel is enhanced. In the ultra-thin body (UTB) structure, only the part of the channel that is close to the gate remains. In the double-gate (DG) structure, the thin channel is controlled by two gates that are located very close to the channel. The gray triangles show the depletion region generated by the

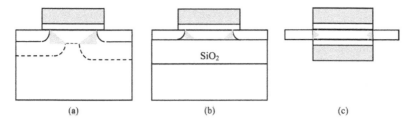

Figure 4.16. Comparison of three device structures: (a) planar structure, (b) ultra-thin body (UTB) structure, and (c) double-gate (DG) structure.

source/drain. We can clearly see the improvement of short-channel effects in the UTB and DG structures since the area of the triangles shrinks as we move from Fig. 4.16(a) to 4.16(c).

The UTB structure is often called a fully depleted silicon-on-insulator (FD-SOI) structure. CMOS devices are fabricated in a crystalline silicon layer on an insulator (usually silicon dioxide) layer. The silicon-on-insulator (SOI) wafers are fabricated by two methods. One method is called separation by implanted oxide (SIMOX), in which oxygen is implanted with an extremely high dose $(1 \times 10^{18} \text{ cm}^{-2})$ and a buried oxide layer is formed with high-temperature annealing. When the buried oxide is formed, the silicon layer on top of it is recrystallized. Another method is wafer bonding and silicon thinning on one side. A very thin buried oxide layer and a high-quality back interface can be created by this method.

In the UTB structure, the thin silicon layer acts as a single unit. The gate affects the back surface potential and the back gate affects the front surface potential. The source/drain junctions are automatically shallow due to the buried oxide. As shown in Fig. 4.16(b), short-channel effects can be suppressed significantly, since charge sharing is reduced in comparison with the planar structure. In order to maintain a good short-channel behavior, however, the buried oxide should be relatively thin. Otherwise, the drain potential can penetrate the buried oxide layer laterally and influence the channel.

The DG structure usually requires a somewhat complicated process sequence, which deviates significantly from the planar process. As will be described in Section 4.3.1, there can be many ways to position the channel, source/drain, and gate, once we make a departure from the planar surface. Despite the variety of device fabrication methods, the operating principles of DG MOSFETs are basically the same. If the body is relatively thick, we can treat the structure as a combination of two UTB structures with a common gate bias. If the body is very thin, bulk inversion occurs and the body should be regarded as one unit. Short-channel effects are well controlled in DG MOSFETs, since the depletion region generated by the source/drain is much smaller in this structure.

4.2.2 Gate Insulator and Stack Engineering

In Section 4.1.7 we found that due to the onset of direct tunneling, the oxide tunneling current can be extremely high for ultra-thin (<3 nm) SiO_2 layers. The tunneling current can be tolerated as long as its value is smaller than the off current of a MOSFET. If the SiO_2 thickness is reduced below 1.5 nm, however, the tunneling is excessively high (> 10 A/cm^2). We definitely need a method to solve this issue.

Since the physical thickness is important in tunneling, we may avoid direct tunneling by using a physically thick material that can still exhibit the electrical property of an ultra-thin silicon oxide layer. What is the electrical property required for appropriate operation of a MOSFET then? It is the large capacitance per unit area. The reason we reduce the oxide thickness is to increase the capacitance, so that the coupling between the gate electrode and the channel is increased. Let us examine the equation for the capacitance of an insulator per unit area. If the insulator has a dielectric constant of κ and a thickness of t_i, the capacitance per unit area is

$$C_i = \frac{\kappa \varepsilon_0}{t_i}, \tag{4.47}$$

where ε_0 is the permittivity of vacuum. In this equation, we can see that the capacitance can be increased by increasing κ instead of decreasing t_i. What we have to do is to use a material with a high dielectric constant (κ) as a gate insulator.

There are many insulators that have a dielectric constant higher than that of SiO_2. Table 4.1 shows the dielectric constant of several materials that have been used or considered as a gate dielectric material. Out of these, hafnium dioxide (HfO_2) has been most

Table 4.1.   Dielectric constants of various insulator materials

Insulator	Dielectric constant
SiO_2	3.9
Si_3N_4	7
Al_2O_3	9
Y_2O_5	15
HfO_2	25
ZrO_2	25
La_2O_3	30

intensively studied as a gate dielectric material and a hafnium-based gate dielectric material is actually used in manufacturing. Depending on the application, Si_3N_4 and Al_2O_3 have also been studied and used.

As these high-κ dielectric materials have been introduced to MOSFET devices only recently, there are still several issues that must be worked out. Most of the high-κ materials have a bandgap smaller than SiO_2, for example, so they cannot be used for high-voltage applications. Since the high-κ materials are deposited on a silicon surface, there may be a large number of interface traps which degrade carrier mobility. Often, an ultra-thin SiO_2 layer is grown to reduce the interface trap density. Many high-κ materials have thermal instability at high temperatures where the dopant activation occurs for poly-silicon gate and source/drain doping. Metal gates with a gate replacement technique are used to avoid the thermal instability issue.

When we fabricate DG MOSFETs, the channel doping concentration is kept very low to minimize the effect of dopant number fluctuation. If we use a channel with a low doping concentration and an n^+ poly-silicon gate, however, we end up with a negative threshold voltage in an n-MOSFET. Since CMOS circuits require a positive threshold voltage for an n-MOSFET, we have to adjust the threshold voltage through work function engineering. By changing the gate electrode material to a metal or a silicide with a larger work function (Eq. (3.1), Section 3.1.1, Chapter 3), we can move the flat-band voltage to a less negative value and, consequently, shift the threshold voltage toward the positive direction.

The use of a metal or silicide gate can solve another problem called **poly depletion**. Even though it is heavily doped, the polysilicon is still a semiconductor, and there is a finite depletion region near the oxide interface. Obviously, this depletion region increases the effective oxide thickness. On the other hand, a metal or silicide has a negligible depletion width near the oxide interface and does not suffer from the poly depletion problem. Another advantage of using metal gates is low resistance. The concerns with metal gates are oxide integrity and compatibility with high-temperature processes.

4.2.3 *Channel and Source/Drain Engineering*

Conventional scaling theory assumes uniform channel doping and requires nothing more than higher doping as the device size is scaled down. There are a number of conflicting requirements, however, if we consider the non-scaling parameters and performance criteria for various applications. In order to reduce the subthreshold swing and the threshold voltage, for example, a low doping concentration is required. We can satisfy these conflicting requirements by utilizing a non-uniform doping profile. We just increase the doping concentration only where it is necessary.

One really useful method is to use the retrograde (low-high) channel profile (Fig. 4.17). With this profile, we can reduce the

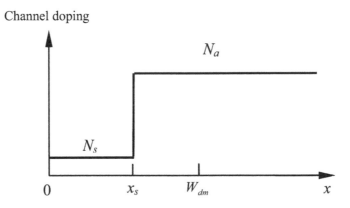

Figure 4.17.   Retrograde (low-high) channel profile.

threshold voltage without having an excessive depletion width. Let us assume $N_s \approx 0$ and find the threshold voltage of the MOSFET with this channel doping profile. Using the fact that all the depletion charge resides in the region between x_s and W_{dm}, we can directly calculate the threshold voltage.

$$V_T = V_{FB} + 2\phi_F + \frac{eN_a}{C_{ox}}(W_{dm} - x_s) \tag{4.48}$$

W_{dm} can be calculated by the strong inversion condition.

$$2\phi_F = \frac{1}{2}\frac{eN_a}{\varepsilon_s}(W_{dm} - x_s)^2 + \frac{eN_a}{\varepsilon_s}(W_{dm} - x_s)x_s \tag{4.49}$$

Solving this quadratic equation, we obtain

$$W_{dm} = \sqrt{\frac{4\varepsilon_s\phi_F}{eN_a} + x_s^2}. \tag{4.50}$$

For sufficiently large N_a, we can approximate Eq. (4.50) by expanding it with a Maclaurin series and retaining the first-order term as follows:

$$W_{dm} \approx x_s + \frac{2\varepsilon_s\phi_F}{eN_a x_s} \tag{4.51}$$

Substituting Eq. (4.51) into Eq. (4.48), we obtain

$$V_T = V_{FB} + 2\phi_F + \frac{2\varepsilon_s\phi_F}{C_{ox}x_s} \tag{4.52}$$

Thus, for sufficiently large N_a we can control the threshold voltage by adjusting the thickness of low-doped region.

There is another advantage of the retrograde channel: the immunity to the dopant number fluctuation. Again, let us assume $N_s = 0$ for simplicity. We can start with Eq. (4.43). This time, however, we should remember that there is no dopant number fluctuation for $0 < x < x_s$.

$$\sigma_{V_T}^2 = \frac{e^2 N_a}{C_{ox}^2 L^2 W^2}\int_0^W\int_0^L\int_{x_s}^{W_{dm}}\left(1 - \frac{x}{W_{dm}}\right)^2 dx\,dy\,dz \tag{4.53}$$

Carrying out the integration, we obtain

$$\sigma_{V_T} = \frac{e}{C_{ox}}\sqrt{\frac{N_a W_{dm}}{3LW}}\left(1 - \frac{x_s}{W_{dm}}\right)^{3/2}. \tag{4.54}$$

If we make N_a sufficiently high, we can make $W_{dm} \approx x_s$. Then, $\sigma_{V_T} \approx 0$. We can suppress the threshold voltage variation caused by dopant number fluctuation almost completely.

Another issue related to dopant number fluctuation is the channel length variation within a device caused by the random positioning of source/drain dopants. Since the average distance between dopants is about 2 nm for 10^{20} cm^{-3} doping, the channel length may vary by 1 nm or more. One way to suppress such fluctuation is to use Schottky junction source/drain. If the source and the drain are made of metal or silicide, we would not have to worry about the dopant fluctuation in source/drain. In addition, the resistance of the source/drain would decrease significantly.

4.3 Double-Gate MOSFETs

4.3.1 *Classification of Double-Gate Structures*

The earlier versions of DG MOSFETs were based on a silicon-on-insulator (SOI) substrate and took the shape shown in Fig. 4.16(c). DG structures are not necessarily implemented on an SOI substrate, however, and can take various three-dimensional forms [6], as shown in Fig. 4.18.

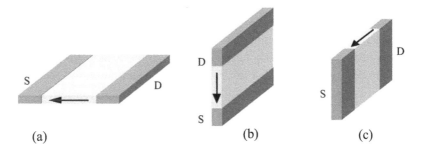

(a)   (b)   (c)

Figure 4.18. Three types of DG structures: (a) Type 1: horizontal length (L), horizontal width (W), (b) Type 2: vertical L, horizontal W, and (c) Type 3: horizontal L, vertical W. The arrow indicates the direction of current. Gates are formed on both sides of the channel, but are omitted in this figure. See also Color Insert.

The structure shown in Fig. 4.18(a) is usually fabricated on SOI and is called Type 1. The channel length and width are defined in the horizontal direction. It has the same structure as that of an SOI MOSFET except that it has another gate under the channel. As is the case with a planar structure, both the length and width of the channel are determined by patterning.

Figure 4.18(b) shows a Type 2 structure, where the current flows in the vertical direction. The channel length is defined in the vertical direction, while the channel width is defined in the horizontal direction. The channel width is determined by patterning, but the channel length is determined by the etch depth and source/drain junctions.

Figure 4.18(c) shows a Type 3 structure, where the channel length is defined in the horizontal direction and the channel width is defined in the vertical direction. In this structure, the channel length is determined by patterning, while the channel width is determined by etch depth. In order to take the advantage of the DG structure, the channel thickness (distance between one surface and the other) should sufficiently small. In the Type 1 structure, the vertical thickness control is important, while, in the Type 2 and Type 3 structures a thin channel should be implemented by patterning.

4.3.2 *Simple Model of Double-Gate MOSFETs*

Let us construct a simple model [7] for an n-channel DG MOSFET and use it to calculate the drain current. In a DG MOSFET, the same bias voltage is applied to the two gates and the front and back oxide thicknesses are identical. Such a symmetry makes the analysis quite simple. Let us first examine the electric field and charge distribution. Because of the symmetry, the front and back surfaces are under the same conditions (accumulation, depletion or inversion). As shown in Fig. 4.19, the electric field and charge distribution reflect the symmetry.

Since the front and back oxide thicknesses are the same, we can write

$$t_{of} = t_{ob} = t_o. \tag{4.55}$$

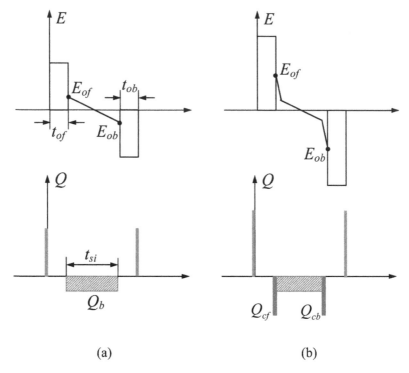

(a)   (b)

Figure 4.19. Electric field and charge distribution of an n-channel DG MOSFET: (a) under the depletion condition and (b) under the inversion condition.

The front and back capacitances are

$$C_{of} = C_{ob} = C_{ox}. \tag{4.56}$$

Now we can calculate the voltage, electric field, and channel charge as follows:

$$V_{of} = E_{of}t_{of} = -E_{ob}t_{ob} = V_{ob} \tag{4.57}$$

$$E_{of} = -E_{ob} = -\frac{1}{2}\frac{\varepsilon_s}{\varepsilon_{ox}}Q_s \tag{4.58}$$

$$Q_s = 2Q_{cf} + Q_b \tag{4.59}$$

The threshold voltage can be calculated as

$$V_T = V_{FB} + 2\phi_F - \frac{1}{2}\frac{Q_b}{C_{ox}}. \tag{4.60}$$

Usually, we leave the channel lightly doped ($Q_b \approx 0$) to suppress the dopant number fluctuation effect. Then, the threshold voltage becomes negative, as is calculated in Example 4.3, if the gate material is n^+ poly-silicon. To make the threshold voltage positive, work function engineering is required as mentioned in Section 4.2.2.

Example 4.3: Threshold voltage of an n-channel DG MOSFET

An n-channel DG MOSFET has a body with 1×10^{15} cm^{-3} p-type doping and 10 nm thickness. The gate oxide thickness is 2 nm. Assuming that the gate material is n^+ poly-silicon, calculate the threshold voltage.

Solution:

For $N_a = 1 \times 10^{15}$ cm^{-3}, $\phi_F = 0.28$ V. The flat-band voltage is $V_{FB} = -0.28 - 0.55 = -0.83$ V. The depletion charge is $Q_b = eN_a t_{Si} = 1.6 \times 10^{-19} \times 1 \times 10^{15} \times 1 \times 10^{-6} = 1.6 \times 10^{-10}$ (C/cm^2) and the oxide capacitance per unit area is $C_{ox} = 3.5 \times 10^{-13}/2 \times 10^{-7} = 1.75 \times 10^{-6}$ (F/cm^2). Q_b/C_{ox} is 9.1×10^{-5} V, which is completely negligible. Thus, $V_T = -0.27$ V.

Finally let us calculate the drain current. Using Q_{cf} and Q_{cb}, we find the drain current in the linear region to be

$$I_D = 2\mu_{eff} C_{ox} \frac{W}{L} \left[(V_G - V_T)V_D - \frac{\gamma V_D^2}{2} \right]. \qquad (4.61)$$

This current is simply twice the value of a single-gate MOSFET.

4.3.3 *Quantum Mechanical Correction to the Threshold Voltage of Double-Gate MOSFETs*

In the previous subsection, we analyzed a DG MOSFET with the assumption that the inversion charge exists is the form of a sheet at the surface. In reality, however, quantum effects push the peak concentration away from the surface and the quantized energy level above the conduction band minimum at the surface. As

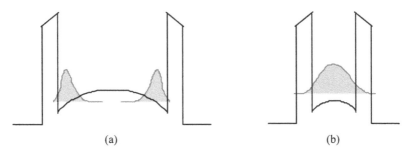

(a)   (b)

Figure 4.20.   Band diagram and charge distribution of an n-channel DG MOSFET: (a) thick channel with surface inversion (the two channels are separated), and (b) thin channel with bulk inversion (the two channels are merged).

discussed in Section 3.1.7, Chapter 3, these effects contribute to the increase of the threshold voltage. If the channel is thick, however, we can still treat the device as a combination of two single gate devices as we can see in Fig. 4.20(a). As long as the two channels are separated, the threshold voltage increase due to the quantum effect will not be affected significantly by the thickness of the channel.

If the channel is sufficiently thin, the ground-state wave function of an electron will have the shape shown in Fig. 4.20(b). The wave function has only one peak, so that there will be a bulk inversion. In addition, the lowest available energy level will be significantly above the conduction band minimum. If the body thickness is just a few nanometers, the ground-state energy level can be tens to hundreds of mV higher than the conduction band minimum as will be seen in Example 4.4.

Example 4.4: Ground-state energy level of an n-channel DG MOSFET with a thin body

An n-channel DG MOSFET has a body with 2 nm thickness. Assuming that the body of the DG MOSFET is a quantum well with infinite potential barriers, calculate the ground-state energy level. The effective mass of electrons is 8.3×10^{-31} kg.

Solution:

From Eq. (1.40c), Section 1.1.5, Chapter 1,

$$E_1 = \frac{\pi^2 \hbar^2}{2m^* t_{si}^2}$$

$$= \frac{3.14^2 \times (1.06 \times 10^{-34})^2}{2 \times 8.3 \times 10^{-31} \times (2 \times 10^{-9})^2} = 1.67 \times 10^{-20} \,(\text{J}).$$

The ground-state energy level is 0.1 eV above the conduction band minimum.

Using a simple quantum well model with a flat bottom and infinite barriers, we can write a quantum mechanical correction term for the threshold voltage of an n-channel DG MOSFET.

$$\Delta V_T^{QM} = \frac{\pi^2 \hbar^2}{2m^* t_{si}^2} \tag{4.62}$$

We have neglected the additional voltage drop across the oxide by assuming that the inversion charge is still negligible at the onset of the strong inversion. Finally, the threshold voltage of an n-channel DG MOSFET can be written as

$$V_T = V_{FB} + 2\phi_F - \frac{1}{2}\frac{Q_b}{C_{ox}} + \frac{\pi^2 \hbar^2}{2m^* t_{si}^2}. \tag{4.63}$$

Figure 4.21 shows the threshold voltage as a function of the body thickness, with the quantum mechanical correction.

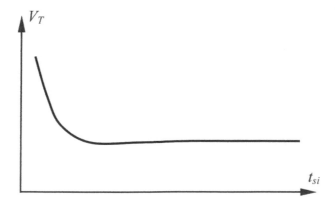

Figure 4.21. Threshold voltage of an n-channel DG MOSFET as a function of the body thickness. The effect of the quantum mechanical correction becomes conspicuous when the body is thinned down to just a few nanometers.

4.4 Tunneling and Resonant Tunneling Devices

4.4.1 *Tunneling Field Effect Transistor*

If we substitute the n^+-source with a p^+-source in an n-channel MOSFET as shown in Fig. 4.22, we obtain a tunneling field effect transistor (TFET). In this device, the carrier injection from the source occurs through an induced tunnel junction between the source and the channel. Unlike normal MOSFETs, this structure has an asymmetric structure.

Figure 4.23 shows how a TFET operates. When the gate bias voltage is below the threshold voltage, electron tunneling from the valence band of the source to the conduction band of the channel is blocked. Holes can be injected into the channel because of the lowered barrier, but they are blocked at the large potential barrier formed at the reverse-biased channel-drain junction. If the gate bias voltage is above the threshold voltage, electrons tunnel from the valence band of the source to the conduction band of the channel. Once injected into the channel, the electrons can be transported as in a normal MOSFET.

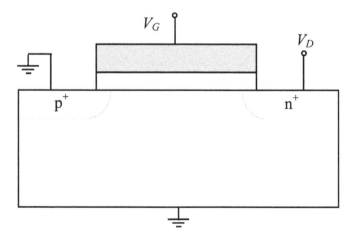

Figure 4.22. Tunneling field effect transistor (TFET). A p^+-source is used instead of the n^+-source in order to use the induced tunnel junction between the source and the channel.

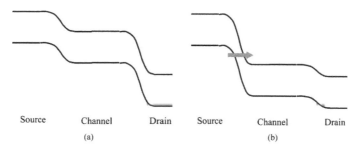

Source   Channel   Drain   Source   Channel   Drain

(a)   (b)

Figure 4.23. Band diagram of a TFET: (a) Off-state: electron tunneling from the valence band of the source to the conduction band of the channel is blocked. (b) On-state: electrons tunnel from the valence band of the source to the conduction band of the channel.

The novelty and advantage of a TFET lies with its carrier injection mechanism. Since the tunneling rate depends on the barrier thickness rather than the barrier height, there is a possibility that the subthreshold swing might be smaller than the thermal injection limit (60 mV/decade at room temperature). Figure 4.24 shows such an example in a fabricated TFET [8].

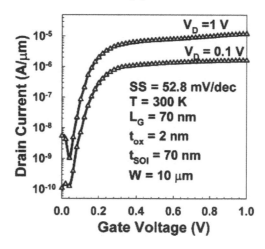

Figure 4.24. Subthreshold characteristics of a TFET. This device shows a subthreshold swing value less than 60 mV/decade at room temperature. Reproduced with permission from W. Y. Choi, B.-G. Park, J.D. Lee, and T.-J. King Liu, "Tunneling Field-Effect Transistors (TFETs) with Subthreshold Swing (SS) Less Than 60 mV/dec," IEEE Electron. Device. Lett., vol. 28, no. 8, pp. 743–745, 2007. © 2007 IEEE.

4.4.2 Resonant Tunneling Diodes

Instead of using a p^+-n^+ junction, we can generate a negative differential resistance by using the resonance of waves between two potential energy barriers. Figure 4.25 shows that such a resonance can actually occur in a double barrier structure. A huge resonance peak can be generated in the quantum well formed by the two potential barriers if the multiply-reflected partial waves interfere constructively within the well. This is analogous to the resonance of optical waves in a Fabry–Perot etalon formed by two partially reflecting mirrors. When such a resonance occurs, the transmission coefficient becomes 1, as shown in Fig. 4.26, and the tunneling current reaches its maximum.

Figure 4.26 shows the transmission coefficients of single and double tunneling barriers. For a single barrier potential, the transmission coefficient increases monotonically as the incident electron energy increases. This reflects the fact that the attenuation of the electron wave function within the barrier weakens exponentially as the incident energy increases. For the double barrier structure, the transmission coefficient generally remains much smaller than that of the single barrier structure because there are two barriers in series. At a few specific incident energies, however, the transmission increases dramatically. These transmission resonances occur when the multiply reflected waves interfere destructively on the cathode side to cancel the reflected incident

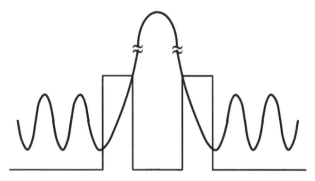

Figure 4.25. Shape of the wave function at the resonance in a double potential barrier structure.

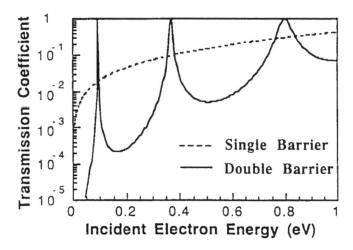

Figure 4.26. Transmission coefficients vs. incident electron energy for a single barrier and a double barrier structure. The barrier is assumed to be 6-monolayer (1.7 nm) AlAs. In the double barrier structure, a 20-monolayer (5.65 nm) GaAs well is assumed to lie between two such barriers. The surrounding material is GaAs. Only the Γ-point minimum in the conduction band is considered.

wave and interfere constructively on the anode side, so that the transmitted waves add up to the value of the incident wave (unity transmission).

Now let us find the resonant tunneling condition with two δ function potentials representing the double potential barrier. Figure 4.27 shows the potential and the propagating waves. Using the continuity of wave functions, we obtain the following relationship between coefficients:

$$\begin{cases} 1 + r_{2B} = A + B \\ -\frac{\hbar^2}{2m}ik[A - B - (1 - r_{2B})] = -E_B b(1 + r_{2B}) \end{cases} \tag{4.64}$$

Integrating the Schrödinger equation at the boundary, we obtain another set of equations:

$$\begin{cases} Ae^{ika} + Be^{-ika} = t_{2B}e^{ika} \\ -\frac{\hbar^2}{2m}ik[t_{2B}e^{ika} - (Ae^{ika} - Be^{-ika})] = -E_B bt_{2B}e^{ika} \end{cases} \tag{4.65}$$

At resonance, the following condition should be satisfied:

$$r_{2B} = 0 \tag{4.66}$$

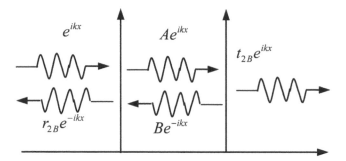

Figure 4.27. Double barrier potential and the propagating waves.

Using this condition, we can find A and B in Eq. (4.64):

$$\left(\begin{array}{l} A = \frac{mE_B b}{ik\hbar^2} + 1 \\ B = -\frac{mE_B b}{ik\hbar^2} \end{array} \right. \tag{4.67}$$

After eliminating t_{2B} in Eq. (4.65), we substitute A and B into Eq. (4.65). Then we can obtain a simple condition for k at resonance:

$$\tan ka = -\frac{\hbar^2 k}{mE_B b} \tag{4.68}$$

We can solve this equation with a graphical method. Figure 4.28 shows the plot of the left-hand-side and right-hand-side function in Eq. (4.68). When $E_B b$ is large, $ka \approx n\pi$. In this case, the resonance energy is close to the bound state energy of an infinitely deep quantum well. When $E_B b$ is small, $ka \approx (n+1/2)\pi$. The resonance energy is located between the bound state energies of an infinitely deep quantum well.

If we change the bias voltage across a resonant tunneling diode, the position of the resonance energy can be changed as shown in Fig. 4.29, so that we can generate negative differential resistance (NDR). The sharply defined resonance can be regarded as an "energy filter" with which we can scan the electron distribution at the cathode side as a function of longitudinal energy. The electron energy distribution at the cathode is determined by the Fermi energy, which is controlled by the doping level. The peak current occurs when the resonance energy is aligned with the bottom of the cathode conduction band; the valley current occurs when the lowest resonance energy is carried below the minimum incident electron

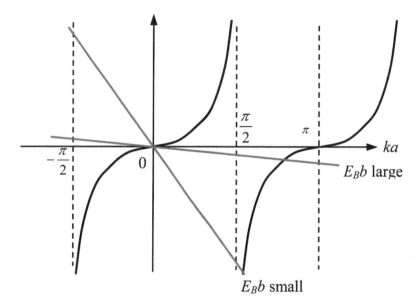

Figure 4.28. Graphical method to solve Eq. (4.68).

energy and the second resonance energy still remains too far above the Fermi level of the cathode side to contribute significantly to the current.

Figure 4.30 shows the current vs. voltage characteristics of a resonant tunneling diode. If we consider the first resonance level only, the characteristic is basically the same as that of a junction tunnel

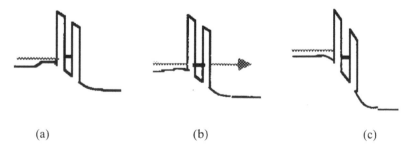

(a)   (b)   (c)

Figure 4.29. Generation of peak and valley current in a resonant tunneling diode: (a) before the peak current, (b) at the peak current, and (c) after the peak current.

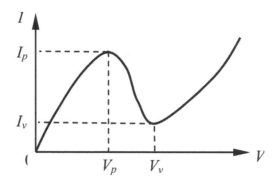

Figure 4.30. Current vs. voltage characteristic of a resonant tunneling diode. V_p and V_v are the peak and valley voltage, respectively. I_p and I_v are the peak and valley currents, respectively.

diode, and it shows an NDR or negative differential conductance (NDC). The NDR is generated when the diode current decreases between the peak and the valley. It is caused by the gradual misalignment of electron energy levels in the cathode side and the resonance energy level in the quantum well.

PROBLEMS

1. **MOSFET scaling and limitations:**

    (a) In constant voltage scaling, what is the scaling factor for the doping concentration when the dimensional scaling factor is $1/\lambda$? Explain why this should be the case.

    (b) According to the 1994 version of the SIA roadmap, the scaling of gate oxide thickness is supposed to be stopped at 3 nm. What was the reason for this?

    (c) In the 1997 version, however, they changed it and predicted gate oxide scaling down to 1 nm. What was the reason for this?

    (d) Explain the method of channel doping engineering used to suppress the dopant fluctuation effect.

2. **Quantum effects in MOSFETs:**

(a) In a MOSFET fabricated on (100) surface, what is the degeneracy of the lowest quantized energy level in the inversion layer? Explain the reason.

(b) What effect does the wave nature of inversion electrons have on the threshold voltage of a MOSFET? State whether the threshold voltage increases or decreases because of the quantum effects and give two reasons for this phenomenon.

(c) Are there any other important quantum effects that are related to a MOS structure?

3. **Drain saturation bias and current with velocity saturation effect:**

(a) Derive Eq. (4.28) from Eq. (4.26) using the condition given by Eq. (4.27).

(b) Derive Eq. (4.29) by substituting Eq. (4.28) into Eq. (4.26).

4. **Output characteristics of a MOSFET:**

When we measured a MOSFET with a channel width of 1 μm, we obtained the characteristic curves as shown below.

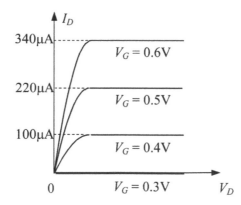

(a) Are these characteristics different from the ideal MOSFET characteristics? What is the difference? Explain the cause of such a characteristic and why we obtain these curves.

(b) From these curves, find the thickness of the gate oxide of this MOSFET.

5. **Double-gate MOSFET:**

   (a) According to the ITRS roadmap, the device structure of a MOSFET will gradually migrate toward a double-gate structure with an ultra-thin body structure in between. Explain why this would be the case.

   (b) In general, the threshold voltage of a double-gate n-MOSFET decreases as the body thickness decreases. Explain why this should be the case.

   (c) When the body of a double-gate n-MOSFET is extremely thin, the trend in (c) is reversed. What is the cause of this phenomenon? Explain the reason briefly.

6. **Threshold Voltage of a Double-Gate MOSFET:**

   (a) In a nanoscale double-gate MOSFET, the channel thickness (the thickness of Si between two gates) should be reduced along with the channel length. If we assume that the threshold voltage is determined mainly by the energy quantization effect in the channel, how would the threshold voltage change as a function of the channel thickness. Why?

   (b) Let us assume that the threshold voltage of a double-gate MOSFET is determined only by the lowest energy level in the channel. If we have a double-gate MOSFET with its channel in the $<100>$ direction and a double-gate MOSFET with its channel in the $<111>$ direction, which one would have a higher threshold voltage? Explain. (We assume that all the other device parameters except the channel direction are the same.)

   (c) In (b), what would be the degeneracy of the lowest energy level in each case (i.e., a device with its channel in the $<100>$ direction and another one with its channel in the $<111>$ direction)?

7. **Subthreshold Swing Less Than 60 mV/decade at Room Temperature:**

   (a) Explain in detail how TFET can achieve a subthreshold swing value less than 60 mV/decade at room temperature. Use the energy band diagram and explain how we can control the barrier thickness rather than the barrier height.

(b) Are there any other carrier injection mechanisms with which we can lower the subthreshold swing value below 60 mV/decade at room temperature? What would be the device structure?

8. JTD and RTD:

(a) In terms of the negative differential resistance formation mechanism, what is the main difference between a junction tunnel diode (JTD) and a resonant tunneling diode (RTD)?

(b) In terms of operation speed, which diode has a potential for higher switching speed? Why?

9. Resonant Tunneling Condition:
Derive Eq. (4.68).

Bibliography

1. R.H. Dennard, F.H. Gaensslen, H.N. Yu, V.L. Rideout, E. Bassous, and A.R. LeBlanc, "Design of ion-implanted MOSFETs with very small physical dimensions," *IEEE J. Solid-State Circuits*, vol. 9, p. 256, 1974.
2. G. Baccarani, M.R. Wordeman, and R.H. Dennard, "Generalized scaling theory and its application to a 1/4 micrometer MOSFET design," *IEEE Trans. Electron. Devices*, vol. 31, p. 452, 1984.
3. R.B. Fair and H.W. Wivell, "Zener and avalanche breakdown in As-implanted low voltage Si n-p junctions," *IEEE Trans. Electron. Devices*, vol. 23, pp. 512–518, 1976.
4. K.F. Schuegraf, D. Park and C. Hu, "Reliability of Thin SiO_2 at Direct Tunneling Voltage," International Electron Devices Meeting, 1994.
5. S.-H. Lo, D.A. Buchanan, Y. Taur, and W. Wang, "Quantum mechanical modeling of electron tunneling current from the inversion layer of ultra-thin oxide nMOSFET's," *IEEE Electron. Device Lett.*, vol. 18, pp. 209–211, 1997.
6. H.-S. P. Wong, K.K. Chan, and Y. Taur, "Self-aligned (top and bottom) double-gate MOSFET with a 25 nm thick silicon channel," International Electron Devices Meeting, pp. 427–430, 1997.

7. K. Suzuki, T. Tanaka, Y. Tosaka, H. Horie, Y. Arimoto, and T. Itoh, "Analytical surface potential expression for thin-film double-gate SOI MOSFETs," *Solid State Electron.,* vol. 37, no. 2, pp. 327–332, 1994.

8. W.Y. Choi, B.-G. Park, J.D. Lee, and T.-J. King Liu, "Tunneling field-effect transistors (TFETs) with subthreshold swing (SS) less than 60 mV/dec," *IEEE Electron. Device Letters*, vol. 28, no. 8, pp. 743–745, 2007.

Chapter 5

Quantum Wire Devices

5.1 Transport in One-Dimensional Electron Systems

5.1.1 *Backgrounds*

As explained in previous chapters, conduction band electrons of n-type semiconductors randomly move around and experience various scattering events. They are scattered either by lattice vibrations or by ionized impurities. The randomness of the electron motion and the scattering results in zero average velocity so that the current through the external contact should be zero. Under external electric fields, these electrons are accelerated in the direction of the electric field in between scattering events. Eventually, the average acceleration process and the average bouncing back due to scattering balance each other, so that the electrons have a non-zero average steady state velocity that is parallel to the direction of the electric field. All of these movements are three-dimensional (3D) because the re-direction of electron motion by random scattering is 3D, and the electrons will go anywhere in the bulk (except the surface where there is a strong depletion electric field, which always kicks electrons back inside). Therefore, the electron statistics are based on the 3D electron gas model [1]. Furthermore, electron transport can be

Nanoelectronic Devices
Byung-Gook Park, Sung Woo Hwang, and Young June Park
Copyright © 2012 Pan Stanford Publishing Pte. Ltd.
www.panstanford.com

modeled by semiclassical drift/diffusion theory, which is based on the random scattering processes.

However, the usual silicon MOSFETs are actually two-dimensional electron systems (2DESs) since the electron motion in the direction perpendicular to the Si/SiO_2 interface is frozen by a steep potential well existing at the interface. The extent of the electronic wave function in the y-direction (the direction perpendicular to the wafer surface) is less than 10 nm, and the electrons in the channel can move freely only in the x- and z-direction. The current flow driven by the electric field between the source and the drain is also two-dimensional, and all the random scattering by phonons and impurities redirects electrons in the x–z plane. Now the surface roughness scattering at the Si/SiO_2 interface becomes important as well. The strong electric field in the y-direction constantly pulls the electrons and makes them bounce backward. With great advances in nano-fabrication technologies, it has become possible to further define a narrow channel from silicon MOSFETs, either by forming narrow-spaced lateral gates or by etching. In those narrow-channel devices, electron motion is further confined in the direction perpendicular to the direction connecting the source and the drain. Since the electrons can move in one direction only, these are called one-dimensional electron systems (1DESs).

Strictly speaking, the above low-dimensional systems are only quasi-2DESs and quasi-1DESs, since they have a finite thickness and width. A quantitative way of describing how much confinement they have is provided by resorting to quantum mechanics. In MOSFETs, the confinement in the y-direction can be modeled as a triangular potential well, resulting in several quantum mechanical subbands. When there is no inter-mixing between these subbands, the electron motion in the y-direction is considered to be frozen out. In 1DESs, because of the confinement in the x and y directions, the momentum in those directions (k_x and k_y) will be quantized. Such quantization will lead to discrete energy levels, which are called 1D subbands. The criterion for the formation of 1DESs is, again, no inter-mixing among these subbands. There are several ways to intermix these subband states. The most common source is through thermal fluctuation. For the formation of 1DESs, the subband energy level spacing is larger than the thermal fluctuation $k_B T$, where k_B is the Boltzmann constant and T is the temperature.

5.1.2 Ideal 1DES

Figure 5.1 shows schematics of a 2DES and a 1DES. While the electrons in 2DESs move around in the x- and z-direction, they move freely only in one direction (z-direction) in the ideal 1DES. Assuming a parabolic band, the kinetic energy of the conduction band electron is only the kinetic energy in the z-direction and is given by

$$E = \frac{p_z^2}{2m^*} = \frac{\hbar^2 k_z^2}{2m^*},\tag{5.1}$$

where p_z, k_z, and m^* are the momentum in the z-direction, the wave vector in the z-direction, and the effective mass, respectively. When L is the length of the 1DES, using the periodic boundary condition, the number of momentum states per unit length is given by $L/(2\pi)$ and the number of allowed states within 1D Fermi strip is given by

$$k_F \frac{L}{2\pi} = \frac{k_F L}{2\pi},\tag{5.2}$$

where k_F is the Fermi wave vector. Since each state accommodates two electrons with opposite spin, the total number of electrons N within the Fermi strip is $k_F L/\pi$. Then the 1D electron density n is

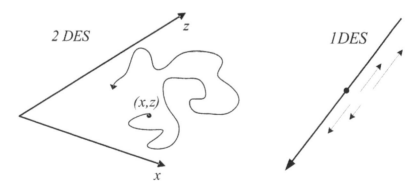

Figure 5.1. Schematics of a 2DES and a 1DES. The electron motion in the y-direction is frozen out in a 2DES. In a 1DES, electrons move freely only in the z-direction.

given by

$$n = \frac{N}{L} = \frac{k_F}{\pi} = \frac{\sqrt{2m^*}}{\pi\hbar}E_F^{1/2} \qquad (5.3)$$

Therefore, the electron density in a 1DES is proportional to the square root of the Fermi energy E_F.

Example 5.1: Calculation of the Fermi energy of a 1DES

A measurement from a GaAs 1DES shows that the average inter-electron spacing r_s is 10 nm. Calculate the Fermi energy E_F.

Solution:

$$n = 1/r_s = 10^8 \text{m}^{-1}$$

$$n = \frac{\sqrt{2m^*}}{\pi\hbar}E_F^{1/2}$$

$$E_F = \frac{\pi^2\hbar^2 n^2}{2 \times m^*} = \frac{\pi^2(6.63 \times 10^{-34}/(2\pi))^2 \times (10^8)^2}{2 \times 9.1 \times 10^{-31} \times 0.068}$$

$$= 5.55 \times 10^{-20}\text{(J)}$$

$$= 3.48 \text{ (meV)}$$

When an ideal 1DES is connected to the source and drain reservoir with a slight difference in the chemical potential ($\Delta\mu$) as shown schematically in Fig. 5.2, the current through the system is given by

$$I = eD(E)v\Delta\mu, \qquad (5.4)$$

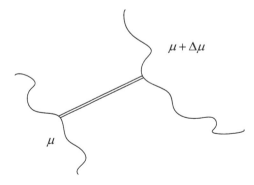

$\mu + \Delta\mu$

μ

Figure 5.2. An ideal 1DES connecting two reservoirs with the chemical potentials μ and $\mu + \Delta\mu$u.

where e is the electron charge, $D(E)$ is the 1D density of states, v is the group velocity. Since $D(E)$ and v are given as follows, the conductance G is expressed as a combination of fundamental constants only.

$$D(E) = \frac{1}{\pi \frac{\partial E}{\partial k}} \qquad (5.5)$$

$$v = \frac{1}{\hbar} \frac{\partial E}{\partial k} \qquad (5.6)$$

$$G = \frac{el}{\Delta \mu} = \frac{2e^2}{h} \qquad (5.7)$$

Here, the quantity $2e^2/h$ is called the conductance quantum. This value does not depend on the actual functional shape of $D(E) = \frac{1}{\pi}\sqrt{\frac{m^*}{2\hbar^2}} E^{-1/2}$ since the energy dependent terms cancel. We can check that the integral of the density of states still gives the same result as Eq. (5.3).

$$n = \int_0^{E_F} D(E)dE = \int_0^{E_F} \frac{1}{\pi}\sqrt{\frac{m^2}{2\hbar^2}} E^{-1/2}dE = \frac{\sqrt{2m^*}}{\pi\hbar} E_F^{1/2} \quad (5.8)$$

Example 5.2: Calculation of the resistance quantum

The 1DES resistance h/e^2 is expressed only in terms of fundamental constants. Calculate the resistance quantum h/e^2.

Solution:

$$\frac{h}{e^2} = \frac{(1.6 \times 10^{-19})^2}{6.63 \times 10^{-34}} = 25898.4 \, (\Omega)$$

5.1.3 Semiconductor 1DESs

The most natural way of realizing semiconductor 1DESs is to etch a 3D semiconductor into a wire-type structure with a small enough cross-sectional area, as shown in Fig. 5.3. As will be shown in Example 5.3, the quantum mechanical energy spacing is inversely proportional to the cross-sectional area. Therefore, it is important to make the cross-sectional area as small as possible. At finite temperatures,

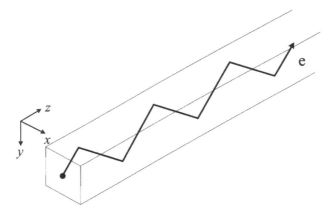

Figure 5.3. Schematic of a semiconductor 1DES. Classically, electrons fly in the z-direction while bouncing off the walls. Quantum mechanically, the small cross section results in subband energy level spacings that are quite large. Each subband corresponds to a quantum state whose wave function is confined inside the cross section. The z-component of the wave function is that of a free electron with the effective mass of the semiconductor.

there is always thermal excitation between the 1D subbands. Classically, this corresponds to a small amount of momentum in the x- and y-direction across the finite area. Therefore, even though most of the electrons' kinetic energy is in the z-direction, the motion of the electrons is a zig-zag type of bouncing along the wall of the 1D tube.

As an example of a realistic 1DES, Fig 5.4 shows a cross-sectional transmission electron micrograph of a recently developed silicon nanowire [2]. The photo shows a silicon cylinder with an almost circular cross section fully surrounded by gate oxide and TiN gate material. The radius of the nanowire and the thickness of the gate oxide are 4 and 3.5 nm, respectively. The bias on the surrounding gate can deplete or induce electrons in the nanowire. To study electron transport, both ends of the nanowire are connected to the source and the drain reservoir so that electrons can be injected into and escape from the nanowire. In the figure, there are two almost identical nanowires between the source and the drain reservoir. The reason for the formation of the twin nanowires is because they had to use a process trick for still using optical lithography techniques.

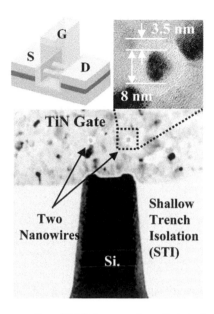

Figure 5.4. Cross-sectional TEM image of a cylindrical silicon 1DES with a radius of 4 nm (from Ref. 2). The wire is surrounded by the gate oxide with a thickness of 3.5 nm and then the TiN gate material. The schematic and the TEM image of the lower panel show that two nanowires bridge the source and drain bulk. There is a silicon plateau underneath the nanowires. Reprinted with permission from Ref. 2. Copyright 2008, American Institute of Physics.

Figure 5.5 shows another way of forming a semiconductor 1DES. These techniques utilize gates to define narrow channels for the electrons and holes [3]. There are two layers of gate structures. The lower gate is identical to the normal MOSFET gate with a thin gate oxide underneath except that it has a long and narrow slit in the middle. The lower gate is negatively biased to deplete the electrons underneath it. In other words, the lower gate is used in a way that is opposite to that of normal MOSFETs. A CVD oxide layer is deposited on top of the lower gate structure, and the upper gate then is deposited on top. The positive bias on the upper gate induces electrons in the long and narrow slit region, forming a 1DES.

1DESs can be fabricated using GaAs/AlGaAs HEMT (high electron mobility transistor) wafers. Figure 5.6 shows a schematic of such a 1DES. Two metal gates with a narrow and long slit are defined on top

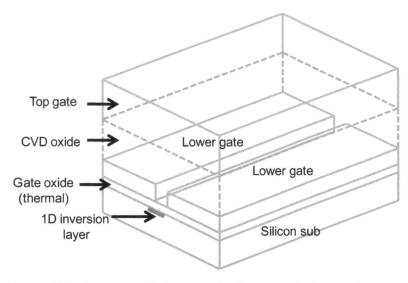

Top gate

CVD oxide

Lower gate

Lower gate

Gate oxide
(thermal)

1D inversion
layer

Silicon sub

Figure 5.5. Formation of silicon 1DES utilizing gate depletion. The positively biased upper gate induces electrons in the silicon region, and the negatively biased lower gate depletes the electrons underneath to form a 1DES. See also Color Insert.

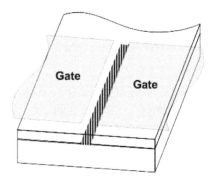

Gate

Gate

Figure 5.6. 1DES fabricated utilizing the split gate technique on GaAs/AlGaAs HEMT wafers. A negative bias on the two gates separated by a narrow gap depletes electrons underneath, forming a narrow wire-type electron layer in the gap region.

of the wafer. (We call these gates as split gates.) A negative bias on these two split gates depletes the 2DES electrons underneath, creating a narrow channel in between.

Example 5.3: Calculation of 1D energy levels

A quasi 1DES is made by forming a rod of GaAs embedded in AlGaAs with a larger bandgap, as shown in Fig. 5.E1(a). The cross section of the rod is a perfect square whose sides are 10 nm across. The length of the rod is much longer than the sides of the square. Calculate the energy difference between the ground state and the first excited state of this 1DES.

Solution:

Since GaAs has a smaller bandgap than AlGaAs, the conduction electrons are confined in the rod, as shown in the schematic conduction band diagram of Fig. 5.E1(b). Here we assume that the GaAs rod is surrounded by infinite potential walls. GaAs has an isotropic effective mass, and we can write down the Hamiltonian for the conduction band electron in the GaAs rod as follows:

$$H = \frac{p_x^2}{2m^*} + \frac{p_y^2}{2m^*} + \frac{p_z^2}{2m^*} + V(x, y)$$

$$V(x, y) = 0, 0 \leq x \leq L \text{ and } 0 \leq y \leq L$$

$$V(x, y) = \infty, \text{ otherwise}$$

Here L is the length of one side of the square (cross-section). The Schrödinger equation is then given by

$$-\frac{\hbar^2}{2m^*} \left(\frac{\partial^2 \Psi}{\partial x^2} + \frac{\partial^2 \Psi}{\partial y^2} + \frac{\partial^2 \Psi}{\partial z^2} \right) = E\Psi,$$

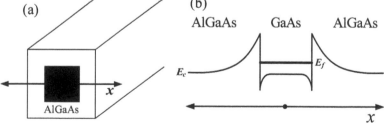

Figure 5.E1. (a) Schematic of a GaAs 1DES embedded in an AlGaAs matrix with larger bandgap. (b) Schematic of the conduction band diagram along the x-axis.

where Ψ is the wave function and E is the energy eigenvalue. The equation can be solved by the standard separation of variables method. We put $\Psi(x, y, z) = X(x)Y(y)Z(z)$ and we obtain the following:

$$\frac{1}{X}\frac{d^2 X}{dx^2} + \frac{1}{Y}\frac{d^2 Y}{dx^2} + \frac{1}{Z}\frac{d^2 Z}{dx^2} = -\frac{2m^* E}{\hbar^2}$$

$$\frac{1}{X}\frac{d^2 X}{dx^2} = -\alpha^2$$

$$\frac{1}{Y}\frac{d^2 Y}{dy^2} = -\beta^2$$

$$\frac{1}{Z}\frac{d^2 Z}{dz^2} = -k_z^2$$

The solution satisfying the boundary condition $(X, Y = 0$ outside of the rod$)$ is given by

$$X = A\sin\alpha x, \alpha = \frac{n\pi}{L}, n = 1, 2, 3\ldots$$

$$Y = B\sin\beta y, \beta = \frac{m\pi}{L}, m = 1, 2, 3\ldots$$

$$Z = C e^{\pm i k_z z}$$

$$\frac{\hbar^2}{2m^*}\left[\left(\frac{n\pi}{L}\right)^2 + \left(\frac{m\pi}{L}\right)^2 + k_z^2\right] = E(n, m)$$

Here, the quantum numbers n and m identify discrete 1D states (we call these states 1D subbands), and k_z denotes the continuous momentum in the z direction. If k_z is the same regardless of the subband, the difference between the ground state and the first excited energy ΔE is given by the following:

$$\Delta E = E(2, 1) - E(1, 1) = E(1, 2) - E(1, 1)$$

$$= \frac{\hbar^2 k_z^2}{2m^*} + \frac{5\hbar^2}{2m^*}\left(\frac{\pi}{L}\right)^2 - \frac{\hbar^2 k_z^2}{2m^*} - \frac{2\hbar^2}{2m^*}\left(\frac{\pi}{L}\right)^2$$

$$= \frac{3\hbar^2}{2m^*}\left(\frac{\pi}{L}\right)^2$$

For $L = 10$ nm,

$$\Delta E = \frac{3 \times (6.63 \times 10^{-34}/(2\pi))^2}{2 \times 9.1 \times 10^{-31} \times 0.067}\left(\frac{\pi}{10 \times 10^{-9}}\right)^2$$

$$= 2.7 \times 10^{-20}\,\text{(J)}$$

$$= 169\,\text{(meV)}$$

This energy is much larger than $k_B T$ (~26 meV) at room temperature.

The 1DESs described above are fabricated by conventional semiconductor processing, such as etching and deposition, starting from wafers. Recently, a new method of forming semiconductor nanowires has been widely studied. Nanowires can be grown in a chamber with the proper combination of catalyst and gas. The most common growth mechanism is the so called VLS (vapor–liquid–solid) method [4]. Vaporized SiH_4 gas reacts in the metal catalyst melting at high temperatures, forming silicon wires. Such self-formed nanowires are cast over a substrate, and contact pads are deposited on both ends of the nanowire. These source drain contact pads, together with the back gate, form a three-terminal nanowire transistor. Figure 5.7 shows an example of such a field effect transistor fabricated from bottom-up synthesized nanowires. The left panel shows a schematic of the device, and the right panel shows an SEM photo of the fabricated device.

All the 1DESs made from semiconductor materials are basically still three-dimensional since they have finite areas in the direction perpendicular to the length direction. The condition under which they behave as a 1DES is that the thermal smearing of the wave function out of the cross-sectional dimension must be negligible so that the electron motion in the cross section is almost frozen.

Figure 5.7. 1DES fabricated from nanowires grown with a bottom-up method. The VLS-grown nanowires are spread over the SiO_2/Si substrate, and two metal contacts are deposited on both sides of the nanowire. The Si substrate acts as a back gate with the insulating layer of SiO_2. See also Color Insert.

(It does have finite quantum energy, but it is actually difficult to find the exact classical counterpart of this energy. Therefore, we cannot say that electrons do not have completely zero kinetic energy in the cross-sectional directions.) This is equivalent to the condition that the inter-mixing between neighboring energy levels is negligible. The criterion $k_B T \ll \Delta E$ gives the same condition. Another condition is that the length of the 1DES should be shorter than the inelastic scattering length, so that the energy level of the electron is not changed by the inelastic scattering. In that case, the electrons move along the 1DES only elastically, slightly bouncing off the wall from time to time (Fig. 5.3). We call this condition "quasi-ballistic." When there are N subbands below the Fermi level (or the chemical potential) and there is no scattering, the conductance of the 1DES is given by

$$G = N \frac{2e^2}{h}, \tag{5.9}$$

since the energy of the electrons in the Nth subband is given by $E_N + \frac{\hbar^2 k^2}{2m^*}$, where E_N is the energy of the Nth subband, and the conductance of that subband is given by Eq. (5.7).

In most of the three-terminal semiconductor 1DESs shown above, a change in the gate bias changes the confinement potential in the cross section of the 1DES. The change in the confinement potential will result in a variation in the 1D energy level spacing. Therefore, when we sweep the gate bias, we can change the number of occupied subbands (subbands below the Fermi energy) consecutively. Whenever the gate bias is the value at which the number of occupied subbands changes by one, the total conductance will change by $2e^2/h$. As a result, as shown in Fig. 5.8, the conductance will show quantized steps as a function of the gate bias. Historically, the first experimental realization of the conductance quantization in a 1DES was done by the Delft group [5]. They fabricated a split-gated GaAs/AlGaAs 1DES and minimized the length of the channel into a sharp "point" to reduce the inelastic scattering length. Therefore, it has been called "quantum point contact". (The electron layer has a finite distance from the surface, and the actual channel becomes longer than the point because of the finite depletion width.) The device structure and the quantized conductance steps are shown in Fig. 5.8. The conductance

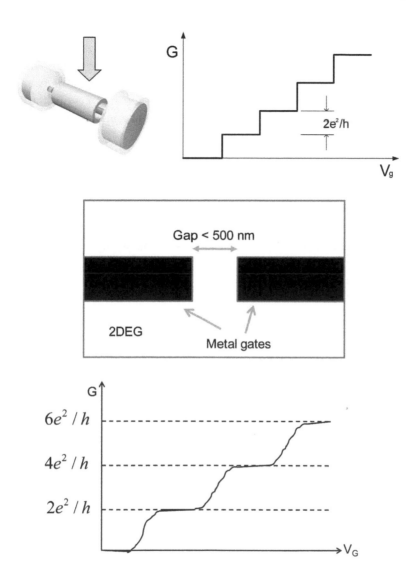

Figure 5.8. The uppermost panel shows the conductance quantization as a function of the gate bias in semiconductor 1DESs. The lower panels show the first experimental observation of conductance quantization. The device structure of the so-called quantum point contact was fabricated by forming the split gates on a GaAs/AlGaAs HEMT. The quantized conductance steps as a function of the gate bias are observed.

quantization was observed at the temperature $T < 4.2$ K, where the inelastic scattering length is much larger than the channel length. As the temperature increases from this cryogenic temperature, the electrons undergo more frequent inelastic scattering and the transition region in the conductance scan between two conductance steps becomes wider. The overall conductance steps smear out eventually.

Example 5.4: Potential calculation of a realistic 1DES

Figure 5.E2 shows the cross section of a 1DES fabricated by deep etching a GaAs/AlGaAs HEMT wafer. The coordinate axes x and z are shown as thick arrows. Here, the y-axis is the transport direction. The potential $\varphi(x, z = l)$ of the 1D electrons at the GaAs/AlGaAs interface can be expressed by the infinite series,

$$\varphi(x, l) = \frac{eN_D w_{dep}^2}{\varepsilon_{GaAs}} \sum_{n=0}^{\infty} f_n(l) \cos(\beta_n x)$$

$$\beta_n = (2n + 1) \frac{\pi}{w_{dep}}$$

$$f_n(l) = (-1)^n \frac{4}{(\beta_n w_{dep})^3} [\cosh(\beta_n d) - 1] e^{-\beta_n l}$$

where e, N_D, w_{dep}, and ε_{GaAs} are the electronic charge, doping concentration of the n-doped AlGaAs region, depletion region width, and dielectric constant of GaAs, respectively. Calculate the potential of a 1DES with $w = 200$ nm and $N_D = 10^{19}$ cm^{-3}. Terminate the series at $n = 15$.

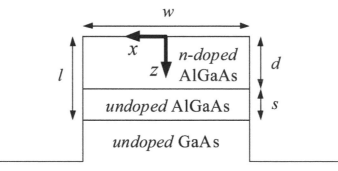

Figure 5.E2. Cross section of a deep etched GaAs 1DES.

Solution:

We note that the surface potential $\varphi(x = w/2, z = l) = 0.8$ V since the Fermi level pinning at the etched surface ($x = w/2$) occurs in the middle of the bandgap. The depletion width w_{dep} can be obtained by iteratively solving the above series for 0.8 V. For the given condition, we obtain $w_{dep} = 73$ nm. Figure 5.E3 shows the calculated potential ($\varphi(x, z = l) = 0$ for $-27 < x < 27$ nm).

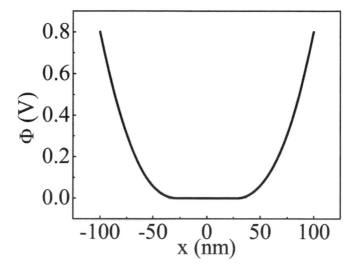

Figure 5.E3. Calculated potential along the width direction of the 1DES

5.1.4 *Silicon 1DESs*

Silicon has an indirect bandgap with complicated valley energies. Figure 5.9 shows six oval shaped valleys along the k-axes. Therefore, a 1DES made of silicon has a variety of subband energy structures and valley degeneracies depending on the crystal axis. For example, a silicon 1D nanowire along the z-direction has simultaneous electron occupation in two constant energy ovals along the k_z-direction. In that case, we might expect conductance steps separated by $4e^2/h$ counting these valley degeneracies. However, there are two different problems with this naïve expectation. The first fact is that, so far,

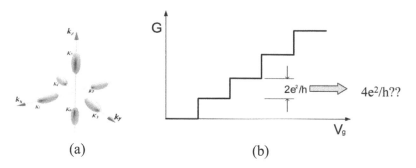

(a)   (b)

Figure 5.9. (a) Valley degeneracy of silicon. There are six oval shaped constant energy surfaces along the *k*-axes. (b) We can expect a different type of conductance quantization due to this valley degeneracy. See also Color Insert.

it has been very rare to observe conductance steps in Si since it is difficult to get a clean system without scattering. Second, there is a possibility that the stress and strain of confined Si structures would change the band structure and lift the spin and valley degeneracies [6]. Recent pioneering works demonstrate such effects, but they are beyond the level of this elementary textbook.

5.1.5 *Wave Nature of Electrons*

Electrons are so small that they obey quantum mechanics. Therefore, they have the well known particle–wave duality. In 3D semiconductors, the conduction band has almost continuous energy levels, and electrons undergo frequent random scattering by phonons (lattice vibrations) and impurities. The quantum mechanical nature is hiding in the band structure and the effective mass, and the semiclassical equation of motion can explain the transport of electrons. The story can be quite different in 1DESs. As shown in Section 5.1.3, the energy level spacing among 1D subbands can be appreciable. The scattering can be much smaller than in 3D semiconductors since electrons lose their random motions in the *x*- and *y*-directions. Then the electrons passing through 1DESs with negligible scattering also have the properties of waves propagating through the channel. The wave function of an electron in each subband in an ideal 1DES described in Sections 5.1.2 and 5.1.3 satisfies the following Schrödinger equation:

$$\frac{d^2\psi_n(z)}{dz^2} = -\frac{2m^*}{\hbar^2}[E_n - V(z)]\psi_n(z) \qquad (5.10)$$

This equation is the same as the equation in Example 5.3, and the solutions to this equation are 1D plane waves propagating either in the $+z$ or the $-z$ direction. These waves are basically the same as light waves, and we have a chance to realize 1DES electron waveguides and other devices incorporating those waveguides.

Example 5.5: An impurity in a semiconductor 1DES

If there is an impurity in a semiconductor 1DES, it can be modeled as a one-dimensional potential well. Set up the Schrödinger equation for this problem.

Solution:

The simplest model potential will be the square well potential. Figure 5.E4 shows such a potential and the corresponding coordinate system. The Schrödinger equation is given by

$$-\frac{\hbar^2}{2m}\frac{d^2\psi(x)}{dx^2} + V(x)\psi(x) = E\psi(x).$$

Here, $V(x) = -V_o$ when $-a < x < a$, and $V(x) = 0$ otherwise. V_o is the effective potential of the impurity, and a is the effective Bohr radius of the impurity atom. Here, the x-axis is the transport direction. The eigenfunctions with $E > 0$ correspond to the electron

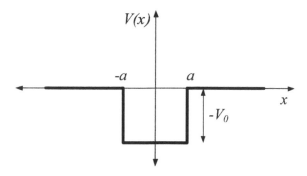

Figure 5.E4.   Potential profile of a 1DES with an impurity at $x = 0$.

waves that are scattered and partially reflected by the impurity. The eigenfunctions with $E < 0$ represent the bound state electrons of the impurity.

These waveguide-based devices can perform all the functions that optical waveguide devices can. Figure 5.10 shows a schematic of an electron Y-branch switch. A single waveguide is divided into two separate waveguides. One of the branches has a gate that controls the local potential. If the value of the gate bias is largely negative, the potential underneath the gate becomes large and the electron wave cannot propagate through that branch, and so the electron wave propagates through only one branch. Eventually, we can have various branch and gate structures and adjust the biases of the gates to control the directional propagation of electron waves.

When two branches are merged again on the opposite side, as shown in Fig. 5.11, the waveguide becomes the Mach–Zehnder interferometer [7]. Again the gate bias changes the local potential and creates a difference in the energies of the electron waves propagating through each waveguide. This will cause a phase difference,

$$\Delta\theta = \Delta E \tau_T / \hbar, \tag{5.11}$$

where $\Delta\theta$, ΔE, and τ_T are the phase difference, energy difference, and transit time, respectively. The value of this phase difference will

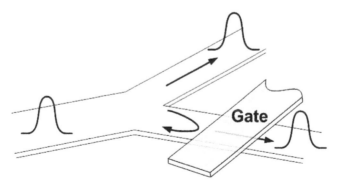

Figure 5.10. Electron Y-branch switch. The waveguide is divided into two branches. The bias of the gate on top of one branch can regulate the passage of electron waves through the branch.

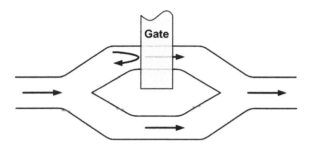

Figure 5.11. Mach–Zehnder interferometer. The waveguide is first divided into two branches, and then they are merged into one. The gate on one of the branches creates a energy difference and thus a difference between the phases of the waves in the two branches. The merging of the two waves with different phases results in the interference pattern.

result in either destructive or constructive interference at the joint branch. This interference is periodic in ΔE since the wave function is basically a sinusoidal function of ΔE. Furthermore, ΔE and thus the phase difference are functions of the gate bias. Therefore, the output current oscillates as a function of the gate bias.

The propagation of an electron wave in a 1D channel is described quantitatively by the Landauer formula. Figure 5.12 schematically shows a 1D channel with a potential barrier in the middle induced by a gate bias. When the transmission probability of this potential barrier is T, the conductance of a single propagating mode (which corresponds to a single subband in Section 5.1.2) becomes the following:

$$ G = \frac{2e^2}{h} T \tag{5.12} $$

The incoming wave from left has a single channel conductance of $2e^2/h$. The transmitted wave to right has the partial conductance of Eq. (5.12), while the conductance of the reflected wave is $(2e^2/h)(1 - T)$. Then the net conductance in the left side of the channel becomes $(2e^2/h)(1 - (1 - T))$ and is the same as Eq. (5.12). The Landauer formula can be extended to the case of multi - channel and multi - port conductors. The book - keeping of the incoming and outgoing waves of each port gives a set of linear equations. By solving the equations, we can obtain the conductance of each port.

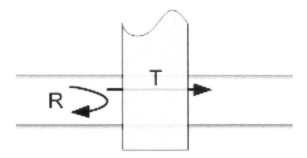

Figure 5.12. A 1D channel with a potential barrier. The conductance through the channel is proportional to the transmission coefficient T of the barrier. The sum of T and the reflection coefficient R is equal to 1.

Example 5.6: Transmission probability of a 1DES with an impurity

Find out the transmission probability of Example 5.5.

Solution:

The 1D Schrödinger equation of the wave function is

$$-\frac{\hbar^2}{2m}\frac{\partial^2\psi(x)}{\partial x^2} + V(x)\psi(x) = E\psi(x).$$

Then

$$\frac{\partial^2\psi(x)}{\partial x^2} + \frac{2m}{\hbar^2}[E - V(x)]\psi(x) = 0.$$

If we write $k^2 = \frac{2mE}{\hbar^2}$, $q^2 = \frac{2m(E+V_0)}{\hbar^2}$, then the Schrödinger equation and its solution in each region are given by the following:

$$x < -a, \quad \frac{\partial^2\psi(x)}{\partial x^2} + k^2\psi(x) = 0$$

$$\psi(x) = e^{ikx} + re^{-ikx}$$

$$-a < x < a, \quad \frac{\partial^2\psi(x)}{\partial x^2} + q^2\psi(x) = 0$$

$$\psi(x) = Ae^{iqx} + Be^{-iqx}$$

$$a < x, \quad \frac{\partial^2\psi(x)}{\partial x^2} + k^2\psi(x) = 0$$

$$\psi(x) = Te^{ikx}$$

The continuity of the probability current $\frac{\hbar}{2im}\left(\psi^*\frac{\partial\psi}{\partial x} - \frac{\partial\psi^*}{\partial x}\psi\right)$ at $x = a$ and $x = -a$ gives

$$\frac{\hbar k}{m}(1 - |r|^2) = \frac{\hbar q}{m}(|A|^2 - |B|^2) = \frac{\hbar k}{m}|t|^2 .$$

The continuity of $\frac{1}{\psi(x)}\frac{\partial\psi(x)}{\partial x}$ again at $x = a$ and $x = -a$ gives

$$x = -a, \quad \frac{ike^{-ika} - ike^{ika}}{e^{-ika} + re^{ika}} = \frac{iqAe^{-iqa} - iqBe^{iqa}}{Ae^{-iqa} + Be^{iqa}}$$

$$x = a, \quad \frac{iqAe^{iqa} - iqBe^{-iqa}}{Ae^{iqa} + Be^{-iqa}} = \frac{ikte^{ika}}{te^{ika}}$$

Getting rid of A and B in the above five equations, we obtain the following:

$$r = ie^{-2ika}\frac{(q^2 - k^2)\sin 2qa}{2kq\cos 2qa - i(q^2 + k^2)\sin 2qa}$$

$$t = e^{-2ika}\frac{2kq}{2kq\cos 2qa - i(q^2 + k^2)\sin 2qa}$$

Finally, we get T.

$$T = |t|^2 = \frac{4k^2q^2}{4k^2q^2\cos^2 2qa + (q^2 + k^2)^2\sin^2 2qa}$$

In Section 5.1.3, we have mentioned that semiconductor 1DESs are quasi-ballistic, with electrons passing through the system without inelastic scattering, bouncing off the walls. That picture is based on the particle nature of the electrons in 1DESs. The conduction band electrons in 1DESs are similar to electrons in vacuum except that they have a different effective mass. Therefore they also exhibit a particle-wave duality, and the wave nature of electrons described in this subsection does not contradict the concept of quasi-ballisticity.

5.1.6 Ballistic Transport in Short-Channel MOSFETs Under High Electric Fields

When the conductance quantization is observed in a 1DES, the transport of the 1DES is called ballistic because there is negligible inelastic scattering. The measurement of the conductance is done by applying the smallest possible voltage difference between two reservoirs on both sides of the 1DES in order to measure the current without noise. The temperature of the system is lowered to minimize inelastic scattering. There is another type of ballistic transport that has quite a different bias condition. When the channel length of a MOSFET is small enough and the electric field across the channel is high enough, the electrons injected from the source have a large chance of passing through the channel without energetic scattering even though there are some scattering centers. This is a type of ballistic transport in a sense that the electron does not undergo inelastic scattering. The quasi ballistic transport introduced in Section 5.1.3 originates from a clean 1DES without any scattering centers. The ballistic transport in this section occurs because the channel is too short and the electric field is too high for the electron to undergo any scattering events.

We will discuss the ballistic transport through planar MOSFETs (two-dimensional electron system). Figure 5.13 shows a schematic band diagram along the short-channel MOSFET. Both the conduction band edge and the valence band edge are sketched. The dotted lines denote the condition when the drain bias is zero. The channel is in the strong inversion condition, and the conduction band edge in the channel region is slightly higher than the source and the drain

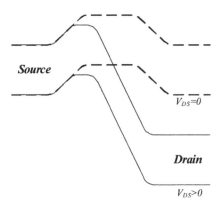

Figure 5.13. Schematic of the band diagram profile of a short-channel MOSFET. The conduction band edge of the channel region is pulled up due to the positive gate bias. When the drain bias is large, the whole band is pulled down from the drain side, leaving a barrier in the source side. The electrons are injected over this source barrier.

contact region. A large drain bias pulls down the drain side of the band diagram. (This condition is denoted by the solid lines in the figure.) The small hill still exists at the source side, acting as a barrier. There is a steep slide in between the source and the drain, suggesting a large electric field. The framework of the theory describing the ballistic transport is two scattering matrices existing at the source barrier and at the channel. An electron reaching the drain has to pass over the source barrier and has to survive in the channel. Then, the electron flux a_D at the drain is related to the source flux a_S [8] by

$$a_D = a_S t_S t_C,\qquad(5.13)$$

where t_S is the transmission probability across the source barrier and t_C is the transmission probability across the channel. The number of electrons in the source region is the sum of electrons just passing over the source barrier and the electrons reflected from anywhere in the channel. The 3D electron density n ($z = 0, y$) at the source and at the depth y is then obtained by dividing the sum of the flux of these electrons by the velocity. It is given by

$$n(0, y) = \frac{t_S a_S + r_C t_S a_S}{v_T} = \frac{t_S a_S(1 + r_C)}{v_T},\qquad(5.14)$$

where v_T is the transit velocity and $r_C = 1 - t_C$. Then t_S from Eq. (5.14) and then a_D from Eq. (5.13) can be written as

$$t_S = \frac{v_T n(0, y)}{a_S(1 + r_C)}$$

$$a_D = v_T n(0, y)\frac{t_C}{1 + r_C}\qquad(5.15)$$

When we integrate both sides of the above equation, keeping in mind the geometry of Fig. 5.13, we obtain the following:

$$\int_0^{y_{max}} a_D\, dy = \int_0^{y_{max}} n(0, y) v_T \frac{t_C}{1 + r_C}\, dy$$

$$= \frac{C_{ox}}{e}(V_{GS} - V_T)v_T \frac{1 - r_C}{1 + r_C}$$

$$eW \int_0^{y_{max}} a_D\, dy = I_{Dsat} = C_{ox} W(V_{GS} - V_T)v_T \frac{1 - r_C}{1 + r_C}\qquad(5.16)$$

We note that as $r_C \to 0$, $I_{Dsat} \to C_{ox} W v_T (V_{GS} - V_T)$, which is the ideal case of no scattering in the channel. The above derivation is for a MOSFET with a large width. Exactly the same formalism can be applied to the case of a 1DES such as the one shown in Fig. 5.3. Now, the 3D electron density at the source is a function of both x and y, $n(z = 0, x, y)$. The integration over the cross section of the nanowire gives the following:

$$\int_0^{x_{max}} \int_0^{y_{max}} a_D \, dy dx = \int_0^{x_{max}} \int_0^{y_{max}} n(0, y) v_T \frac{t_C}{1 + r_C} dy dx$$

$$= \frac{C_{ox}}{e} (V_{GS} - V_T) v_T \frac{1 - r_C}{1 + r_C}$$

$$e \int_0^{x_{max}} \int_0^{y_{max}} a_D \, dy dx = I_{Dsat} = W^2 C_{ox} (V_{GS} - V_T) v_T \frac{1 - r_C}{1 + r_C} \quad (5.17)$$

Here, C_{ox} is now the gate capacitance per unit length, and W is both the width and the height of the nanowire. Except for this C_{ox} and the W^2 dependence, the shape of the formula is similar to Eq. (5.16), suggesting that the physics of high field ballistic transport are the same in both the 2D and 1D cases.

The linear dependence on $(V_{GS} - V_T)$ of I_{Dsat} in the ballistic regime shown in Eq. (5.17) is similar to the case of short-channel transistors. Even though the bias dependence is the same, the transport physics of the short-channel transistor is velocity saturation, and it is totally different from ballistic transport. Because the compact modeling should be based on the physics of the transport, it is important to know the exact transport mechanism. Recent nanowire devices usually exhibit the cross-over from the $(V_{GS} - V_T)^2$ dependence of the long channel transistor (which will be shown in the next section) to the $(V_{GS} - V_T)$ dependence. It still is a difficult task to tell whether this $(V_{GS} - V_T)$ dependence originates from the velocity saturation or from ballistic transport, even though some experimental evidences have been reported to show that a few nanowire transistors are in the ballistic regime. Much more work is needed to clarify this subject.

5.2 Nanowire MOSFETs

5.2.1 *Evolution of MOSFETs*

As was explained in Section 5.1.1, the MOSFET is basically a 2D device in which the electrons are confined in the potential well at the Si/SiO$_2$ interface. Here, the electric field perpendicular to the gate

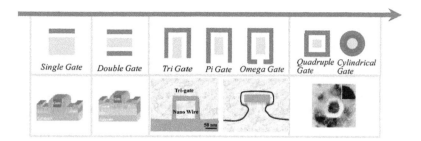

Figure 5.14. Evolution of silicon 3D transistors. The names of the devices reflect the number and shape of the gates. (Tri-gate and Omega-gate device in the lower row are schematically redrawn from Refs. 9 and 10. The photo of the cylindrical-gate device is from Ref. 2.) See also Color Insert.

plane either induces the inversion layer or turns off the transistor. As the MOSFET scales down into the deep nanometer regime, this electric field is not overwhelmingly stronger than the electric field in the direction from the source to the drain. In that case, it is difficult to completely turn off the channel. Such an incomplete turn-off combined with weaker control of the channel by the gate bias is called the short-channel effect.

Recently, there has been much effort to reduce the short-channel effect by changing the structure of the MOSFET from its simple 2D one to a 3D structure. This corresponds to creating gate electric field that can squeeze the channel off in an additional direction. Figure 5.14 summarizes the development of such 3D transistors. The first row contains schematic cross sections of the various 3D MOSFETs. The red regions denote the gate, and the yellow regions denote the semiconductor. The double gate transistor simply increases the number of gates by sandwiching the silicon layer between the top and the bottom gate. The π-gate and Ω-gate transistors have two additional gates on the two sides of the silicon channel. Of course the shape of the two side gates in the Ω-gate device also intrudes into the bottom. Even though it is still difficult to fabricate, a gate-all-around MOSFET is the most efficient form because the electric field that controls the channel formation comes from all angles. A MOSFET with a surrounding gate that has a circular cross section will be even better since the electric field in the radial direction is uniform and there will be no corner effects (crowding of electric fields at the corners).

These gate-all-around circular MOSFETs with a small radius have the perfect shape of nanowire MOSFETs.

5.2.2 *Analytical Model of Nanowire MOSFETs*

Figure 5.15 shows a schematic of the gate-all-around nanowire MOS-FET [11]. A cylindrical silicon pillar with radius R is surrounded by wraparound SiO_2 and then a metal (poly-Si) gate material. The top and bottom parts of the silicon are heavily doped and make contacts with metal so that they form the drain and the source. The cross section and the coordinate system are shown in the right panel of the figure.

In this section, we offer a simple analysis of the gate-all-around nanowire MOSFET shown in Fig. 5.15. It is generally difficult to obtain the current–voltage characteristics of a nanowire MOSFET analytically because of its 3D geometry. A gate-all-around nanowire MOSFET is the only exception with enough symmetry. Here, we rework the derivation of Ref. 11. We use the cylindrical coordinate

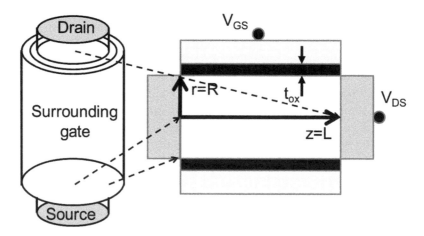

Figure 5.15. Schematic of a gate-all-around silicon nanowire field effect transistor. A silicon cylinder is surrounded by the gate oxide and the poly-Si gate. The cylinder is undoped except for both ends, which are heavily n doped to form the source/drain contacts. The right panel shows the cross section of the cylinder and the axes of the cylindrical coordinate system used to mathematically model the physics of the device. (The figure was schematically redrawn from Ref. 11.)

system in which the z-axis lies along the center of the Si cylinder, and r is the radial direction. If the cylinder is undoped, and we neglect holes, Poisson's equation in the cylinder is written as

$$\nabla^2 \phi(r, z) = -\frac{-en}{\varepsilon_{Si}}$$

$$= \frac{1}{r}\frac{\partial}{\partial r}\left(r\frac{\partial}{\partial r}\phi(r, z)\right) + \frac{\partial^2}{\partial z^2}\phi(r, z)$$

$$= \frac{e}{\varepsilon_{Si}}n_i \exp\left\{\frac{e[\phi(r, z) - \phi_F]}{k_B T}\right\}. \tag{5.18}$$

Here, $\phi(r, z)$, $e\ (> 0)$, n, n_i, ε_{Si}, and ϕ_F are the potential, electronic charge, electron density, intrinsic carrier concentration, dielectric constant of Si, and electron quasi Fermi potential, respectively. This equation is considered only in the white region of Fig. 5.15. The heavily doped contact regions simply serve as ideal reservoirs. In the gradual channel approximation, where the variation of $\phi(r, z)$ along the z direction is much smaller than the variation of $\phi(r, z)$ in the r direction, we can convert Eq. (5.18) into an ordinary differential equation with only the variable r.

$$\frac{\partial^2 \phi(r, z)}{\partial z^2} \approx 0$$

$$\nabla^2 \phi(r, z) \approx \frac{d^2\phi(r)}{dr^2} + \frac{1}{r}\frac{d\phi(r)}{dr}$$

$$= \frac{k_B T}{e}\delta \exp\left\{\frac{e[\phi(r) - \phi_F]}{k_B T}\right\}$$

$$\delta = \frac{e^2 n_i}{k_B T \varepsilon_{Si}} \tag{5.19}$$

The boundary conditions for this ordinary differential equation are given by the following:

$$\left.\frac{d\phi}{dr}\right|_{r=0} = 0$$

$$\phi|_{r=R} = \phi_s. \tag{5.20}$$

The first condition states that the radial electric field at the center of the cylinder should be zero. The existence of a radial electric field at the center would suggest an infinite charge density at $r = 0$. The quantity ϕ_s is the surface potential at the Si/SiO$_2$ interface.

Since the current flows mainly along the z direction, we can assume that ϕ_F is constant in r. Then Eq. (5.19) can be analytically solved, and the solution is as follows:

$$\phi(r) = \phi_F + \frac{k_B T}{e} \ln \left(\frac{-8B}{\delta(1 + Br^2)^2} \right) \qquad (5.21)$$

Another boundary condition that $\phi(r)$ should satisfy is Gauss's law. The unknown constant B in Eq. (5.21) is determined by this boundary condition. Figure 5.16 shows the Gaussian surface surrounding a cylinder with radius R and length L in the middle of the cylinder. Then Gauss's law gives the following:

$$\oint D \cdot dS = Q$$

$$\varepsilon_{Si} \left(-\frac{d\phi}{dr}_{r=R} \right)(2\pi RL) = 2\pi RL\sigma_s$$

$$\varepsilon_{Si} \left(-\frac{d\phi}{dr}_{r=R} \right) = -C_{ox}(V_{GS} - \phi_{ms} - \phi_s)$$

$$\varepsilon_{Si} \left(\frac{d\phi}{dr}_{r=R} \right) = C_{ox}(V_{GS} - \phi_{ms} - \phi_s) \qquad (5.22)$$

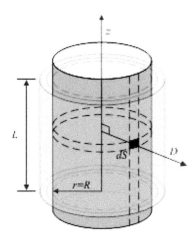

Figure 5.16. A Gaussian surface inside the silicon cylinder. The Gaussian surface is part of the dark cylinder with radius R and length L. The source and drain are away from the top and bottom of the Gauss surface. Only the oxide and the gate within L are shown. The electric field perpendicular to the surface exists only at the sidewall of the surface.

Here, σ_s is the surface charge density, C_{ox} is the gate capacitance, ϕ_{ms} is the work function difference, and V_{GS} is the gate bias. The third row of Eq. (5.22) is obtained by expressing σ_s with the MOS capacitor formula. The negative sign comes from the fact that there are electrons at the interface. By inserting Eq. (5.21) into Eq. (5.22), we obtain

$$\frac{q}{k_B T}(V_{GS} - \phi_{ms} - \phi_F) - \log\left(\frac{8}{\delta R^2}\right) = \log\left(\frac{1-\beta}{\beta^2}\right) + \eta\left(\frac{1-\beta}{\beta}\right)$$

$$\eta = \frac{4\varepsilon_{Si}}{C_{ox}R}$$

$$B = \frac{\beta - 1}{R^2}. \tag{5.23}$$

When V_{GS} is given, β can be obtained as a function of ϕ_F by using Eq. (5.23).

The drain current I_{DS} is flowing along a thin hollow cylinder with thickness Δt as shown in Fig. 5.17, and it is given by the following equation:

$$I_{DS} = SJ = \Delta t(2\pi R)e\mu n E_z$$

$$= 2\pi \mu RQE_z = 2\pi \mu RQ \frac{d\phi_F}{dz}, \tag{5.24}$$

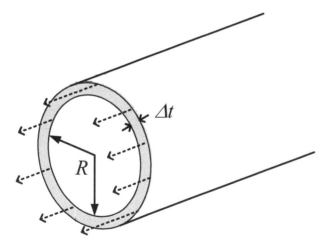

Figure 5.17. Calculation of the current distributed within Δt from the surface.

where μ is the electron mobility, E_z is the electric field along the z direction, and Q is the sheet charge density. By integrating both sides of Eq. (5.24), we obtain the following:

$$I_{DS} = \mu \frac{2\pi R}{L} \int_0^{V_{DS}} Q(\phi_F) d\phi_F = \mu \frac{2\pi R}{L} \int_{\beta_S}^{\beta_D} Q(\beta) \frac{d\phi_F}{d\beta} d\beta \qquad (5.25)$$

By differentiating Eq. (5.23) we obtain the following:

$$-\frac{e}{k_B T} d\phi_F = \frac{-d\beta}{1-\beta} - \frac{2\beta}{\beta^2} d\beta + \eta \frac{-d\beta}{\beta} - \eta \frac{1-\beta}{\beta^2} d\beta$$

$$\frac{d\phi_F}{d\beta} = \frac{k_B T}{e} \left[\frac{1}{1-\beta} + \frac{2}{\beta} + \eta \frac{1}{\beta} + \eta \frac{1-\beta}{\beta^2} \right]$$

$$= \frac{k_B T}{e} \left[\frac{2-\beta}{\beta(1-\beta)} + \eta \frac{1}{\beta^2} \right] \qquad (5.26)$$

And the surface charge density is given by Eq. (5.22).

$$Q_s = \varepsilon_{Si} \frac{d\phi(r)}{dr}\bigg|_{r=R} = \frac{4\varepsilon_{Si}}{R} \frac{k_B T}{e} \frac{1-\beta}{\beta} \qquad (5.27)$$

We obtain I_{DS} by using Eqs. (5.26) and (5.27).

$$I_{DS} = \mu \frac{4\pi \varepsilon_{Si}}{L} \left(\frac{2 k_B T}{e} \right)^2 \int_{\beta_S}^{\beta_D} \left[\frac{1 - \beta/2}{\beta^2} + \frac{\eta}{2} \frac{1-\beta}{\beta^3} \right] d\beta$$

$$= \mu \frac{4\pi \varepsilon_{Si}}{L} \left(\frac{2 k_B T}{e} \right)^2 \left[\frac{\eta}{4\beta^2} + \frac{1 - \eta/2}{\beta} + \frac{1}{2} \ln \beta \right]_{\beta_D}^{\beta_S} \qquad (5.28)$$

We define the following functions:

$$f(\beta) = \ln(1 - \beta) - \ln \beta^2 + \eta \left(\frac{1-\beta}{\beta} \right)$$

$$g(\beta) = \frac{\eta}{4\beta^2} + \frac{1 - \eta/2}{\beta} + \frac{1}{2} \ln \beta \qquad (5.29)$$

Then, from Eq. (5.23),

$$f(\beta_D) = \frac{e}{k_B T} (V_{GS} - V_o - V_{DS})$$

$$f(\beta_S) = \frac{e}{k_B T} (V_{GS} - V_o)$$

$$V_o = \phi_{ms} + \frac{k_B T}{e} \ln \left(\frac{8}{\delta R^2} \right). \qquad (5.30)$$

Then I_{DS} is given by

$$I_{DS} = \mu \frac{4\pi \varepsilon_{Si}}{L} \left(\frac{2k_B T}{e}\right)^2 [g(\beta_S) - g(\beta_D)]. \tag{5.31}$$

In the linear regime, both β_S and $\beta_D \to 0$ and then

$$f(\beta) \approx \eta \left(\frac{1-\beta}{\beta}\right), \quad g(\beta) \approx \frac{\eta}{4\beta^2}$$

$$I_{DS} = \mu \frac{4\pi \varepsilon_{Si}}{L} \left(\frac{2k_B T}{e}\right)^2 \left[\frac{\eta}{4\beta_S^2} - \frac{\eta}{4\beta_D^2}\right]. \tag{5.32}$$

Finally, we obtain

$$I_{DS} = 2\mu C_{ox} \frac{\pi R}{L} \left(V_{GS} - V_T - \frac{V_{DS}}{2}\right) V_{DS}$$

$$V_T = \phi_{ms} + \frac{k_B T}{e} \ln\left(\frac{8}{\delta}\right) - \frac{2k_B T}{e} \ln\left[R\left(1 + \frac{t_{ox}}{R}\right)^{\frac{2\varepsilon_{Si}}{\varepsilon_{ox}}}\right] \tag{5.33}$$

Here we use $C_{ox} = \frac{\varepsilon_{ox}}{R \ln\left(1 + \frac{t_{ox}}{R}\right)}$, where C_{ox} is the gate capacitance per unit area, t_{ox} is the thickness of the oxide, and ε_{ox} is the dielectric constant of the oxide.

In the saturation regime, $\beta_S \to 0$ and $\beta_D \to 1$. Then

$$f(\beta_S) \approx \eta \frac{1 - \beta_S}{\beta_S}, \quad f(\beta_D) \approx \ln(1 - \beta_D)$$

$$g(\beta_S) \approx \frac{\eta}{4\beta_S^2}, \quad g(\beta_D) \approx \frac{\eta}{4\beta_D^2} + (1 - \eta/2)\frac{1}{\beta_D} \tag{5.34}$$

$$I_{DS} = \mu \frac{4\pi \varepsilon_{Si}}{L} \left(\frac{2k_B T}{e}\right)^2 \left[\frac{\eta}{4\beta_S^2} - \frac{\eta}{4\beta_D^2} - (1 - \eta/2)\frac{1}{\beta_D}\right].$$

The drain current I_{DS} becomes

$$I_{DS} = \mu \frac{4\pi \varepsilon_{Si}}{L} \left(\frac{2k_B T}{e}\right)^2 \left[\frac{\eta}{4}\left(1 + \frac{e}{k_B T \eta}(V_{GS} - V_o)\right)^2\right.$$

$$- \frac{\eta}{4} \frac{1}{\left(1 - \exp\frac{[e(V_{GS} - V_0 - V_{DS})]}{k_B T}\right)^2}$$

$$\left. - (1 - \eta/2)\frac{1}{\left(1 - \exp\frac{[e(V_{GS} - V_o - V_{DS})]}{k_B T}\right)}\right] \tag{5.35}$$

Example 5.7: Comparison of planar and nanowire MOSFETs

A nanowire MOSFET has radius $R = 5$ nm and a gate oxide thickness of $t_{ox} = 3.5$ nm. What should be the width of the planar MOSFET if its gate capacitance is the same as that of the nanowire MOSFET with the same and length L?

Solution:

Let the gate capacitances of the nanowire MOSFET and planar MOSFET be C^w_{ox} and C^p_{ox}, respectively. Then

$$C^w_{ox} = C^p_{ox}$$

$$\frac{2\pi\varepsilon_{ox}L}{\ln\left(\frac{R+t_{ox}}{R}\right)} = \frac{WL\varepsilon_{ox}}{t_{ox}}$$

$$W = \frac{2\pi t_{ox}}{\ln\left(\frac{R+t_{ox}}{R}\right)}$$

$$= \frac{2\pi \times 3.5}{\ln(8.5/5)} = 41.4 \text{(nm)}$$

Example 5.8: Current–voltage characteristics of a nanowire MOSFET

Plot I_{DS} as a function of V_{GS} at various values of V_{DS} (in the linear region) of a nanowire MOSFET with radius $R = 10$ nm and gate oxide thickness of $t_{ox} = 2$ nm.

Solution:

We use Eq. (5.33) to calculate I_{DS} in the linear region. First, C_{ox} and δ are given by the following:

$$C_{ox} = \frac{\varepsilon_{ox}}{R\ln(1 + t_{ox}/R)}$$

$$= \frac{3.9 \times 8.854 \times 10^{-12}}{10 \times 10^{-9}\ln(1 + 2 \times 10^{-9}/10 \times 10^{-9})} = 0.0189[\text{F}/\text{m}^2]$$

$$\delta = \frac{e^2 n_i}{k_B T \varepsilon_{Si}} = \frac{1.6 \times 10^{-19} \times 10^{10}}{0.0259 \times 11.7 \times 8.854 \times 10^{-12}} = 596.3[\text{m}^{-2}]$$

Then V_T is given by

$$V_T = \phi_{ms} + \frac{k_B T}{e} \ln\left(\frac{8}{\delta}\right) - \frac{2k_B T}{e} \ln\left[R\left(1+\frac{t_{ox}}{R}\right)^{2\varepsilon_{Si}/\varepsilon_{ox}}\right]$$

$$= 0 + 0.0259 \times \ln\left(\frac{8}{596.3}\right) - 2 \times 0.0259$$

$$\times \ln\left[10 \times 10^{-9} \times \left(1 + \frac{2 \times 10^{-9}}{10 \times 10^{-9}}\right)^{\frac{2 \times 11.7}{3.9}}\right] \doteq 0.785 \text{ [V]}$$

We input these values into Eq. (5.33) and calculate I_{DS} as a function of V_{GS} in the range $0 < V_{GS} < 2$ V, at $V_{DS} = 0.1, 0.2, 0.4, 0.6, 0.8,$ and 1 V. Figure 5.E5 shows the calculated results. We assume a mobility of 340 cm^2/Vs. The results calculated from a three-dimensional device simulation with the same geometry are also shown in the figure.

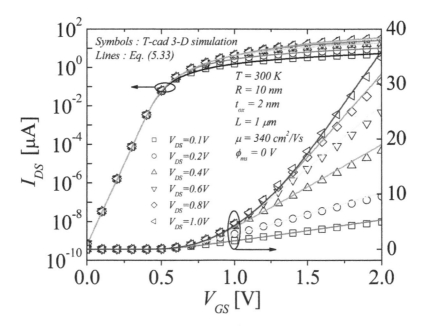

Figure 5.E5. I_{DS} as a function of V_{GS} at various values of V_{DS} in the linear regime. The results are calculated from Eq. (5.33). The results of numerical simulations are also shown. See also Color Insert.

PROBLEMS

1. **Electrons confined in a silicon nanowire:**
   A more realistic potential profile for the electrons confined in the semiconductor 1DES than the abrupt potential of Example 5.3 is a harmonic potential. Electrons confined in a silicon cylinder experience a harmonic potential in the radial direction. Another interesting aspect of this setup is that the conduction band of silicon has valleys. Write down the Hamiltonian.

2. **Quantized conductance steps:**

   (a) Find the value of quantized conductance steps of the quantum point contact fabricated on GaAs/AlGaAs HEMT when there is no magnetic field.
   (b) Find the value of quantized conductance steps of the same quantum point contact when there is strong magnetic field perpendicular to the wafer plane.

3. **Effect of finite contact resistance:**
   Realistic quantum point contact devices usually have finite contact resistances. Find the value of the first and the second conductance step of a quantum point contact when the contact resistance is 2 kΩ.

4. **Condition for observing quantized conductance steps:**
   Using the same structure of Example 5.3, design 1DES that will exhibit at least three quantized conductance steps. Use the criterion $\Delta E \approx 10\, k_B T$.

5. **Pinch-off voltage of a split gate:**
   The depletion region extending underneath the split gate of Fig. 5.8 is 20 nm when the gate voltage $V_G = 0$ V. If the depletion region increases by 10 nm for every 0.1 V of the negative gate bias, find the pinch-off voltage of the 200 nm-wide quantum point contact.

6. **Transmission probability:**
   Obtain the transmission probability for the one-dimensional potential barrier with the width a and the height V_o.

7. Derive Eq. (5.21).

8. Derive Eq. (5.23).

9. Derive Eq. (5.27).

10. Calculation of I_{Dsat} of a ballistic MOSFET:
A MOSFET is in the ballistic regime. It has planar type gates. The length (L), width (W), oxide thickness (t_{ox}), threshold voltage (V_T), gate bias (V_{GS}), and saturation velocity (v_T) are 0.1 μm, 10 μm, 1 nm, 0.2 V, 2 V, and 10^7 cm/sec, respectively. Calculate the saturation current I_{Dsat}.

Bibliography

1. N.W. Ashcroft and N.D. Mermin, *Solid State Physics*, Holt Rinehart Winston (1976).

2. K.H. Cho, K.H. Yeo, Y.Y. Yeoh, S.D. Suk, M. Li, J. M. Lee, *et al.*, "Experimental evidence of ballistic transport in cylindrical gate-all-around twin silicon nanowire metal-oxide-semiconductor field-effect transistors," *Appl. Phys. Lett.*, **92**, 052102 (2008).

3. J.H.F. Scott-Thomas, S.B. Field, M.A. Kastner, H.I. Smith, and D.A. Antoniadis, "Conductance oscillations periodic in the density of a one-dimensional electron gas," *Phys. Rev. Lett.*, **62**, 583–586 (1989).

4. G. Liang, J. Xiang, N. Kharche, G. Klimeck, C.M. Lieber, and M. Lundstrom, "Performance analysis of a Ge/Si core/shell nanowire field-effect transistor," *Nano Lett.*, **7**, 642 (2007).

5. B.J. van Wees, H. van Houten, C.W.J. Beenakker, J.G. Willianson, L.P. Kouwenhoven, D. van der Marel, *et al.*, "Quantized conductance of point contacts in a two-dimensional electron gas," *Phys. Rev. Lett.*, **60**, 848 (1988).

6. K.H. Cho, Y.C. Jung, B.H. Hong, S.W. Hwang, J.H. Oh, D. Ahn, *et al.*, "Observation of three-dimensional shell filling in cylindrical silicon nanowire single electron transistors," *Appl. Phys. Lett.*, **90**, 182102 (2007).

7. S. Datta, M.R. Melloch, S. Bandyopadhyay, and R. Noren, "Novel interference effects between parallel quantum wells," *Phys. Rev. Lett.*, **55**, 2344–2347 (1985).

8. M. Lundstrom, "Elementary scattering theory of Si MOSFET," *IEEE Elec. Dev. Lett.* **18**, 361–363 (1997).

9. C. Jahan, O. Faynot, M. Cassé, R. Ritzenthaler, L. Brévard, L. Tosti, *et al.*, "ΩFETs transistors with TiN metal gate and HfO$_2$ down to

10 nm," Symposium on VLSI Technology Digest of Technical Papers, 112 (2005).

10. Y.-C. Wu, P.-W. Su, C.-W. Chang, and M.-F. Hung, "Novel twin poly-Si thin-film transistors EEPROM with trigate nanowire structure," *IEEE Elect. Dev. Lett.*, **29**, 1226 (2008).

11. H.A. El Hamid, B. Iniguez,, and J.R. Guitart, "Analytical model of the threshold voltage and subthreshold swing of undoped cylindrical gate-all-around-based MOSFETs," *IEEE Trans. Elect. Dev.* **54**, 572–579 (2007).

Chapter 6

Quantum Dot Devices

6.1 Zero-Dimensional Electron Systems

6.1.1 Semiconductor Quantum Dot as 0DESs

In Section 5.1, we defined the ideal 1DES and introduced realistic 1DESs that can be fabricated from various types of semiconductors. If the free electron motion in a 1DES is frozen by further confinement in the z-direction (the direction in which electrons freely move in the 1DES), then the system is called a zero-dimensional electron system (0DES). Figure 6.1 shows a conceptual diagram of a 0DES. It is basically the ideal 1DES of Fig. 5.1 with two walls added to the figure to denote the potential confinement, blocking electron motion in the z-direction. Electrons are captured between the potential walls and cannot move freely even in the z-direction. The only way for the electrons between the two walls to be released is either by quantum mechanical tunneling through the walls or by jumping over the walls. Classically, the most important energy that these 0DES electrons have is the electrostatic energy determined by the total charge (the self potential energy) and the electrostatic potential of the system (the external potential energy).

Nanoelectronic Devices
Byung-Gook Park, Sung Woo Hwang, and Young June Park
Copyright © 2012 Pan Stanford Publishing Pte. Ltd.
www.panstanford.com

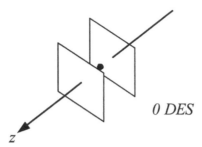

Figure 6.1. Schematic of 0DES. The electrons would have moved freely in the z-direction without two walls (1DES). The walls create the potential barriers, and the electrons can go across the walls either by quantum mechanical tunneling or by jumping over them. The electrostatic energy of such a 0DES is a strong function of the number of captured electrons when the distance between the two walls is small enough.

In the early 1900s, many physicists, including Millikan, showed that the minimum unit of charge was the electron charge e of -1.6×10^{-19} C (with the famous Millikan's oil drop experiment). In spite of this fundamental discovery, we consider the charge and thus the electrostatic energy continuous physical quantities in most macroscopic systems. This is mainly because the number of electrons in these systems is large, and so the change of one electron only gives mathematically infinitesimal change of electrostatic energy. However, the size of the 0DES is small enough so that the system has only a small number of electrons, and the electrostatic energy of the 0DES changes greatly even for a "one-electronful" variation in the total charge. Therefore, as will be shown shortly, the electrostatic energy of the system is a very sensitive function of the number of electrons, and we must take into account the discreteness of the electron charge. Thus, the charge of the 0DES is an integer multiple of e, and the electrostatic energy of 0DES is basically discrete too.

The left panel of Fig. 6.2 shows a more realistic schematic of 0DES. The 0DES is made by creating a small piece of semiconductor with the smaller bandgap (the short cylinder in the figure) surrounded by an insulator or other semiconductor (the large box in the figure) with the larger bandgap. The right panel of Fig. 6.2 shows a schematic band diagram (conduction band edge) along the small semiconductor cylinder and into the insulator box. Since the

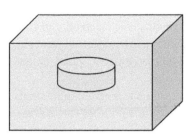 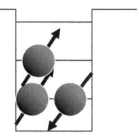

Figure 6.2. Semiconductor 0DES and schematic band diagram. A small piece of semiconductor is surrounded by an insulator with a larger bandgap. The right figure shows a sketch of the conduction band with a potential well (small piece of semiconductor) and quantum mechanical energy levels occupied with a few electrons with different spins.

bandgap of the semiconductor is smaller than that of the insulator, there are band discontinuities along the surface of the semiconductor, and, in that case, the electrons in the semiconductor are confined in the potential well created by this band discontinuity. Let us assume that there is one conduction band electron in the potential well. Elementary quantum mechanics tells us that the conduction band electron in such a potential well has discrete quantum energy levels. Therefore, we call such 0DESs "quantum dots." Here, the term "dot" suggests a 0DES. The term "quantum" means that the size of the potential well is so small that the electron energies are determined by the quantum mechanics. The figure also indicates how electrons with different spin orientations fill the potential well. A quantum dot is also called an "artificial atom" since it has discrete energy levels and a finite number of electrons consecutively fill in those energy levels. As will be shown later, the energy level spacing usually increases as the size of the quantum dot decreases, approaching the size of a real atom. Of course, how "quantum" the quantum dot really is, is determined by how much larger the quantum energy spacing is than the thermal fluctuation. This criterion is similar to the case of 1DES in which the condition for resolving the conductance quantization is that the 1D subband energy level spacing is larger than the thermal fluctuations. If we try to put additional electrons in a quantum dot that is already occupied by one electron, we need to supply the electrostatic charging energy of a single electron mentioned above in addition to the energy difference between the quantum energy level of the original electron and that of the added electron. Usually, it is a good approximation to treat the electrostatic energy of adding a single electron and the quantum energy levels independently.

There are many types of interesting materials other than semiconductors that can be used to form the quantum dots (or 0DESs). Recently, there have been a number of advances in the fabrication of metal nano-particles with sizes smaller than 10 nm, and these are also called quantum dots. However, strictly speaking, their sizes should be really small in order to be called quantum dots. The number of conduction electrons is usually quite large in metals, and therefore, it is difficult to have only a small number of electrons in a metal dot unless its size is extremely small. On the other hand, the number of electrons in semiconductor quantum dots can be easily reduced, and the electronic properties are usually quantum mechanical in most of the semiconductor quantum dots. Semiconductor quantum dots can be made by various techniques. Figure 6.3(a) shows a transmission electron microscope image of a quantum dot created by the self-assembled growth. The small mountain in the photo is an InAs quantum dot grown in a metal organic chemical vapor deposition (MOCVD) chamber. The diameter of the quantum dot is smaller than 20–30 nm, so we can even count the finite number of lattice sites. A slow epitaxial growth at a particular temperature results in the nucleation of such quantum dots, rather than growing flat layers. This growth mode is known as the Stranski–Krastanov (S–K) growth mode [1]. Figure 6.3(b) shows a scanning electron micrograph image of a vertical quantum dot. A small metal disc is deposited on top of a GaAs/AlGaAs quantum well wafer by electron beam lithography, metal evaporation, and lift-off. Slow wet etching creates a vertical pillar underneath the metal mask. The pillar then has a small volume of GaAs (or InGaAs) surrounded on the top and bottom by AlGaAs barriers and on the side by the etched surface as shown in the schematic in the figure. This technique has the advantage of creating atomically precise quantum dots and barriers. However, the sidewalls still have randomness because of the wet etching process. Figure 6.3(c) shows a quantum dot defined by split gates. In Chapter 5, we introduced the split-gate technique for

(a)

(b)

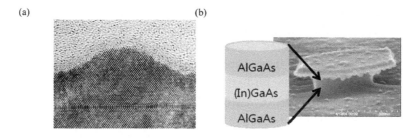

(c)

(d)

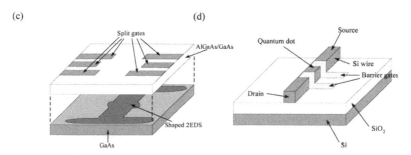

Figure 6.3. Examples of semiconductor 0DES: (a) a TEM photo of an InAs self-assembled quantum dot embedded in a GaAs matrix; (b) 0DES formed by the etching of a vertical pillar using a GaAs/AlGaAs quantum well wafer; (c) split gates on top of a GaAs/AlGaAs HEMT wafer; and (d) a 1D nanowire etched on an SOI wafer with two metal gates acting as barriers. See also Color Insert.

fabricating 1DESs. After the split gates were fabricated on top of a GaAs/AlGaAs HEMT wafer, a negative bias was applied to these gates to deplete the electrons underneath and form a narrow electron channel. A combination of several split gates with particular shapes can now define a quantum dot instead of the nanowire 1DES. Since there is a finite depletion layer from the edge of the gate, these gates can be separated and biased by different voltages while still forming the barrier region in between. This biasing technique can create a tunnel barrier connecting the quantum dot and the reservoir. Figure 6.3(d) shows a quantum dot fabricated by the combined process of etching and gate definition. This technique represents the realization of the conceptual diagram of Fig. 6.1. First, a long, narrow wire is defined by wet etching of a GaAs/AlGaAs HEMT wafer or a silicon-on-insulator (SOI) wafer. Then two sets of narrow metal gates are patterned and deposited across the channel. When negative biases are applied on these gates, they can deplete the electrons underneath them and form potential barriers isolating the region between them from other parts of the nanowire. The etching and gate definition used in creating a 0DES are similar to the methods for fabricating 1DESs. The only difference is that there is an additional dimension of confinement in the fabrication of 0DESs.

6.1.2 *Coulomb Blockade*

In this subsection, we will forget about the quantum energy level of the electron in a quantum dot for a while, and we will discuss how to calculate the electrostatic energy of a single electron added to the quantum dot. In the early 1950s, Gorter found that conduction through granular metals was suppressed at low temperatures and at small applied voltages [2]. It was found that the tunneling of a single electron into the grain required an increase in the electrostatic energy by e^2/C, where C is the capacitance of the grain. Therefore, when $e^2/C \gg k_B T$, the tunneling was difficult, and the conductance became very small. This phenomenon nowadays is called Coulomb Blockade (CB). Exactly the same phenomenon occurs in the semiconductor quantum dots introduced in the previous section. As was already discussed in the previous section, the only way of changing the number of electrons in the quantum dot is having them tunnel in and out of it through the barriers that define the quantum dot. The left panel of Fig. 6.4 schematically shows a quantum dot connected to the source and the drain reservoir by two tunnel barriers. The right panel shows an example of the current–voltage characteristic of such a quantum dot. (Such characteristic is what Gorter observed in early days.) The I–V data show the calculation results using the so called orthodox theory, which will be introduced in later sections. It shows a clear suppression of the current at low source-drain bias (Coulomb blockade), and this suppression disappears when the temperature increases. The tunnel barriers in Fig. 6.4 could be Al-GaAs barriers in Fig. 6.3(b) or gate-bias-induced depletion regions

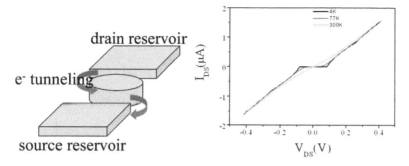

Figure 6.4. (Left) A schematic of the quantum dot connected to the source and the drain reservoir by tunnel barriers. (Right) A calculated current–voltage characteristic of such a quantum dot at three different temperatures. At the lowest temperature, a suppression of the current near zero bias can be observed. This is the manifestation of the Coulomb blockade phenomenon. See also Color Insert.

in Fig. 6.3(c) and (d). If we want to form two tunnel barriers for the quantum dot of Fig. 6.3(a), two small electrodes with a gap whose size is comparable to the size of the quantum dot should be formed on either side of the quantum dot. This requires electron beam lithography and lift-off procedures.

To observe single-electron tunneling behavior experimentally, there is another important requirement in addition to the condition of $e^2/C \gg k_B T$. The average tunneling time Δt of a single electron is given by $-e/I$. The difference in energy, ΔE, before and after tunneling is $-eV$. From the uncertainty principle, $\Delta E \Delta t \geq \hbar$, and the resistance of the tunnel barrier $R = V/I > h/e^2$. Therefore, the resistance of the tunnel barrier should be larger than the resistance quantum ($h/e^2 \sim 25.9$ kΩ). Figure 6.5 shows the I–V characteristics of the quantum dot at $C = 0.5$, 1, and 5 aF and $T = 4.2$ K. Very clear gaps near zero bias can be seen in Fig. 6.5, and the size of the gap is proportional to $1/C$. At this temperature, $e^2/C \gg k_B T$, and the resistance, which is 5 MΩ, is $\gg h/e^2$. One more interesting observation is that the size of the gap region is larger for smaller values of C. The size of the gap is proportional to the single-electron charging energy e^2/C, and this energy gets larger as the C value gets smaller.

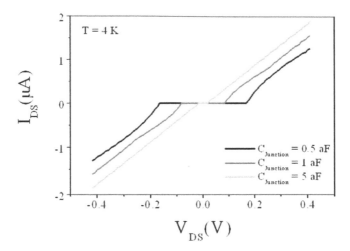

Figure 6.5. *I–V* characteristics of the quantum dot with values of $C = 0.5$, 1, and 5 aF and $T = 4.2$ K. All of them show clear gap structures near zero bias, suggesting that the single-electron charging energy is larger than $k_B T$ in all three quantum dots at this temperature. See also Color Insert.

Example 6.1: Classical charging energy of a silicon nano-sphere

Calculate the total electrostatic energy of a metallic sphere with radius 5 nm surrounded by silicon and containing two electrons.

Solution:

The self capacitance of the sphere is $C = 4\pi \varepsilon a$, where ε is the dielectric constant of the silicon and a is the radius of the sphere. Bringing the first electron into the empty sphere does not require energy. The potential V of the sphere with the first electron then is given by $V = q/C = q/4\pi \varepsilon a$ ($q = -e$), and the energy required to bring the second electron is given by

$$E = qV = \frac{q^2}{4\pi \varepsilon a}$$

$$= \frac{(1.6 \times 10^{-19} \text{C})^2}{4\pi \times 11.7 \times 8.854 \times 10^{-12} \text{F/m} \times 5 \times 10^{-9} \text{m}}$$

$$= 24.6 (\text{meV})$$

Example 6.2: Charging energy of silicon spheres

To observe the discreteness of the electron charge, the charging energy should be larger than the thermal fluctuation. Calculate the charging energy of above metal spheres with various radii a.

Solution:

The following table summarizes the formulae of the self capacitance and single-electron charging energy, together with their values at various radii a.

a (nm)	$C = 4\pi\varepsilon a$ (aF)	$E_C = q^2/C$ (meV)
1	1.3	123
5	6.5	24.6
10	13	12.3
50	65	2.46
100	130	1.23

6.1.3 *Single-Electron Transistors*

More interesting phenomena can occur when there is a gate that changes the potential of the quantum dot continuously. Figure 6.6 shows the schematic of such a three terminal device. It consists of a quantum dot connected to the source and the drain reservoir by tunnel barriers and the gate. The tunneling between the quantum dot and the gate electrode is forbidden, and the only action of the gate is changing the potential of the quantum dot continuously. This device is called a single-electron transistor, since it can control the transport of a single electron through the source/drain and the quantum dot by changing the gate bias. Of course, to observe such transistor action, the same condition for observing single-electron tunneling discussed in the previous section should be satisfied here. There can be multiple gates that can control the potential of the quantum dot independently.

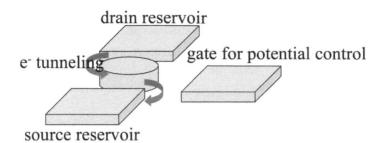

drain reservoir

e⁻ tunneling   gate for potential control

source reservoir

Figure 6.6. Schematic of a single-electron transistor. There are two tunnel barriers for the electron tunneling to and from the source and drain reservoir and the gate that can control the potential of the quantum dot continuously.

When there are N electrons in the quantum dot, the total electrostatic energy is given by

$$E = \frac{(Ne)^2}{2C} + Ne\phi_{ext} = \frac{e^2}{2C}\left(N + \frac{C\phi_{ext}}{e}\right)^2 - \frac{1}{2}C\phi_{ext}^2, e < 0 \quad (6.1)$$

where C is the total capacitance of the quantum dot and ϕ_{ext} is the electrostatic potential that is determined by the gate bias. Here the total capacitance of the quantum dot is simply the self-capacitance of the quantum dot. For example, if the shape of the quantum dot is a sphere with a radius r and it is surrounded by a dielectric with dielectric constant ε_R, $C = 4\pi r \varepsilon_R$.

Now, when ϕ_{ext} is such that E has its minimum at integer $N = i$ ($\phi_{ext} = -ei/C$), the nearby states at $N = i+1$ and at $N = i-1$ have energies larger than that of the state at $N = i$, by $e^2/(2C)$. In this case, the conduction of an electron through the dot is suppressed at low T because adding or subtracting one electron to or from the dot requires a finite energy of $e^2/(2C)$. On the other hand, when ϕ_{ext} is such that E has its minimum at a half integer value $i- 1/2$ ($\phi_{ext} = -e(i - 1/2)/C$), the states at $N = i- 1$ and at $N = i$ have the same energy, and no extra Coulomb energy is needed for the transport. As ϕ_{ext} is continuously varied by the gate bias, these two cases occur in turn, and the conductance G shows a series of peaks as a function of ϕ_{ext}. These are called "Coulomb oscillations." Each peak in the Coulomb oscillations represents the change of one electron in the QD population. Figure 6.7 summarizes this

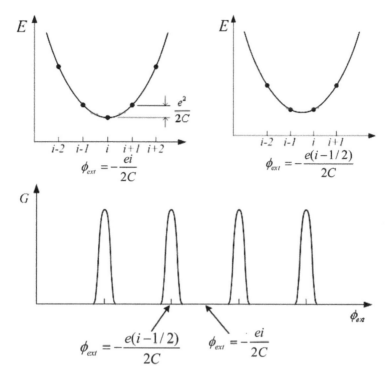

Figure 6.7. (Upper left) Electrostatic energy of the quantum dot when ϕ_{ext} is an integer multiple of $-e/2C$, (Upper right) electrostatic energy of the quantum dot when ϕ_{ext} is a half integer multiple of $-e/2C$, (Lower) gate bias dependence of the conductance (or current) as a function of ϕ_{ext}.

gate bias dependence. It shows E as a function of electron number when $\phi_{ext} = -ei/C$ and $\phi_{ext} = -e(i-1/2)/C$. It also shows the conductance G as a function of ϕ_{ext}, exhibiting the periodic on/off behavior.

6.1.4 Fock–Darwin States

Now we will focus on the quantum mechanical energy levels that the electron occupies in a quantum dot. The electrostatic energy calculated in the previous subsections is determined by the overall geometry of the single-electron transistor and the total number of electrons in it. (It is basically self capacitance of the quantum dot.)

The quantum mechanical energies are determined by Schrödinger equation with a confining potential term. This confining potential is a different type of Coulomb potential an electron feels. (The single-electron charging energy is the result of Coulomb interaction among the electrons in a quantum dot.) The calculation of quantum energy levels of an electron in a quantum dot is similar to the case for the electron in a hydrogen atom. In semiconductor quantum dots, band discontinuity creates a potential well for the electrons. In real atoms, the Coulomb potential of the nucleus confines electrons. As was mentioned in Section 6.1.1, it is usually a good approximation that these quantum mechanical energy levels can simply be added to the single-electron charging energy:

$$E = \sum_{i=1}^{N} E_p + \frac{(Nq)^2}{2C} - Nq\phi_{ext}$$

$$= \sum_{i=1}^{N} E_p + \frac{q^2}{2C}\left(N - \frac{C\phi_{ext}}{q}\right)^2 - \frac{1}{2}C\phi_{ext}^2, q > 0, \quad (6.2)$$

where E_p is the quantum energy level of the quantum dot. In this case, all the features of single-electron transport are the same except there are shifts proportional to E_p's in the spacings between current peaks. One of the most realistic but analytically solvable confinement potentials is a harmonic potential. In 1928, Fock and Darwin [3, 4] had already solved the quantum mechanical energy levels of a single electron residing in a 2D harmonic potential where the magnetic field is applied on the quantum dot. (These energy levels are called "Fock–Darwin" states.) The energy E_{nl} is given by the following formula:

$$E_{nl} = (2n + |l| + 1)\hbar\left(\frac{1}{4}\omega_c^2 + \omega_o^2\right)^{1/2} - \frac{1}{2}l\hbar\omega_c \quad (6.3)$$

Here n ($= 0, 1, 2,...$) is the radial quantum number, l ($= 0, \pm1, \pm2, ...$) is the angular momentum quantum number, ω_o is the electrostatic confinement strength, and ω_c is the cyclotron energy given by qB/m^*, where B is the magnetic field perpendicular to the disc and m^* is the effective mass of the electron. In semiconductors, they were observed in a vertically defined, disc-shaped GaAs quantum dot [5]. The quantum dot is similar to the one in the photo in Fig. 6.3(b), except it has the gate surrounding the side of the pillar.

The way of observing these energies is behind Eq. (6.2). The spacing between two current peaks in the current oscillation as a function of the gate potential is proportional to the energy in Eq. (6.2). (It is called the addition energy [5].) Since the addition energy has two additive terms of Coulomb energy and the quantum energy, subtraction of the Coulomb energy from the addition energy gives the quantum energy. Figure 6.8 shows the current peak position as a function of B. Except for the gaps between each of the lines (which correspond to the Coulomb energy), the movement of the peak position is similar to the calculated E_{nl} in the lower panel of Fig. 6.8. Note that

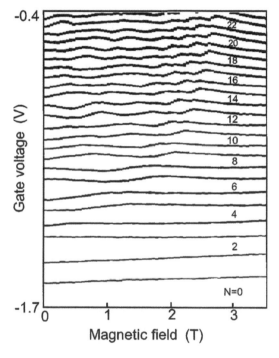

Figure 6.8. Experimental observation of Fock–Darwin states in GaAs quantum dot. (Upper) the current peak position of the Coulomb blockade oscillations in the gate bias as a function of the magnetic field. (Lower) Calculated Fock–Darwin states as a function of the magnetic field. Both diagrams have a good similarity. Reprinted with permission from S. Tarucha, D.G. Austing, T. Honda, R.J. van der Hage, and L.P. Kouwenhoven, "Shell filling and spin effects in a few electron quantum dot," *Phys. Rev. Lett.* **77**, 3613–3616 (1996). Copyright (1996) by the American Physical Society.

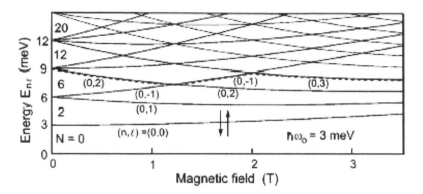

Figure 6.8. (Continued).

each energy level in the calculated diagram can be occupied with two electrons with opposite spin. Zeeman splitting is neglected since the direction of B is perpendicular to the radius of the disc.

6.1.5 *Spin States and Quantum Computing*

The conduction electrons in semiconductors have Zeeman splitting when a strong magnetic field is applied. In the expression of Fock–Darwin states shown in Eq. (6.3), only the orbital energy change due to the magnetic field was considered, and the Zeeman splitting of spin-up and spin-down electron was neglected. When a large magnetic field is applied in the plane with the quantum dot, the orbital energy of Eq. (6.3) will be small, and the main contribution from the magnetic field will be Zeeman splitting differentiating the spin-up and spin-down electron states.

In such a situation, we can realize a bi-level system consisting of spin-up and spin-down states of a single electron. For example, we can fill the quantum dot with exactly one electron under the in-plane magnetic field. The ground state of a GaAs quantum dot is a spin-up state. By applying high-speed electrical pulse on the gate, we can create a superposed state of the spin-up and spin-down states with tight control of the relative phase between them. Of course, the inverse of the pulse duration should be comparable to the Zeeman energy so that mixing between the two states can occur. This one electron spin state is a spin qubit [6]. We can also fabricate neighboring quantum dots with quantum mechanical coupling so that one qubit in a quantum dot can interact with another qubit in a second quantum dot. This type of coupling can form quantum gates. By extending these coupling schemes into N qubits in N quantum dots, we can perform quantum computing operations.

6.1.6 *Example of Semiconductor Single-Electron Transistors*

Coulomb blockade oscillations as a function of the gate bias were first observed in an etched silicon nanowire [7]. Scott-Thomas *et al.* observed periodic current oscillations as a function of the gate in a narrow Si MOSFET with a width of 25 nm. This device actually had the geometry of a 1DES as shown in Fig. 5.5. The researchers were apparently trying to fabricate an ideal 1DES, and they initially interpreted these oscillations as an evidence of charge density wave that was expected to be observed in a correlated 1DES. Delft people [8] right away pointed out that the data could be explained by a Coulomb blockade originating from a quantum dot defined by random potential fluctuations in a nanowire near threshold. Since then, many types of semiconductor single-electron transistors (SETs) have been reported. One of the most popular types is a quantum dot defined by the so-called split-gate technique (shown in Fig. 6.3(c)). Figure 6.9 illustrates the formation of a split-gate quantum dot where a set of gates on top of GaAs/AlGaAs HEMT wafer depletes the electrons underneath them and defines a small electron-holding area. Figure 6.10 shows two more examples of GaAs [9] and silicon SETs [10]. Electron beam lithography and wet etching of a HEMT wafer were used to define a quantum dot connected to a large source and drain reservoir by narrow necks. A metal layer was deposited on the top of the device, and the bias on it modulated the potential of the quantum dot. The silicon SET was fabricated with modern silicon VLSI processing technologies. This device was fabricated by first forming a 1D wire from an SOI wafer and then depositing two narrow depletion gates perpendicular to the wire. The width of these gates could be smaller than the lithographic limit since the sidewalls of the CMOS were used. Finally, a top gate was deposited on the whole device in order to

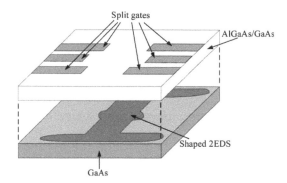

Split gates

AlGaAs/GaAs

Shaped 2EDS

GaAs

Figure 6.9. Spilt gate single-electron transistor (same as Fig. 6.3(c)). The negative biases on six split gates deplete the electrons underneath. Finite depletion layers create a disc-shaped electron region connected to the source and the drain reservoir by two bottlenecks. The biases on the gates also change the potential of the quantum dot.

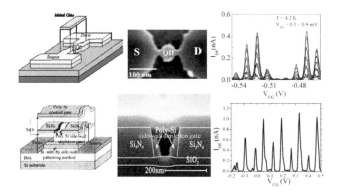

Figure 6.10. (Upper) Schematic, SEM photo, and the $I-V$ characteristic of a GaAs SET The quantum dot, tunnel barriers, and the reservoir are defined simultaneously by a single etching step (reprinted with permission from S.H. Son, K.H. Cho, S.W. Hwang, K.M. Kim, Y.J. Park, Y.S. Yu, and D. Ahn, "Fabrication and characterization of metal-semiconductor field-effecttransistor-type quantum devices," *J. Appl. Phys.* **96**, 1, 704–708, copyright 2004, American Institute of Physics). (Lower) Schematic, SEM photo, and the $I-V$ characteristic of a silicon SET. This SET was fabricated by full silicon VLSI processing technology (reprinted from *Microelectronic Eng.*, **63** (1–3), B.H. Choi, S.H. Son, K.H. Cho, S.W. Hwang, D. Ahn, D.H. Kim, J.D. Lee, B.G. Park, "Direct observation of excited states in double quantum dot silicon single electron transistor," 129–133, 2002, with permission from Elsevier).

induce the electron layer. Figure 6.10 also shows the current as a function of the gate bias. Both devices showed sharp current oscillations as a function of the gate bias, which were the strong evidence of single-electron tunneling. The temperature of the measurement was kept low, so that the $e^2/C \gg k_B T$ condition was satisfied. The current level was less than 10 nA in those devices, automatically satisfying another condition, $R = V/I \gg h/e^2$.

6.2 Modeling of Single-Electron Transistors

6.2.1 *Effect of the Drain Bias*

A quantum dot connected to the source and drain reservoir by tunnel barriers as shown in the left panel of Fig. 6.4 can be described by the equivalent circuit shown in Fig. 6.11. The quantum dot is a space between two tunnel capacitors C_1 and C_2. The resistances R_1 and R_2 denote the tunnel resistances, the inverses of which are proportional to the electron tunneling rates of the tunnel barriers. The external bias V is divided into V_1 and V_2 across the tunnel barrier 1 and the tunnel barrier 2. Then the net charge δQ in the quantum dot

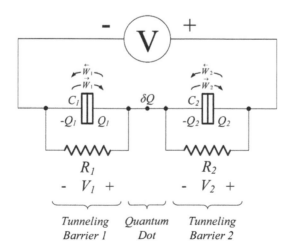

Figure 6.11. Equivalent circuit of the double junction corresponding to the schematic in the left panel of Fig. 6.4.

is expressed by

$$\delta Q = Q_1 - Q_2 = ne \, (e < 0)$$
$$Q_1 = C_1 V_1$$
$$Q_2 = C_2 V_2$$
$$V = V_1 + V_2. \tag{6.4}$$

Here, Q_1 and Q_2 are the charges of the tunnel capacitors C_1 and C_2, respectively.

In the case when one electron jumps from the quantum dot across the tunnel junction 2, then the charge states change as follows:

$$Q_1 \rightarrow Q_1$$
$$Q_2 \rightarrow Q_2 + e$$
$$\delta Q \rightarrow \delta Q - e \tag{6.5}$$

Now the new voltage $\left(V_1' + V_2' \right)$ drop becomes

$$V_1' + V_2' = \frac{Q_1}{C_1} + \frac{Q_2 + e}{C_2} = V_1 + V_2 + \frac{e}{C_2} = V + \frac{e}{C_2} \neq V \tag{6.6}$$

To keep $V_1 + V_2$ constant, an external source has to move ΔQ from the negative side of the junction 1 to the positive side of the junction 2 (through the source itself). Then

$$V_1'' + V_2'' = \frac{Q_1 + \Delta Q}{C_1} + \frac{Q_2 + e + \Delta Q}{C_2} \equiv \frac{Q_1}{C_1} + \frac{Q_2}{C_2}$$
$$\therefore \Delta Q = -\frac{e C_1}{C_1 + C_2} \tag{6.7}$$

Eventually,

$$Q_1 \rightarrow Q_1 + \Delta Q = Q_1 - \frac{e C_1}{C_1 + C_2}$$
$$Q_2 \rightarrow Q_2 + e + \Delta Q = Q_2 + \frac{e C_2}{C_1 + C_2} \tag{6.8}$$

The increase in the Coulomb energy after this tunneling event is

$$\frac{1}{2} C_1 V_1''^2 + \frac{1}{2} C_2 V_2''^2 - \frac{1}{2} C_1 V_1^2 - \frac{1}{2} C_2 V_2^2$$
$$= \frac{-e(Q_1 - Q_2)}{C_1 + C_2} + \frac{e^2}{2(C_1 + C_2)}. \tag{6.9}$$

The work provided by the external source is

$$-\Delta Q V = \frac{e C_1 V}{C_1 + C_2}. \tag{6.10}$$

And the total energy increase is

$$\Delta \vec{E}_2 = \frac{e^2}{C_1 + C_2} \left[\frac{1}{2} - n + \frac{C_1 V}{e} \right]. \tag{6.11}$$

This results in a tunneling rate for this particular tunneling event at zero temperature of

$$\vec{W}_2(n) = \frac{-\Delta \vec{E}_2}{e^2 R_2} \theta(-\Delta \vec{E}_2)$$

$$= \frac{n - \frac{1}{2} - \frac{C_1 V}{e}}{R_2(C_1 + C_2)} \theta \left(n - \frac{1}{2} - \frac{C_1 V}{e} \right). \tag{6.12}$$

Here the function θ is the unit step function and suggests that the tunneling occurs only when $\Delta \vec{E}_2 < 0$.

Let P_n be the probability of finding charge state en in the dot. Then P_n satisfies the time-dependent master equation (with the other tunneling events as denoted in Fig. 6.11):

$$\frac{d P_n}{dt} = [\vec{W}_2(n+1) + \bar{W}_1(n+1)] P_{n+1}$$

$$+ [\vec{W}_1(n-1) + \bar{W}_2(n-1)] P_{n-1}$$

$$- [\vec{W}_2(n) + \bar{W}_1(n)] P_n - [\vec{W}_1(n) + \bar{W}_2(n)] P_n \tag{6.13}$$

To calculate DC I–V, we must find a stationary solution for the P_n's. In the stationary state, $d P_n / t = 0$, and the right-hand side of Eq. (6.13) being equal to 0 gives us the detailed balance for in and out processes across each barrier.

$$[\vec{W}_2(n+1) + \bar{W}_1(n+1)] P_{n+1} = [\vec{W}_1(n) + \bar{W}_2(n)] P_n \tag{6.14}$$

This equation provides a recursion formula for obtaining all the P_n values from P_0. The unknown can be obtained from the normalization condition, $\Sigma P_n = 1$. The DC current is given by

$$I = -e \sum_n [\vec{W}_1(n) - \bar{W}_1(n)] P_n \tag{6.15}$$

Example 6.3: Electron tunneling rates

Determine three other tunneling rates for a single electron besides the one given by Eq. (6.12).

Solution:

We can repeat the same derivation procedure (Eqs. (6.5–6.12)) for the other tunneling processes. The charge configurations for other three single-electron tunneling processes are as follows:

$$\overleftarrow{W_2}, Q_1 \rightarrow Q_1, Q_2 \rightarrow Q_2 - e, \delta Q \rightarrow \delta Q + e$$
$$\overrightarrow{W_1}, Q_1 \rightarrow Q_1 + e, Q_2 \rightarrow Q_2, \delta Q \rightarrow \delta Q + e$$
$$\overleftarrow{W_1}, Q_1 \rightarrow Q_1 - e, Q_2 \rightarrow Q_2, \delta Q \rightarrow \delta Q - e$$

Then the changes of electrostatic energies are as follows:

$$\Delta \overleftarrow{E_2} = \frac{e^2}{C_1 + C_2} \left[\frac{1}{2} + n - \frac{C_1 V}{e} \right]$$

$$\Delta \overrightarrow{E_1} = \frac{e^2}{C_1 + C_2} \left[\frac{1}{2} + n + \frac{C_2 V}{e} \right]$$

$$\Delta \overleftarrow{E_1} = \frac{e^2}{C_1 + C_2} \left[\frac{1}{2} - n - \frac{C_2 V}{e} \right]$$

The resulting tunneling rates are given by

$$\overleftarrow{W_2}(n) = \frac{-\Delta \overleftarrow{E_2}}{e^2 R_2} \theta(-\Delta \overleftarrow{E_2}) = \frac{-n - \frac{1}{2} + \frac{C_1 V}{e}}{R_2(C_1 + C_2)} \theta \left(-n - \frac{1}{2} + \frac{C_1 V}{e} \right)$$

$$\overrightarrow{W_1}(n) = \frac{-\Delta \overrightarrow{E_1}}{e^2 R_1} \theta(-\Delta \overrightarrow{E_1}) = \frac{-n - \frac{1}{2} - \frac{C_2 V}{e}}{R_1(C_1 + C_2)} \theta \left(-n - \frac{1}{2} - \frac{C_2 V}{e} \right)$$

$$\overleftarrow{W_1}(n) = \frac{-\Delta \overleftarrow{E_1}}{e^2 R_1} \theta(-\Delta \overleftarrow{E_1}) = \frac{n - \frac{1}{2} + \frac{C_2 V}{e}}{R_1(C_1 + C_2)} \theta \left(n - \frac{1}{2} + \frac{C_2 V}{e} \right).$$

Example 6.4: Calculation of the probabilities

Find out the stationary probabilities P_n of the double junction using the tunneling rates and the detailed balance of Eq. (6.14). The parameters of the double junction are $R_1 = R_2 = 1 \text{ M}\Omega, C_1 = C_2 = 1 \text{ aF}$, and the bias $V = 0.2 \text{ V}$.

Solution:

$$\frac{C_1 V}{e} = \frac{C_2 V}{e} = \frac{1.6 \times 10^{-18} \times 0.2}{-1.6 \times 10^{-19}} = -2$$

Using this and the rate equations above, the following table gives the tunneling rates in units of $1/(2 R_1 C_1)$ for various n values.

N	$\overrightarrow{W_2}(n)$	$\overleftarrow{W_2}(n)$	$\overrightarrow{W_1}(n)$	$\overleftarrow{W_1}(n)$
−4	0	3/2	11/2	0
−3	0	1/2	9/2	0
−2	0	0	7/2	0
−1	1/2	0	5/2	0
0	3/2	0	3/2	0
1	5/2	0	1/2	0
2	7/2	0	0	0
3	9/2	0	0	1/2
4	11/2	0	0	3/2

We set $p_0 = c$. Then from Eq. (6.14),

$$p_1 = \frac{\overrightarrow{W_1}(0) + \overleftarrow{W_2}(0)}{\overrightarrow{W_2}(1) + \overleftarrow{W_1}(1)} p_0 = \frac{3/2 + 0}{5/2 + 0} c = \frac{3}{5} c$$

$$p_2 = \frac{\overrightarrow{W_1}(1) + \overleftarrow{W_2}(1)}{\overrightarrow{W_2}(2) + \overleftarrow{W_1}(2)} p_1 = \frac{1/2 + 0}{7/2 + 0} p_1 = \frac{1}{7} p_1 = \frac{3}{35} c$$

$$p_3 = \frac{\overrightarrow{W_1}(2) + \overleftarrow{W_2}(2)}{\overrightarrow{W_2}(3) + \overleftarrow{W_1}(3)} p_2 = \frac{0 + 0}{9/2 + 1/2} p_2 = 0.$$

The hole occupation probabilities p_{-n} are calculated as follows:

$$p_{-1} = \frac{\overrightarrow{W_2}(0) + \overleftarrow{W_1}(0)}{\overrightarrow{W_1}(-1) + \overleftarrow{W_2}(-1)} p_0 = \frac{3/2 + 0}{5/2 + 0} c = \frac{3}{5} c$$

$$p_{-2} = \frac{\overrightarrow{W_2}(-1) + \overleftarrow{W_1}(-1)}{\overrightarrow{W_1}(-2) + \overleftarrow{W_2}(-2)} p_{-1} = \frac{1/2 + 0}{7/2 + 0} p_{-1} = \frac{1}{7} p_{-1} = \frac{3}{35} c$$

$$p_{-3} = \frac{\overrightarrow{W_2}(-2) + \overleftarrow{W_1}(-2)}{\overrightarrow{W_1}(-3) + \overleftarrow{W_2}(-3)} p_{-2} = \frac{0 + 0}{9/2 + 1/2} p_{-2} = 0$$

The normalization condition gives

$$p_{-2} + p_{-1} + p_0 + p_1 + p_2 = 1$$
$$\left(\frac{6}{35} + \frac{6}{5} + 1 \right) c = 1, c = \frac{35}{83}.$$

Therefore, we get

$$p_0 = \frac{35}{83}$$

$$p_1 = p_{-1} = \frac{21}{83}$$

$$p_2 = p_{-2} = \frac{3}{83}$$

$$\text{others} = 0$$

6.2.2 Effect of Finite Temperature

The derivation in Section 6.2.1 was restricted to the case of $T = 0$. At finite temperature, however, Fermi-Dirac statistics must be considered, and this will change the tunneling rates. For example, it can be shown that the $T = 0$ tunneling rate of Eq. (6.12) is modified to the following formula at finite T [11].

$$\vec{W}_2(n) = \frac{-\Delta \vec{E}_2}{e^2 R_2} f(-\Delta \vec{E}_2)$$

$$f(z) = \frac{1}{1 - \exp(-z)} \tag{6.16}$$

Figure 6.12 shows the current calculated for various finite values of T using this formula. The parameters used in the calculations are $R_1 = 0.9$ MΩ, $R_2 = 0.1$ MΩ, and $C_1 = C_2 = 1.6$ aF. The figure gives the I–V curves at $T = 10, 30, 77, 200$, and 300 K. The staircases that can be seen clearly at low T smear out as T increases. The $I - V$ relationship is almost linear when $T = 300$ K.

6.2.3 Effect of the Gate Bias

Figure 6.13 shows the circuit diagram of a single-electron transistor consisting of a double junction and a gate that was shown in Fig. 6.6. By superposition, the external voltage on junction 2 is a simple sum of the voltage across the junction 2 when $V_g = 0$ and the voltage across the junction 2 when $V_b = 0$. Therefore, we obtain the following:

$$V_2^{\text{ext}} = \frac{C_1 + C_g}{C_1 + C_g + C_2} V_b + \frac{-C_g}{C_1 + C_g + C_2} V_g \tag{6.17}$$

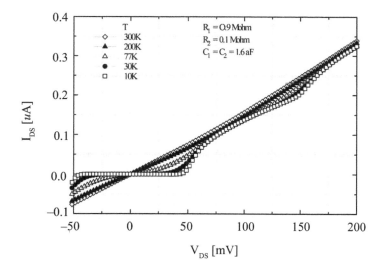

Figure 6.12. Typical temperature dependence of a single-electron transistor. It shows a Coulomb gap near zero bias ($V_{DS} = 0$) at low temperatures. The Coulomb gap smears out as T increases, finally becoming linear.

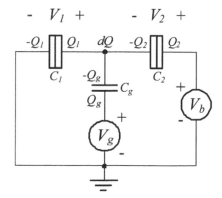

Figure 6.13. Circuit diagram of a single-electron transistor consisting of a double junction and a gate. The drain bias and the gate bias are denoted as V_b and V_g, respectively. The capacitance C_1 and C_2 are tunnel capacitances. The capacitance C_g is the normal capacitance.

When an electron tunnels our of the quantum dot across junction 2, both the sum of the voltage drop across junction 1 and across junction 2 and the sum of the voltage drop across junction 1 and across the gate capacitor before and after the tunneling event should be constant. For this, the external source V_g has to move ΔQ_g from the negative side of junction 1 to the positive side of the normal capacitance C_g. On the other hand, the external source V_b has to move ΔQ from the negative side of junction 1 to the positive side of junction 2.

$$\frac{Q_1 + \Delta Q + \Delta Q_g}{C_1} + \frac{Q_2 + \Delta Q + e}{C_2} = \frac{Q_1}{C_1} + \frac{Q_2}{C_2} \qquad (6.18)$$

$$\frac{Q_1 + \Delta Q + \Delta Q_g}{C_1} + \frac{Q_g + \Delta Q_g}{C_g} = \frac{Q_1}{C_1} + \frac{Q_g}{C_g} \qquad (6.19)$$

From the above two equations, we obtain the following:

$$\Delta Q = -\frac{(C_1 + C_g)e}{C_T}$$

$$\Delta Q_g = \frac{C_g e}{C_T}$$

$$C_T = C_1 + C_2 + C_g \qquad (6.20)$$

Then the charge variations can be expressed as follows:

$$Q_1 \rightarrow Q_1 - \frac{C_1 e}{C_T}$$

$$Q_2 \rightarrow Q_2 + \frac{C_2 e}{C_T}$$

$$Q_g \rightarrow Q_g + \frac{C_g e}{C_T}. \qquad (6.21)$$

The change of the Coulomb energy after the tunneling event is

$$\frac{\left(Q_1 - \frac{C_1 e}{C_T}\right)^2}{2C_1} + \frac{\left(Q_2 + \frac{C_2 e}{C_T}\right)^2}{2C_2} + \frac{\left(Q_g + \frac{C_g e}{C_T}\right)^2}{2C_g}$$

$$-\frac{Q_1^2}{2C_1} - \frac{Q_2^2}{2C_2} - \frac{Q_g^2}{2C_g}$$

$$= \frac{(Q_2 - Q_1 + Q_e)e}{C_T} + \frac{e^2}{C_T}. \qquad (6.22)$$

The work provided by external sources is

$$
\begin{aligned}
&- \Delta Q\, V_b - \Delta Q_g V_g \\
&= \frac{(C_1 + C_g)e}{C_T} V_b - \frac{C_g e}{C_T} V_g.
\end{aligned}
\tag{6.23}
$$

Therefore, the total energy increase is

$$
\Delta \overrightarrow{E_2} = \frac{e^2}{C_T}\left[\frac{1}{2} - n + \frac{C_1 + C_g}{e}V_b - \frac{C_g}{e}V_g\right].
\tag{6.24}
$$

Here we use the definition of the charge $ne = Q_1 - Q_2 - Q_g$. Note that the last two terms corresponds to V_2^{ext} in Eq. (6.17). Again from superposition principle, we obtain the effective bias across the junction 1.

$$
V_1^{\text{ext}} = \frac{C_2}{C_1 + C_2 + C_g} V_b + \frac{C_g}{C_1 + C_2 + C_g} V_g
\tag{6.25}
$$

Then, using similar procedures, we can obtain energy differences for other tunneling events.

$$
\begin{aligned}
\Delta \overleftarrow{E_2} &= \frac{e^2}{C_T}\left[n + \frac{1}{2} - \frac{C_T V_2^{\text{ext}}}{e}\right] \\
&= \frac{e^2}{C_T}\left[n + \frac{1}{2} - \frac{C_1 + C_g}{e}V_b + \frac{C_g}{e}V_g\right] \\
\Delta \overrightarrow{E_1} &= \frac{e^2}{C_T}\left[n + \frac{1}{2} + \frac{C_T V_1^{\text{ext}}}{e}\right] = \frac{e^2}{C_T}\left[n + \frac{1}{2} + \frac{C_2}{e}V_b + \frac{C_g}{e}V_g\right] \\
\Delta \overleftarrow{E_1} &= \frac{e^2}{C_T}\left[-n + \frac{1}{2} - \frac{C_T V_1^{\text{ext}}}{e}\right] \\
&= \frac{e^2}{C_T}\left[-n + \frac{1}{2} - \frac{C_2}{e}V_b - \frac{C_g}{e}V_g\right]
\end{aligned}
\tag{6.26}
$$

When these energy differences are smaller than zero, the corresponding tunneling events are possible. From those, the bias conditions for a specific tunneling event can be obtained. For example, when $n = 0$, the conditions for various tunneling events are as follows:

$$\Delta\overrightarrow{E}_2(n=0) < 0, V_b < \frac{C_g}{C_1+C_g}V_g - \frac{e}{2}\frac{1}{C_1+C_g}$$

$$= \frac{C_g}{C_1+C_g}V_g + \frac{|e|}{2}\frac{1}{C_1+C_g}$$

$$\Delta\overleftarrow{E}_2(n=0) < 0, V_b < \frac{C_g}{C_1+C_g}V_g - \frac{|e|}{2}\frac{1}{C_1+C_g}$$

$$\Delta\overrightarrow{E}_1(n=0) < 0, V_b < -\frac{C_g}{C_2}V_g + \frac{|e|}{2C_2}$$

$$\Delta\overleftarrow{E}_1(n=0) < 0, V_b < -\frac{C_g}{C_2}V_g - \frac{|e|}{2C_2} \tag{6.27}$$

Figure 6.14 shows these conditions. For a positive V_b, one of the main current flow is single-electron tunneling across the junction 1 into the Coulomb island ($\Delta\overrightarrow{E}_1(n=0) < 0$), and subsequent

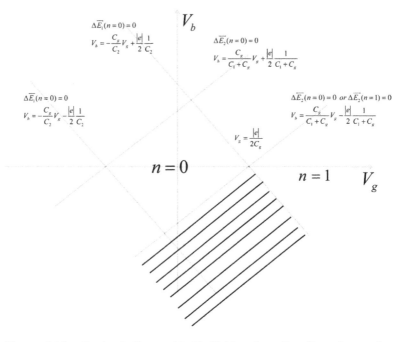

Figure 6.14. Coulomb diamond in V_g–V_b bias plane. Four lines denote the boundary delineating the region of each tunneling condition. The hatched region is where both $\Delta\overrightarrow{E}_1(n=0) < 0$ and $\Delta\overrightarrow{E}_2(n=1) < 0$ are possible so that the current flow from junction 1 to junction 2 is possible.

tunneling across the junction 2 out of the Coulomb island ($\Delta\overrightarrow{E_2}(n = 1) < 0$). The overlapped region of these two conditions is hatched in Fig. 6.14.

6.2.4 Brief Review of Single-Electron Circuit Simulation

In this section, we will briefly review how to carry out single-electron circuit calculations. Figure 6.15(a) shows an example of a single-electron circuit consisting of a few single-electron transistors and biases. Each transistor has a quantum dot confining electrons and it is represented by a node in the figure. Each node is connected to the source and drain through tunneling capacitances, and to the gate through a normal capacitance. If the size of the connecting capacitance is small enough, the junction between the two single-electron transistors again becomes a single-electron node. The most difficult problem in simulating a single-electron circuit is that the tunneling rate of a particular tunneling capacitance is a function of the charge configurations of all the nodes in the circuit. Therefore, we have to set up a master equation describing the charge states of all the nodes altogether.

We will consider the particular case shown in Fig. 6.15(a) and concentrate on one quantum dot denoted as node 5. Nodes 3 and 4 are connected to the external biases E_3 and E_4. Nodes 1 and 2 are other quantum dots. The circuit can be visualized as a linear superposition of a circuit considering external biases only (Fig. 6.15(b)) and one with internal voltages only (Fig. 6.15(c)). The charge at node 5, q_5, is expressed as follows:

$$
\begin{aligned}
q_5 &= Q_1 + Q_2 + Q_3 + Q_4 \\
&= C_1 \left(V_5^{\text{int}} - V_1^{\text{int}} \right) + C_2 \left(V_5^{\text{int}} - V_2^{\text{int}} \right) \\
&\quad + C_3 \left(V_5^{\text{int}} - 0 \right) + C_4 \left(V_5^{\text{int}} - 0 \right) \\
&= (C_1 + C_2 + C_3 + C_4) V_5^{\text{int}} - C_1 V_1^{\text{int}} - C_2 V_2^{\text{int}} \quad (6.28)
\end{aligned}
$$

Here, Q_1, Q_2, Q_3, and Q_4 are the charges on the capacitances. V_1^{int}, V_2^{int}, V_3^{int}, and V_4^{int} are the internal node voltages. This can be extended into a general matrix form

$$
\mathbf{C}\mathbf{V}^{\text{int}} = \mathbf{q}, \quad (6.29)
$$

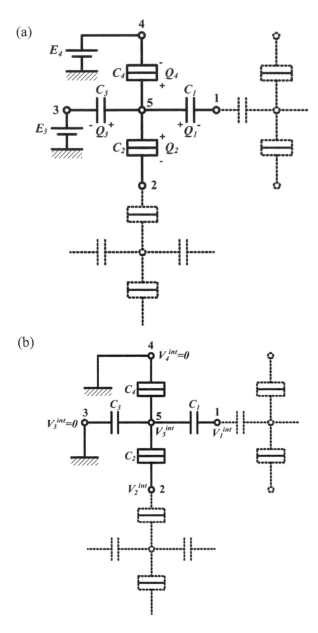

Figure 6.15. (a) A part of a single-electron circuit showing three single-electron transistors, (b) equivalent circuit when external biases are set to zero, and (c) equivalent circuit when only external biases are considered.

(c)

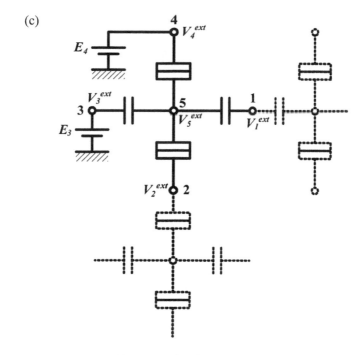

Figure 6.15. (Continued).

where C_{ij} = sum of the capacitances connected to the ith node, when $i = j$

 −(sum of the capacitances in between the ith node and the jth node), when $i \neq j$ and there are no nodes in between

 0, when $i \neq j$ and the ith node and the jth node are not directly connected.

We can utilize Kirchhoff's current law in the time domain (Laplace transform) for node 5.

$$sC_1\left(V_5^{\text{ext}} - V_1^{\text{ext}}\right) + sC_2\left(V_5^{\text{ext}} - V_2^{\text{ext}}\right) + sC_3\left(V_5^{\text{ext}} - V_3^{\text{ext}}\right)$$
$$+sC_4\left(V_5^{\text{ext}} - V_4^{\text{ext}}\right) = 0$$
$$(C_1 + C_2 + C_3 + C_4)V_5^{\text{ext}} - C_1V_1^{\text{ext}} - C_2V_2^{\text{ext}} = C_3V_3^{\text{ext}}$$
$$+C_4V_4^{\text{ext}} = C_3E_3 + C_4E_4 \tag{6.30}$$

This result can also be generalized into a matrix form

$$\mathbf{CV^{\text{ext}}} = \mathbf{EV_e} \tag{6.31}$$

Here, the matrix $\mathbf{C}$ is the same as in Eq. (6.18), and $\mathbf{E}$ is the $N \times M$ matrix whose elements are expressed as follows:

E_{ij} = sum of the capacitances between the ith node and the jth node, when they are directly connected

0, when the ith node and the jth node are not directly connected.

$\mathbf{V}_e$ is the $M \times 1$ vector representing the external biases (E_3, E_4, etc.).

The tunneling rate of the ith tunneling barrier is similar to the case of a single-electron transistor and is given by

$$\Gamma_i = \frac{1}{e^2 R_t(i)} \frac{\Delta E_i}{1 - \exp\left(-\frac{\Delta E_i}{k_B T}\right)}$$

$$\Delta E_i = eV_{eff}(i) + \sum_{n=1}^{N} \frac{Q_n(\mathbf{q_k})}{2C_n} - \sum_{n=1}^{N} \frac{Q_n(\mathbf{q'_k})}{2C_n} \qquad (6.32)$$

Here, $V_{eff}(i)$ is the effective voltage drop across the ith tunnel barrier, which is the sum of internal external voltages calculated using Eqs. (6.29) and (6.31), respectively. The vectors $\mathbf{q}_k$ and $\mathbf{q}'_k$ are the electron configurations of the old and the new state (before and after tunneling through the ith barrier), respectively. The symbols C_n and Q_n are the nth capacitance and its charge. Finally, the electron configuration of the quantum dot for obtaining the charge state $\mathbf{q}$ is determined by the master equation.

$$\dot{P}_i = \sum_j (\Gamma_{ji} P_j - \Gamma_{ij} P_i) \qquad (6.33)$$

Here, $P_i(n)$ is the probability of filling n electrons in the ith quantum dot. One efficient way of solving the master equation is using the Monte Carlo method. Figure 6.16 shows an algorithm for solving Eq. (6.33) with the Monte Carlo method. When a single-electron circuit is given, the initial electron occupation configuration of all the quantum dots is set as an initial guess. Then the tunneling rates of all the tunnel junctions are obtained by Eq. (6.33). A set of random numbers is generated to give the next electron configuration after a certain amount of time. After the next electron configuration is determined, the tunneling rates of the tunnel junctions are calculated again. This process is repeated until enough iterations have been done, and the average (steady state) electron configuration is obtained from the different electron configurations of all the steps.

Figure 6.17 shows a single-electron inverter consisting of two single-electron transistors in series [12]. The gates of the SETs are connected and form the input. The load capacitor C_L is connected in between the two SETs. When the input becomes high and the load capacitor is in the logic high, the upper SET becomes less on. The lower SET is more on and the load capacitor is discharged through it, and the output voltage (the voltage across the load capacitor) decreases, inverting the input. As the voltage of the load capacitor decreases, the upper SET becomes a little more on and tries to re-charge the load capacitor. Therefore, the voltage of the load capacitor cannot drop completely to zero in this inverter. When the input becomes low and the output is in the logic low state, the upper SET turns more on and the lower SET turns more off. Then the load capacitor is charged more through the upper SET, resulting in the logic high. Figure 6.18 shows the calculated transfer characteristics when $C_G = 1.6$ aF, $C = 3.2$ aF, $R_t = 100$ MΩ, $C_L = 5.12$ aF, and $V_B = 0.03$ V. Even though the logic swing is smaller than the bias window, there clearly is an inverting characteristic. Figure 6.19 shows the T dependence of the transfer characteristics. As T increases, the logic swing decreases, and, finally, the inverter does not work at high T.

6.2.5 *Examples of Single-Electron Circuits*

Single-electron transistors are a basic building block of single-electron circuits. Single-electron circuits (SECs) have been widely studied since the advent of single-electron transistors, since they are considered to be an ultimate limit of electronic circuits in which the change of one electron can represent a logic level. There have been many proposals for realizing single-electron circuits, and in this section we will describe several categories of them. They are full SET logic circuits with one or a few gates, SET-FET (field effect transistor) hybrid logic circuits, and SET-FET hybrid memory circuits.

The way of configuring single-electron logic circuits is similar to how MOS circuits are configured, except that on and off repeatedly show up as functions of the gate bias in SETs. Therefore, utilizing this repeating on-offs, it is possible to realize a multi-valued logic. The first type of SE logic was proposed by Chen [13] and consists of

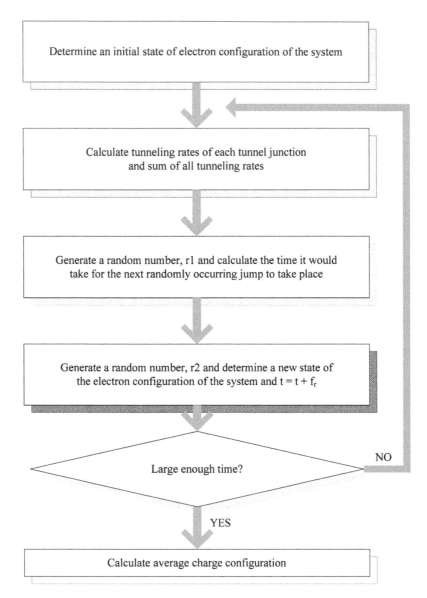

Figure 6.16. Flow chart for obtaining the solution to the master equation of Eq. (6.33).

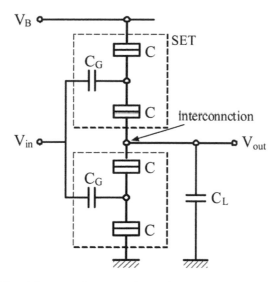

Figure 6.17. Schematic diagram of a single-electron inverter consisting of two single-electron transistors in series. Both gates of the SETs are hooked together, forming an input terminal. The load capacitor C_L is connected in the junction between the two SETs. The output voltage is the voltage drop across C_L.

SETs with a single gate terminal. Figure 6.20(a), (b), and (c) show, respectively, an inverter [12], a two-input NOR [13], and two-input OR logic gates [14] consisting of single gate SETs. These logic gates utilize only one threshold and represent binary logic. The unit SET is similar to the MOSFET and has three terminals (gate, source, and drain). It consists of five components, $C_g, C_s, C_d, R_s,$ and $R_d,$ denoting the gate normal capacitance, source tunnel capacitance, drain tunnel capacitance, source tunnel resistance, and drain tunnel resistance. The symbols V_B and C_L denote the bias voltage and the load capacitance, respectively. The operation principle of these logic gates is similar to that of MOSFET logic gates. When at least one of the inputs is switched on in the 2-input OR gate shown in Fig. 6.20(c), that input turns on the SET connected to it. Then the on SET charges the load until the voltage of the load capacitor (V_{out}) becomes comparable to V_B, decreasing the charging current. The load capacitor is always discharged through the bottom tunnel junction, and the on SET supplies the decreased charge. When both inputs go to the low

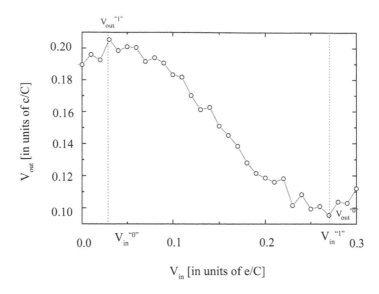

Figure 6.18. Calculated transfer characteristics of the single-electron inverter shown in Fig. 6.15. The output voltage swing of this inverter is smaller than the bias voltage, and the logic low does not approach zero in this type of inverter.

states, both SETs turns off, and the load capacitor is fully discharged through the bottom tunnel junction. Then the final output voltage is low.

The major difference between SETs and MOSFETs is that SETs can have multiple gates that control the potential of the quantum dot independently. The second type of logic circuit consists of SETs with two gates [15]. Figures 6.21(a), (b), (c), and (d) show, respectively, an inverter, a two-input XOR, a two-input NAND, and a two-input NOR logic gate. Here, all the symbols have the same meanings as in Fig. 6.20. The additional gate capacitance of each SET is denoted as C_b. The role of the additional gate is similar to the back gate of the MOSFET. This additional gate is pre-biased to give a shift in the transfer characteristic of each SET. For example, the second gate of the upper SET of the inverter circuit shown in Fig. 6.21(a) is grounded, while the second gate of the lower SET is biased high (V_B). When V_{in} (denoted as a logic state A in the figure) is high, the upper SET is turned off, and the lower SET is turned on. Then the load

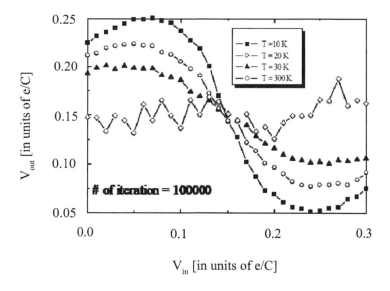

Figure 6.19. T-dependence of the transfer characteristics of the inverter shown in Fig. 6.15. The logic swing is a strong function of T, approaching zero at $T = 300$ K, at which temperature the Coulomb blockade is not effective because of the thermal fluctuation.

capacitance is discharged through the lower SET until V_{out} becomes low. Once V_{out} becomes low, the current through the lower SET becomes negligibly small. When V_{in} is low, the upper SET is turned on, and the load capacitor is charged through the upper SET up to high voltage. The current through the upper SET is again negligible once the load capacitor is fully charged. The lower SET is turned off, and there is no current through the lower SET. By having additional gates, we can realize CMOS-type gates. The current flow occurs only during transients.

The action of a two-input NOR shown in Fig. 6.21(d) can be explained as follows. When both inputs (A and B) are low, two upper SETs in series are turned on, and two parallel lower SETs are turned off. Then the upper load capacitor in between the two series SETs is charged up first, then this charge flows into the lower load capacitor, resulting in the output being high. When the input A is high and the input B is low, the uppermost SET is turned on and the second series SET is turned off. On the other hand, the left lower SET is

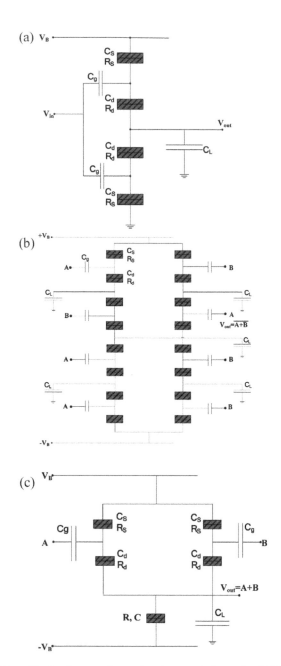

Figure 6.20. Single-electron logic circuits consisting only of SETs with one gate: (a) inverter, (b) NOR, and (c) OR gate.

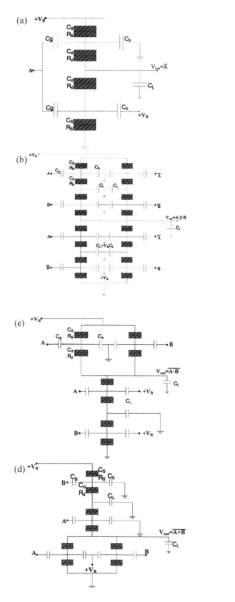

Figure 6.21. Single-electron circuits consisting of SETs with two gates: (a) inverter, (b) XOR, (c) NAND, and (d) NOR gate.

turned on, and the right lower SET is turned off. The upper load capacitor is charged up, but it does not affect the output since the second series SET is turned off. The lower capacitor is discharged through the left lower SET, resulting in a low state in V_{out}. When the input A is low and the input B is high, the uppermost SET is turned off and the second series SET is turned on. On the other hand, the left lower SET is turned off, and the right lower SET is turned on. The upper load capacitor is discharged through the second series SET into the lower load capacitor, but all these charges again are discharged through the right lower SET, again resulting in a low state in V_{out}. When both inputs are high, both upper SETs are turned off and both lower SETs are turned on. Then the lower load capacitor is discharged through the lower SETs, and the logic output becomes low. Other logic gates (two-input NAND and two-input XOR) can be explained in a similar way.

The final type of single-electron logic consists of circuits that utilize multi-gate SETs. Figures 6.22(a) and (b) show a multi-input NAND [15] and multi-input XOR [16], respectively. The upper part of the multi-input NAND is similar to the two-input NAND shown in Fig. 6.21(c) except that there are N two-gate SETs. On the other hand, the lower SET has N gates and therefore N inputs and one back gate. The back gates of upper SETs are grounded, and the back gate of the multi-input lower SET is biased high. When all the inputs are high, all the upper SETs are turned off, and the lower SET is turned on. Then the load capacitance is discharged until it becomes low. When at least one input of the upper SETs is low, that SET is turned on, and the lower SET is turned off. The on SET then charges the load capacitor up to a high state.

One serious problem of SECs is the small current driving capabilities. To overcome this problem, many researchers have proposed hybrid SECs in which SETs and MOSFETs are used together. Fig. 6.23(a) show a cascade type circuit consisting of an SET in series with an MOSFET. The circuit is driven by a constant current load. It is called the universal literal gate and was proposed by Inokawa [17]. When the input of the SET gate is high, the SET is turned on, and the constant current load drives the SET current in the rising edge of the Coulomb oscillation. In this case, to regulate the current increase, the voltage drop across the MOSFET becomes small, and the output is

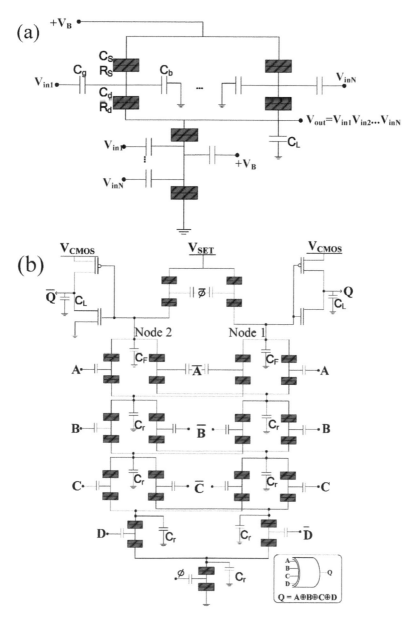

Figure 6.22. Multi-input SECs: (a) NAND and (b) XOR gate.

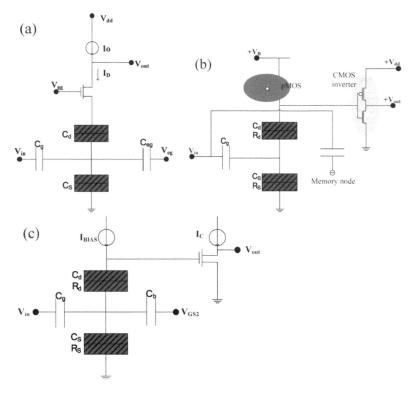

Figure 6.23. Single-electron FET logic circuits: (a) universal literal gate (ULG) proposed by Inokawa, (b) ULG by Uchida, and (c) ULG by Mahapatra.

almost shorted to the ground (low state). When the input of the SET becomes low, the SET tries to be turned off, and the constant current load drives the SET current in the falling edge of the Coulomb oscillation. Then the voltage drop across the MOSFET becomes larger, and the output goes to the high state. The output voltage swing in this universal literal gate is almost the same as the MOSFET bias voltage V_{dd}, which is much larger than the input swing of the SET.

Figure 6.23(b) shows a converter consisting of an SET in series with a pMOS. This circuit was proposed by Uchida [18]. The output of the converter is fed into a CMOS so that the overall circuit acts as an inverter with amplified output. The pMOS is always in a low current state, since only positive biases are used. However, it works as a load to feedback the action of the SET. When input is high, the SET is turned on, and the voltage drop across the pMOS decreases to regulate the current. This action leads to the high state in the output of the pMOS-SET series circuit. When the input is low, the SET tries to be turned off. Then the voltage drop across the pMOS increases in order to increase the current, resulting in a low voltage in the output. Therefore, the pMOS-SET series circuit works as a converter, but with an amplified output. The output of this series circuit is eventually inverted by the CMOS inverter connected to it.

Figure 6.23(c) shows a two-stage circuit consisting of a current biased SET and a MOSFET [19]. When the gate bias of the SET increases from an on state to an off state (this can, for example, be caused by an increase of the voltage since the SET has periodic on/off oscillations), the drain voltage should increase to keep the bias current constant. This increased drain bias turns on the second stage MOSFET, increasing the drain current of the MOSFET. Therefore, this circuit can convert an input with a small voltage swing to a large current. When the MOSFET stage is current biased as shown in the figure, the increased gate bias of the MOSFET (drain bias of the SET) decreases the drain bias of the MOSFET (output of the circuit) to keep the current constant. This is an inverting characteristic. (Other characteristics are also possible depending on the gate bias window.) Of course, the output voltage swing is that of the MOSFET and will be much larger than the input swing.

The multiple on/off characteristics of SETs can be utilized naturally to produce a multi-valued memory. However, the state of the single-electron memory is represented by the change of a single electron so that the logic levels are usually very small. To overcome this problem, many memory circuits consisting of SETs and MOSFETs have been proposed. Here again, the SET determines the logic state by the change of a single electron, and MOSFET increases the voltage (or sometimes the current) level. Most of the proposals are basically conventional MOSFET memory circuit, such as SRAMs (static random access memories) or latches, except for having one of the MOSFETs replaced with an SET. Figure 6.24(a) is an SET-MOS SRAM cell [17]. The bit line and the word line are connected to the drain and the gate of the FET switch M_3, respectively. When both the word line and the bit line are selected (the bias of the word line is high, and the bit line is connected to the writing or reading circuit), M_3 is on

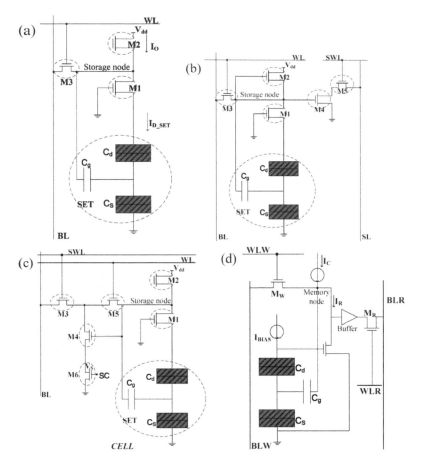

Figure 6.24. Single-electron-FET memory circuits: (a) SRAM cell by Inokawa, (b) modified SRAM cell, (c) SRAM cell by Yu, and (d) SRAM cell by Mahapatra.

and transfers the charge in and out of the memory node. This SRAM is based on the universal literal gate of Fig. 6.23(a), except that it is designed to have negative differential resistance characteristics for SRAM action. This structure has only 4 transistors. Furthermore, the capacitance of the memory node is less than a few fF, so that the writing speed is relatively fast. However, it has static power consumption due to the current biasing, and the read speed is slow because the small current has to charge up the bit line capacitance (usually a few tens of fF). Figure 6.24(b) is a modification of Fig. 6.24(a), utilizing the current sensing scheme [17] to increase the read speed. Figure 6.24(c) is a similar version of Fig. 6.24(b), but the number of lines is reduced to three [20]. Figure 6.24(d) is an interesting latch structure in which the gate of the SET is connected to the drain of the MOSFET, and the gate of the MOSFET is connected to the drain of the SET [19]. When the logic high is written at the memory node, the gate of the SET becomes high, and it turns on the SET. Then it pulls down the current and lowers the voltage of the gate of the MOSFET. This turns off the MOSFET, and the memory node becomes a stable high level. This memory scheme uses double bit line and double word line architecture, so that the writing and reading operations occur using separate lines. For the writing operation, BLW and WLW are biased and the FET M_W is turned on. For the reading operation, BLR and WLR are biased and the FET M_R is turned on. When the logic low is written on the memory node, it turns off the SET, and the drain of the SET (gate of the MOSFET) goes high. This turns on the MOSFET, and the memory node is in a stable low condition. The read operation is performed by switching on the word line and connecting the bit line to the sense amplifier input. Since there is a buffer between the memory node and the sense amplifier, the logic level of the memory node is conserved during and after the read operation.

Finally, regarding the simulation of these single-electron circuits, it was emphasized in the previous subsection that the junction between two SETs should be treated as another quantum dot, and the electron population has to be kept track of together with other quantum dots in other SETs. Yu *et al.* found that if the total capacitance of such interconnections is much larger than the total capacitances of other normal SETs, then the electron populations of these interconnections can be neglected [12]. In this case, each SET in the circuit can be treated independently, and compact modeling techniques, which are widely used in conventional circuit simulations, can be used. Usually, compact modeling uses a combination of diodes to simulate the characteristics of SETs. Sometimes built-in routines in SPICE-type simulators are used. Many of single-electron circuits introduced in this subsection have SETs embedded in CMOS circuits. That kind of circuit can be simulated using these SPICE-type simulators [21].

PROBLEMS

1. **Electrons confined in a silicon cylinder:**
   The electrons in a gate-all-around silicon nanowire field effect transistor such as the one shown in Fig. 6.P1 feel an oval-shaped confinement potential. Write down the Hamiltonian of the silicon quantum dot with oval potential confinement.

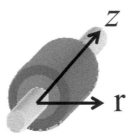

Figure 6.P1 A gate-all-around silicon nanowire field effect transistor. The electrons confined inside the nanowire feed an oval-shaped potential. The oval has the short axis in the r-direction of the cylindrical coordinate, and the long axis along the z-direction. See also Color Insert.

2. **DC current of a double junction:**
   Calculate the DC current of the double junction with parameters $R_1 = R_2 = 1$ MΩ, $C_1 = C_2 = 1$ aF, and the bias $V = 0.2$ V.

3. **Energy of a quantum box:**
   (a) Find out the energy difference between the ground state and the first excited state of a rectangular quantum box with length L.
   (b) Find the value of L when this energy is the same as the Coulomb charging energy of a sphere with the same radius.

4. **Various characteristics of a single-electron transistor:**
   (a) The following figure shows a schematic of a single-electron transistor consisting of a double junction and a gate (same as Fig. 6.13). When $C_1 = 2$ aF, $C_2 = 3$ aF, $C_g = 2$ aF, $V_b = 0.5$ V, and $V_g = 1$ V, find the external voltage on junction 2.
   (b) Calculate the total energy increase for the tunneling of a single electron across the junction 2 when $n = 0$.

(c) Calculate the total energy increase for the tunneling of a single electron across the junction 1 when $n = 0$.
(d) Obtain the slopes of the Coulomb diamond.

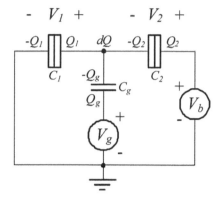

Figure 6.P2 Schematic of a single-electron transistor consisting of a double junction and a gate. The drain bias and the gate bias are denoted as V_b and V_g, respectively. The capacitance C_1 and C_2 are tunnel capacitances. The capacitance C_g is the normal capacitance.

5. **Analysis of the measured Coulomb diamond:**
   Figure 6P3 shows a current measured from a single-electron transistor. Find out the capacitances.

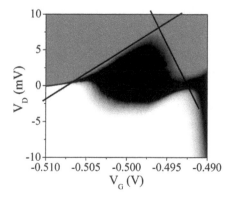

Figure 6.P3 Example of measured Coulomb diamond. V_{DS} and V_{GS} correspond to V_b and V_g, respectively. The solid lines denote the slopes of the diamond.

6. **Calculation of Coulomb charging energy:**
Calculate the Coulomb charging energy of the single-electron transistor of Fig. 6.P2 when $C_1 = C_2 = 1$ aF and $C_g = 5$ aF.

Bibliography

1. K.P. Leonard, K. Pond, and P.M. Petroff, "Critical layer thickness for self-assembled InAs islands on GaAs" *Phys. Rev. B* **50**, 11687 (1994).

2. C.J. Gorter, "A possible explanation of the increase of the electrical resistance of thin metal films at low temperatures and small field strengths," *Physica* **17** (8), 777–780 (1951).

3. C.G. Darwin, "The diamagnetism of the free electron," *Proc. Camb. Philos. Soc.* **27**, 86 (1930).

4. V. Fock, "Bemerkung zur Quantelung des larmonischen Oszillators im Magnetfeld," *Z. Phys.* **47**, 446–448 (1928).

5. S. Tarucha, D.G. Austing, T. Honda, R.J. van der Hage, and L.P. Kouwenhoven, "Shell filling and spin effects in a few electron quantum dot," *Phys. Rev. Lett.* **77**, 3613–3616 (1996).

6. R. Hanson, L.P. Kouwenhoven, J.R. Petta, S. Tarucha, and L.M.K. Vandersypen, "Spins in few-electron quantum dots," *Rev. Mod. Phys.* **79**, 1217 (2007).

7. J.H.F. Scott-Thomas, Stuart B. Field, M.A. Kastner, Henry I. Smith, and D.A. Antoniadis, "Conductance oscillations Periodic in the density of a one-dimensional electron gas," *Phys. Rev. Lett.* **62**, 583–586 (1989).

8. H. van Houten and C.W.J. Beenakker, "Conductance oscillations periodic in the density of a one-dimensional electron gas," *Phys. Rev. Lett.* **63**, 1893 (1989).

9. S.H. Son, K.H. Cho, S.W. Hwang, K.M. Kim, Y.J. Park, Y.S. Yu, and D. Ahn, "Fabrication and characterization of metal-semiconductor field-effect-transistor-type quantum devices," *J. Appl. Phys.* **96**, 1, 704–708 (2004).

10. B.H. Choi, S.H. Son, K.H. Cho, S.W. Hwang, D. Ahn, D.H. Kim, J.D. Lee, B.G. Park, "Direct observation of excited states in double quantum dot silicon single electron transistor," *Microelectronic Eng.* **63**(1–3), 129–133 (2002).

11. G.-L. Ingold and Y.V. Nazarov, in *Single Charge Tunneling* (eds. by H. Gravert and M. Devoret), Plenum, New York 1992, Chapter 2, "Charge tunneling rates in ultrasmall junctions" p. 21.

12. Y.S. Yu, S.W. Hwang, and D. Ahn, "Macromodeling of single-electron transistors for efficient circuit simulation," *IEEE Tran. Elect. Dev.*, **46**(8), pp. 1667–1671 (1999).

13. R. H. Chen, A.N. Korotkov, and K.K. Likharev, "Single-electron transistor logic," *Appl. Phys. Lett.* **68**, 1954 (1996).

14. I. Karafyllidis "Single-electron OR gate," *Elect. Lett.*, **36**, 407 (2000).

15. M.-Y. Jeong, Y.-H. Jeong, S.-W. Hwang, and D. M. Kim, "Performance of single-electron transistor logic composed of multi-gate single-electron transistors," *Jpn. J . Appl. Phys.* **36**, 6706 (1997).

16. K. Uchida, K. Matsuzawa, and A. Toriumi, "A new design scheme for logic circuits with single electron transistors," *Jpn. J . Appl. Phys.* **38**, 4027 (1999).

17. H. Inokawa, A. Fujiwara, and Y. Takahashi, "A multiple-valued logic and memory with combined single-electron and metal-oxide-semiconductor transistors," *IEEE Trans. Elec. Dev.* **50**, 462 (2003).

18. K. Uchida, J. Koga, R. Ohba, A. Toriumi, "Programmable single-electron transistor logic for future low-power intelligent LSI: proposal and room-temperature operation," *IEEE Trans. Elec. Dev.*, **50**, 1623 (2003).

19. S. Mahapatra and A.M. Ionescu, "Realization of multiple valued logic and memory by hybrid SETMOS architecture," *IEEE Trans. Nanotechnol.* **4**, 705 (2005).

20. Y.S. Yu, H.W. Kye, B.N. Song, S.-J. Kim, and J.-B. Choi, "Multi-valued static random access memory (SRAM) cell with single-electron and MOSFET hybrid circuit," *Elec. Lett.*, **41**, 3134 (2005).

21. YunSeop Yu, SungWoo Hwang, and David Ahn, "Modeling of single-electron transistors for efficient circuit simulation and design," in *Handbook of Theoretical and Computational Nanotechnology* (ed. M. Rieth and W. Schommers, Vol. X , pp. 319–362 (2006).

Chapter 7

MOSFET as a Molecular Sensor

7.1 Introduction

In this chapter, we explore the potential applications of MOSFET structures to the sensing of molecular reactions and explain the operating principles behind these applications. First, for the sake of completeness, we briefly describe the principles of potentiometry and amperometry. The MOSFET devices, which depend on potentiometry, are then extensively explained with the ISFET structure as a model device.

7.1.1 *Potentiometry vs. Amperometry*

The (bio)chemical sensors [1, 2] used for electrical signal detection either detect the electric charges directly or else detect the electric field through electric charges. The former type is called the "amperometric sensor," and the latter is the "potentiometric sensor." The electric charges have different forms depending on the (bio)chemical reactions involved in the specific sensing. The charges can be electrons, protons (H^+), or the charges from the backbone of DNA (nucleotide) or proteins. In the electrometric sensor, electrons resulting

Nanoelectronic Devices
Byung-Gook Park, Sung Woo Hwang, and Young June Park
Copyright © 2012 Pan Stanford Publishing Pte. Ltd.
www.panstanford.com

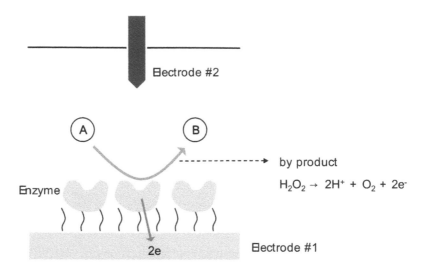

Figure 7.1. Operational principles of the amperometric bio sensor. The target molecule A is changed to B with the help of the enzyme coated on the working electrode (Electrode #1), and the resulting electrons are absorbed by the working electrode to form the electric current. By measuring the electric current through Electrode #1, the existence of the molecule A is sensed.

from the (bio)chemical reactions are measured directly by a metal electrode in the form of the electrical current.

Figure 7.1 shows a typical example of a sensor based on amperometry. Two electrodes, the working and the reference electrodes, are immersed in an aqueous environment, and an appropriate voltage is applied. An enzyme material is coated on the working electrode and works as a catalyst for the (bio)reactions of interest.

Let us consider lactate sensing as an example. Lactate is a metabolite material formed from pyruvate in muscles and liver when the oxygen supply is not sufficient. The pattern of change or the increasing trend of lactate in blood gives a sensitive indicator of a patient's likelihood of survival. A material called LOD (lactate oxidase) is used as the catalytic material specific to lactate. Various polymers [3] have been introduced to function both as the chemically stable ground on which LOD is coated and as an efficient conducting film to carry the electrons resulting from the biochemical reactions to the working electrode. The following equations describe a typical chemical reaction involved in sensing the lactate level.

$$L - \text{Lactic acid} + O_2 \xrightarrow{LOD} \text{Pyruvic acid} + H_2O_2$$

$$H_2O_2 \rightarrow 2H^+ + O_2 + 2e^-$$

$$\text{Poly(An-Co-FAn)}_{(oxi)} + 2e^- \rightarrow \text{Poly(An-Co-FAn)}_{(red)}$$

Here lactic acid is transformed to pyruvic acid with H_2O_2 as a byproduct. H_2O_2 is transformed to two protons, an oxygen molecule, and two electrons. These electrons are then absorbed by the working material, which gives rise to an increase in the electric current in the circuit. The working electrode is "reduced" in this case. Notice that the material is oxidized when electrons are given out from the material and reduced if electrons are absorbed after the chemical reactions.

Even though the sensing principle is simple and many catalyst materials have been developed for the sensing specific target molecules of interest, the sensors based on "amperometry" may not be appropriate for integration with the semiconductor chips because the sensing devices are not based on MOSFETs (or MOSFET-like structures). A semiconductor chip is necessary for the signal processing of the electrical signal obtained from the sensor devices, and semiconductor chips are mostly composed of MOSFET structures. So it may be useful to build the sensor devices based on the FET principle. Therefore, this chapter is mostly dedicated to the sensor devices based on "potentiometry" where the electrical charges that result from the (bio)chemical reactions are detected by the sensor devices by way of the "electric field."

7.1.2 FET-Based (Bio)Chemical Sensor

In the previous chapters, we explained the operational principles of semiconductor devices with a particular emphasis on devices based on the FET (field effect transistor). In the MOSFET structures, the channel conductance between two electrodes (source and drain) is controlled by the electric field coming from the gate electrode. The intensity of the electric field is controlled by the electric charges in the gate material, and the amount of the charge is proportional to the intensity of the signal. In another words, the signal is converted

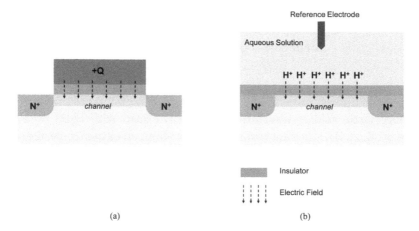

Figure 7.2. A schematic diagram illustrating the similarity between N-type (a) MOSFET device and (b) FET-based (bio)chemical sensor. Notice that the charges are the signals for both cases and are converted into the current between source and drain.

into the MOSFET current between the source and the drain by way of the field effect.

In the chemical sensor application of FETs, the electrical charges are generated by (bio)chemical reactions rather than by the electrical signals generated by the electrical circuits. However, the rest of operation, which is the conversion of the electrical charges to the source/drain current by way of the field effects originating from the electrical charges is same for the MOSFET and the FET-based (bio)chemical sensors.

Figure 7.2 shows a schematic diagram illustrating the similarity between the MOSFET device and the FET-based (bio)chemical sensor device. In the figure, both structures consist of the gate electrode (named the reference electrode in (b)), aqueous solution in the case of the FET-based (bio)chemical sensor, insulator, and semiconductor. The vertical electric field originates from the gate electrode in the MOSFET device, whereas the electric field comes from the H^+ ions residing in the aqueous solution adjacent to the gate insulator in the case of the chemical FET devices. Even though the H^+ ions are shown in the figure as a typical example, any change in the charges in the molecules, which are usually bound to the gate insulator as the result of the biomolecule binding activity, will result in a modulation of the vertical field, thereby changing the conductance of the electrical channel between the source and drain regions.

The FET structure shown in 7.2(b) is called the ISFET (ion-sensitive FET) and was first introduced by Bergveld [4, 5]. The aim of the ISFET is to sense the H^+ concentration of the aqueous solution. A simple MOSFET theory can be applied to find out the relationship between the sheet charge (H^+ concentration at the surface of the gate insulator) and the change in the MOSFET current. Once the relationship between the H^+ concentration of the aqueous solution and sheet charge is understood, pH of the aqueous solution is known by monitoring the MOSFET current. After the ISFET structure was introduced, a number of alternatives were introduced that modified the insulator interface by the receptor molecules which may match with the target molecules of interest. They are called the enzyme FET, chemical FET, and the immunoFET where the gate insulator is attached to an enzyme, chemicals, and protein molecules, respectively. The operation principle of various biochemical FETs is more or less similar to the principle of the ISFET device. Hence, we will explain the theory of ISFET device in detail in Section 7.3.

7.1.3 *Two Types of (Bio)Chemical FET Sensor: ISFET and Affinity-Based FET*

The FET sensor can be categorized into two types according to the states of the charges generated as a result of biochemical reactions. In the first type, the charges in the EI (electrolyte–insulator) interface are changed by the molecules generated by the (bio)chemical reactions. The typical example of these byproducts are H^+ ions, and they can be sensed by the ISFET, which will be studied in Section 7.3. In the second type, the "probe" molecules, which have a specific binding affinity to the target molecules to be sensed, are attached to the insulator surface of the FET. The charge states of the probe molecules are changed if they become bound to the target molecules. The typical example in this category is the DNA FET, where the probe DNA's with a specific affinity with the target DNA's are attached to the surface in the affinity-based FET to be studied in Section 7.4.

7.2 Some Basics of Chemistry and the Importance of H^+

Usually, a biochemical sensor works in an aqueous environment since the biochemical reactions take place in an aqueous environment. In some cases, an aqueous environment is the ideal buffer solution. In other cases, it may be the actual human serum, blood, or other types of secretion from the body. Non-ideal conditions in which the sensor works hinder the stable operation of the sensors because there are also a number of other molecules coexisting with the target molecules in the aqueous solution (hereafter simply called either the "solution" or the "electrolyte") and impurity ions such as Na^+, Cl^-, K^+, and Mg^+, to name a few. So it is important to understand the interface between the solution and the insulator and between the solution and the semiconductor/conductor, which are the places where most of the biochemical reactions take place. Sections 7.2.1 and 7.2.2 are dedicated to the fundamentals of the ion states of the solutions and some of the molecules.

7.2.1 *Some Basics of Chemistry*

7.2.1.1 Aqueous solution

Aqueous solution containing ions may be considered a semiconductor material containing dopant impurities. In semiconductor, carriers are electrons and holes, and dopant ions are immobile. However, in the aqueous solution, the impurities such as NaCl are ionized to ions (Na^+, Cl^-) and contribute to electric conduction together with the H^+ and OH^-, which corresponds to electrons and holes in the semiconductor since they are mobile in the water solution.

Also, the transport of the ions may cause the motions of the water molecules due to viscosity of the water molecules. The effect is called the electro-osmotic flow. The transport of the positive ions will accompany the drift of the water molecules nearby and influence the drift of the negative ions. However, net effect to the total electric current may be zero since same influence will be given to the adjacent positive ions. The electro-osmotic flow may give an important effect to the transport of the biological molecules in the solution, which may be used to manipulate the molecules of interest [6].

7.2.1.2 pH

Since the characteristics of the solution in which most of the bio-chemical sensing takes place can be best described by the concentration of the H^+ ions and OH^- ions, the definition of pH denoting the concentrations of those ions is briefly explained in this section. pH is defined as the negative logarithm of the hydrogen ion concentration in moles per liter (molar concentration) [mol/L], or

$$pH = -\log[H^+] \tag{7.1}$$

The concentration of H^+ ions, $[H^+]$, in pure water is 10^{-7} moles per liter, and since the number of molecules per mole is 6.02×10^{23} (Avogadro's number), the concentration of $[H^+]$ is around $6.02 \times 10^{13}/cm^3$. OH^- concentration in pure water is the same as $[H^+]$. The mass action law states that $[OH^-][H^+]$ in pure water is 3–$4 \times 10^{26}/cm^3$ at room temperature, and the number is preserved if impurity molecules are added to change either $[H^+]$ or $[OH^-]$.

Notice that the mass action law is similar to the mass action law in semiconductor where n·p product corresponds to $[OH^-][H^+]$; n·p in silicon is $2.2 \times 10^{20}/cm^3$ at room temperature.

If $[H^+]$ is larger (smaller) than $[OH^-]$, the solution is called an acid (base). Note the similarity in the mass action law between the electron and hole concentrations in semiconductor materials and the H^+ and OH^- concentrations in a solution. Conventionally, the concentrations of electrons and holes are denoted by number/cm^3 in semiconductor science and engineering, whereas the ion concentration in an electrolyte is expressed by mol/L. For conversion of mol/L to number/cm^3 is simply to multiply $6.02 \times 10^{20}/cm^3/(mol/L)$ to the chemistry scale.

It is interesting to note that electrons and holes determine the many physical properties of the semiconductor, including the electrical conductivity and optical properties. Similarly, $[H^+]$ and $[OH^-]$ determine the properties of the water (and electrolyte) since many of the chemical (and biochemical) reactions are determined by $[H^+]$ and $[OH^-]$. Since the direct measurement of $[H^+]$ is impossible, the definition in Eq. (7.1) is only ideal. The more practical definition of pH comes from the hydrogen ion activity rather than $[H^+]$, which may be measured by measuring the electromotive force taken place between the solution containing $[H^+]$ and the electrode. (See, for example, http://en.wikipedia.org/wiki/PH and Ref. 7 therein.)

7.2.1.3 Buffer solution

A buffer solution is an aqueous solution whose pH value does not change, or changes very little, when acid or base is added to the solution. Buffer solutions are used as a means of maintaining pH value at a constant value in a wide variety of biochemical applications. For example, many biological enzymes are active only in a solution within a certain range of pH values, so it is important for the biological system to have a certain buffer function in the biological solution, such as blood. Actually, in blood plasma, buffers of carbonic acid (H_2CO_3) and bicarbonate (HCO_3^-) are present to maintain a pH in blood between 7.35 and 7.45. It is beyond the scope of this book to explain all the details of the buffer solution, but it may be worthwhile to introduce the basic principles of buffer activity and a couple of important parameters related to the buffer solution.

In a simple buffer solution with a mixture of a weak acid HA, there is an equilibrium between HA and its conjugate base A^- as

$$HA + H_2O \leftrightarrow H_3O^+ + A^-. \tag{7.2}$$

According to Le Chatelier's principle, the chemical reaction from the right-hand side to the left-hand side is enhanced if H^+ is added to the solution. If OH^- is added to the solution, the activity on the right-hand side is enhanced as OH^- removes H^+ through the reaction $H^+ + OH^- \rightarrow H_2O$. K_a is defined as

$$K_a = [H^+]_{eq}[A^-]_{eq}/[HA]_{eq} \tag{7.3}$$

In terms of K_a, the logarithmic expression of the above equation gives

$$pH = pK_a + \log_{10}[A^-]/[HA] \tag{7.4}$$

If one knows K_a, the pH value of the buffer solution is easily found by solving

$$K_a = \frac{x^2}{([HA]_i - x)} \tag{7.5}$$

where x is the concentration of $[H^+]$(or H_3O^+) and $[OH^-]$ in Eq. (7.2). By solving (7.5) for x for a given K_a and $[HA]_i$, the added concentration of the acid in the buffer solution, $pH = -\log_{10}(x)$, can be easily calculated (see Problem 4).

In actual buffer solutions, the pK_a values for each buffer additive (some are weak acid, some are weak base), the temperature coefficients defined as $d(pH)/dT$ are varied. Also, it should be noticed that the solutions are not ideal because the buffer range is not zero but rather a finite value, showing a finite buffer capability.

It may be interesting to note the equivalence of the ionization in Eq. (7.2) to the ionization of the donor type dopant atom from N_D^0 to N_D^+.

$$[N_D^0] + Si = [N_D^+] + n, \qquad (7.6a)$$

and

$$K_a = [N_D^+]n/[N_D^0] = n(1 - f_t)/f_t, \qquad (7.6b)$$

where f_t is the occupation probability of the donor level in the energy gap of the semiconductor. Condition for full ionization corresponds to $f_t = 0$, or K_a is infinity. Figure 7.3 shows the situations describing the equations in (7.2) and (7.6). In the case of Eq. (7.2), HA is dissociated to H^+ and A^-, whereas the impurity atoms are dissociated to N_D^+ and e^- in Eq. (7.6). In the figure, the energy of the two states (dissociated and associated) is schematically shown.

One thing to notice with regard to sensing applications is the role of the secondary ions—such as A^- in Eq. (7.2)—which may play an

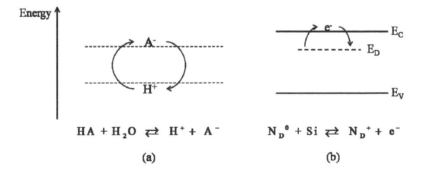

Figure 7.3. A schematic diagram showing (a) the dissociation and recombination of HA molecule and (b) donor atom in the semiconductor. The energy is not to scale, but the energies of the ions will be different after dissociation in water.

important role by way of the electric double layer at the interface between the electrolyte solution and the solid material. This point will
be revisited in Section 7.3.2, where the theory of the EDL is explained
in detail.

7.2.1.4 Many biological reactions involve H$^+$

Many biochemical reactions generate or consume the proton, H$^+$,
in the solution. As the result, the concentration of the proton ions
in the aqueous solution changes, subject to the molecules to be
probed. That is why many biochemical sensors are based on a pH
sensor.

In Fig 7.4, typical examples of the ionic exchange through
the membranes of the biological cell are denoted. Two important
metabolic activities, glycolysis and respiration, are shown in the figure. When the cell is triggered by the ligands, which are protein

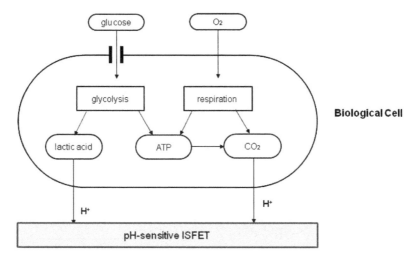

Figure 7.4. Schematic representation of the biochemical reactions related
to glycolysis and the respiration taking place in the cell. The outputs of
glycolysis and respiration include ATP, lactic acid, and CO$_2$. Many of them
generate protons as a byproduct. Also shown in the figure is the ISFET structure, which senses the change in [H$^+$]. By sensing the change in [H$^+$], the
metabolism around the cell can be sensed. For example, in the glycolysis cycle, glucose is supplied from the blood as the result of the intake of nutrition,
and ATP is generated with some H$^+$ generated as the by product.

molecules [7], the cell membrane is open to the glucose molecules. After a lengthy process, the glucose produce ATP (adenosine triphosphate) molecules, which serve as the energy source for cell operations, with lactic acid as the byproduct. Also shown in the figure is the respiration process in which an O_2 molecule is received by the cell membrane and becomes CO_2 with H_2 as the byproduct. Now, by measuring [H^+] around the cell, information related to the metabolism can be known. One simple way of measuring [H^+] is to use ISFET devices. Even though detailed explanations of each reaction taking place in the cell are beyond the scope of this book [7], it should be noted that the biological reactions related to the metabolism of the biological cell give out the [H^+] as the output of the reactions.

7.2.2 *Charges of Important Biomolecules*

Many biochemical sensors are designed to detect specific molecules in solution. The most widely used biosensors are the DNA and protein sensors, where the unknown DNA molecules and protein molecules are detected by attaching "probe" molecules whose molecular structures are designed in such way that the probe molecules may form pairs with the unknown molecules of interest. Those processes are called "bio conjugation" or "hybridization." In a DNA sensor, the probe molecule is single–strand (ss) DNA with a genome sequence that fits with the sequence of the target DNA. In the case of a protein chip, the probe molecule may be an antibody [8] or DNA (or RNA), called aptamer [9], which can be paired with the target protein by way of the protein binding for the case of antibody-antigen binding [10], or the DNA (or RNA)-protein binding in the case of the aptamer–protein binding [11].

In either case, DNA or protein sensors, there should be appreciable changes in the quantity of electrical charges during the binding action so that the change in the charges can affect the FET device in the form of changing electrical conductance. Therefore, it is important to understand the charge states of the molecules (DNA and protein) in the solution. Figure 7.5 illustrates the electrical charges present in the DNA structures and protein structures. In DNA, the charges come from the phosphate acid, which connects

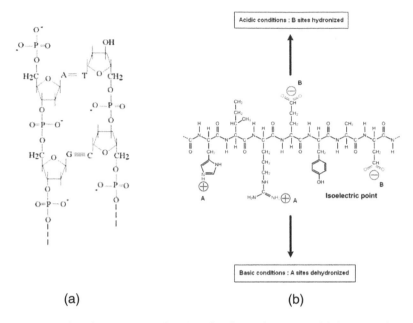

(a)   (b)

Figure 7.5. Charge states of DNA molecules and proteins: (a) A DNA molecule is negatively charged in the solution as the proton in the phosphor backbone is emitted; (b) a protein molecule is positively charged or negatively charged according to emission and absorption of the proton in NH (A sites) and COOH (B sites) radicals.

with the dixoyribose to form the phosphate backbone. The oxygen, which is not saturated with other atoms (either the carbon or oxygen of the neighbor dioxyribose), is negatively charged. From Fig. 7.5(a), it can be seen that $-1e$ is charged for each nucleotide and $-2e$ if a double strand is formed.

The charge states of the protein are determined by the amino acids that constitute the protein. As shown in Fig. 7.5(b), amino acids have four different groups attached to a central carbon atom: a hydrogen atom, an amino group (NH_3^+), a carboxyl group (COO^-), and a side chain (R group in the figure). In addition to the important function of the side chain in determining the structure and function of the protein when the amino acids are folded to form a three-dimensional structure, some side chains are electrically positive, some are negative, and some are electrically neutral. The charge states of the protein are determined mainly by the dissociable groups on the surface of protein after the polymerization (protein folding). For example, dissociable groups such as $-NH_3^+$ and $-COOH$ may be deprotonated to $-NH_2 + H^+$ and $-COO^- + H^+$, respectively. Usually, the charge states of protein molecules are influenced by the pH value since the surface reaction will be influenced by $[H^+]$ concentration in the solution [12].

The detailed description of the charge states and their locations in the DNA molecules and protein molecules are available for obtaining the charge states of DNA and proteins [12] with the help of the molecule dynamic simulations [13].

7.3 EISFET

As pointed out in Section 7.2, ISFET has been used to sense the pH value of the solution. A basic operation principle is that the surface potential of the FET is modulated by the charge states in the EI interface, which in turn is dependent on the $[H^+]$ concentration in the electrolyte. In this section, the structure of ISFET sensors and their operating principles will be explained.

7.3.1 EISFET Structures

When the ISFET structure was first introduced [14, 15], the device was operated without any reference electrode. However, it was demonstrated that [16, 17] a reference electrode is necessary to stabilize the electric potential of the electrolyte for reliable operation of the sensor devices. The sensor device structure with the reference electrode, which may be regarded as the gate electrode in a conventional MOSFET with fixed potential, will be referred to as the electrolyte–insulator–semiconductor field-effect transistor (EISFET).

As shown in Fig. 7.6, the EISFET device may be divided into three regions according to the roles as electrochemical sensor. The first region is the reference electrode and electrolyte solution region, where

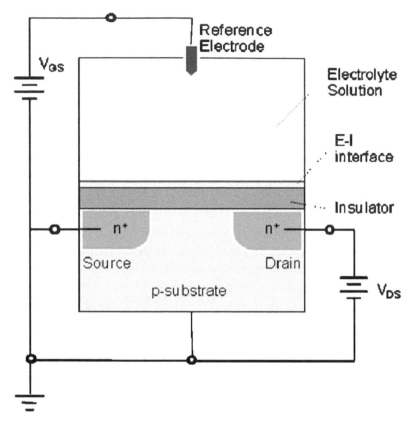

Figure 7.6. EISFET structure with the n-type FET. The device can be divided into three important regions: the gate-electrolyte region, the FET region, and the EI interface region. The FET device is an n-channel device, and V_{GS} is applied to the electrode in the electrolyte with respect to the source voltage.

the unknown concentration of [H$^+$] to be sensed is contained in the solution. The second region is the FET region, where the semiconductor device with the source and drain contacts is covered with the dielectric material, which is in contact with the solution. The third region, probably the most important region, is the interface region, called the EI (electrolyte–insulator) interface, where the charge states of the interface region vary with [H$^+$]. At the interface between the electrolyte (liquid) and insulator (solid), there exists an electric double layer, which is abbreviated as EDL.

It is interesting to notice that the EISFET may be viewed as a structure where the gate insulator is sandwiched by two semiconductor materials; one is the ordinary semiconductor in the FET and the other is the electrolyte solution containing various ions as the electrical carriers. Naturally, two interfaces, one between the aqueous solution and the gate insulator and the other at the semiconductor surface, determine the electrical characteristics of the EISFET. The charge–potential relationships in the two interfaces are the key theories to be studied in this section. The theory for the charge potential relationship at the EI interface is called the theory of EDL (electric double layer), whereas the theory at the semiconductor surface is called the MOS surface theory.

The relationship between $[H^+]$ and the charge states in the EI interface can be derived from the kinetic theory at the insulator surface coupled with the theory of the electric double layer. We will devote the majority of Section 7.3.2 to explaining the EDL. The charge states in the EI interface in turn change the carrier concentration in the FET surface region, thereby modulating the source/drain current of the FET device.

7.3.1.1 Reference electrode

The reference electrode is introduced to stabilize the electrostatic potential of the electrolyte. Usually the reference electrode is chosen as an inert material in the electrolyte solution, such as Pt or Ag/AgCl. In this book, the reference electrode is called the gate electrode even though electric signal is not applied to the electrode.

7.3.1.2 Electrolyte

An electrolyte is the aqueous solution containing a certain concentration of ions, which may exist in the natural solution or may be intentionally introduced to mimic a real physiological environment. As mentioned above, it may be interesting to view the electrolyte as a semiconductor material with the ions as the carriers of the electrical transport.

7.3.1.3 Insulator film

EISFETs can have different materials between the gate insulator and the electrolytes, and they are divided into categories according to the material. The first group consists of device structures adopting an ion-selective membrane on top of the gate insulator [18]. The second consists of those adopting an inorganic insulator film. The typical example of the first group is the Enzyme ISFET where the EI interface is modified by enzymes that catalyze the biochemical reaction consuming or producing an H^+ ion. Based on this idea, many kinds of ISFET enzyme sensors have been developed [19]. In general, enzymes are attached to the gate insulator of ISFET by the use of organic thin membranes as the interface film. The organic membranes where the enzyme is immobilized should be an important factor in determining the performance of the sensors. Even though a detailed study of the Enzyme ISFET is beyond the scope of this book, it should be mentioned that the choice of proper organic membrane with the combination of the enzyme material is crucially important because the signal, usually an H^+ ion, is produced as a result of an enzyme-catalyzed reaction of analyte on and/or in the membrane. Also the enzyme activity and the diffusion rates of analytes and reaction products are dependent on the chemical and physical properties of the membrane used to modify the EI interface.

Various inorganic films have been used as the insulator material for the EI interface. The first dielectric used was SiO_2 [15], followed by Si_3N_4 [20], Al_2O_3 and Ta_2O_5 [21, 22], and mixed dielectrics such as borosilicate glass (BSG) [23]. So far, Si_3N_4 has been used most frequently because there exist two different types of sites on the interface, ionizable sites that directly interact with the electrolyte to either release or bind hydrogen ions according to the concentration of the hydrogen ions in the electrolyte.

7.3.1.4 EI (Electrolyte insulator) interface

Usually, at the interface between the electrolyte and the gate insulator, there exist space charge region and the sheet charges with negligible thickness. Because of electrostatic interactions with the water molecules and the ions in the electrolyte, a very thin layer forms, called the EDL (electric double layer). The EDL is composed of two layers; inner layer with the sheet charges, Q_{ei}, and outer layer with the space charge region, Q_{dif}. The theory of the EDL, which will be the topic of the next section, deals with the relationship between the charges (both the space charges and sheet charges shown in Fig. 7.6) and the potential drop across the EDL.

7.3.2 Theory of EISFET [24]

The purpose of EISFET theory is to obtain a relationship between the concentration of the ions in the electrolyte and the electrical current of the FET. As we learned from MOSFET theory in Section 3.2, the source drain current of the MOSFET device can be rewritten as

$$I_{DS} = \beta \left[\left(V_{GS} - V_{TH(MOS)} \right) V_{DS} - \frac{1}{2} V_{DS}^2 \right] \text{ (linear region)} \qquad (7.7a)$$

$$I_{DS} = \frac{1}{2} \beta \left(V_{GS} - V_{TH(MOS)} \right)^2 \text{ (saturation region)} \qquad (7.7b)$$

$$\beta = C_{ox} \mu_n \frac{W}{L}, \qquad (7.7c)$$

where $V_{TH(MOS)}$ can be written as

$$V_{TH(MOS)} = V_{FB} + Q_b / C_{ox} + 2\varphi_f. \qquad (7.8)$$

Equivalently, the I_{DS} of the EISFET can be written in the same way if C_{ox} and V_{TH} are substituted for by C_{in} and V_{TH} (EISFET), respectively.

The best way understand the situation is to draw the energy band diagrams (or energy diagram) for the two cases, MOSFET and EISFET, as shown in Fig. 7.7, when $V_G (V_{GS,S}) = V_{TH}$ is applied. In the figure, the charge distributions in both cases are also shown. In the case of the MOSFET, the induced charge in the gate electrode, Q_G, is assumed to be like a sheet charge so that there exists negligible potential drop across the gate to insulator interface. The situation is different for the case of EISFET, where the potential drop across the EDL is not negligible.

As shown in Chapter 3 and Eq. (7.8), the threshold voltage of MOSFET is the gate voltage needed to induce the surface potential

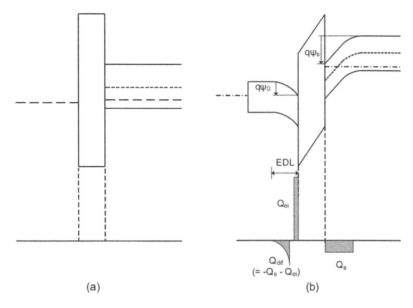

(a) (b)

Figure 7.7. Energy diagram of (a) MOS and (b) EISFET when $V_{GS} = V_{TH}$ is applied. V_{FB} for MOS is the difference between the work function of the metal and semiconductor if there exist no charges in the insulator and semiconductor surface. In EISFET, even though V_{FB} of MOS is applied, the semiconductor band is not flattened because of the charges (Q_{ei}) at the EI interface.

of semiconductor channel to $2\phi_F$. The flat-band voltage (V_{FB}) for the MOS device is the difference between the work function of the metal and semiconductor if there exist no non-ideal charges in the insulator and semiconductor surface. In the case of EISFET, even though the V_{FB} of the MOS is applied, the semiconductor band is not flattened because of the charges at the EI interface. In Fig. 7.7 (b), we consider the case when a sheet charge, Q_{ei}, exists at the EI interface. Q_{ei} may be due to the molecular charges of the DNA in the case of DNA sensor or due to the chemical interaction of protons (H^+) in the solution and the insulator surface in the case of the ISFET. This situation is analogous to the case when a non-ideal sheet charge, Q_{ox}[C/cm^2], exists at the oxide with a centroid form or Q_{ss} at the insulator–semiconductor interface (see Problem 5). As Q_{ox} changes the threshold voltage through modifying the flat-band voltage of the MOSFET, Q_{ei} changes the threshold voltage of the EISFET.

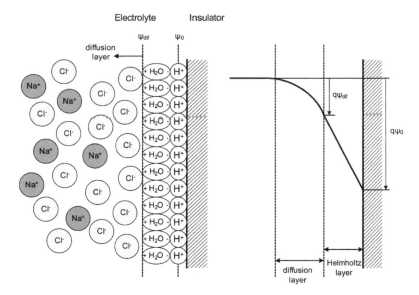

Figure 7.8. Schematic diagrams of (a) the EDL (electric double layer), and (b) the energy diagram across the EDL for the electrolyte containing N_a^+ and Cl^- ions. Notice that the EDL consists of two layers, the diffusion and Helmholtz layers.

Let us consider the potential drop and charges in the EDL. In Fig. 7.7, the potential across the EDL is denoted as ψ_0. It is the goal of EDL theory to develop the relationship between the Q_{ei} and ψ_0, which in turn are function of $[H^+]$ in the electrolyte solution. The Q_{ei} gives effect to the threshold voltage, thereby modulating the EISFET current.

The best way of describing the EISFET structure is writing the potential equation similar to the MOS equation in chapter 3. Referring to Fig. 7.8 ,

$$V_G - V_{FB} = \psi_{dif} + V_h + V_{ox} + \psi_s, \qquad (7.9)$$

where ψ_{dif} and V_h are the potential drops across the outer layer, called the diffusion layer, and the inner layer, called the Helmholtz layer, of the EDL, respectively. Eq. (7.9) simply states that $V_G - V_{FB}$ is applied across the EDL (ψ_{dif} and V_h), gate insulator (V_{ox}), and the semiconductor (ψ_s). The Helmholtz layer can be regarded as the thin insulator made of the water molecules attaching to the insulator surface. It is the inner layer where the sheet charges may exist. If the sheet charge, Q_{ei}, is located at the insulator–inner layer interface, $V_{ox} = -Q_s/C_{in}$ and $V_h = -(Q_s + Q_{ei})/C_h$ (see Problem 6). From the charge neutrality, $Q_s = -(Q_{dif} + Q_{ei})$.

If $Q_s(\psi_s)$, $Q_{dif}(\psi_{dif})$, and $Q_{ei}(\psi_0)$ are known, Eq. (7.9) can be solved to obtain Q_s and V_G relationship. Or, from

$$V_G - V_{FB} = Q_{dif}^{-1}(\psi_{dif}) - (Q_s + Q_{ei})/C_h - Q_s/C_{in} + \psi_s,$$

$$(7.10a)$$

the threshold voltage may be written as

$$V_{TH} = V_{FB} + Q_{dif}^{-1}(\psi_{dif})$$
$$-(Q_{d,max} + Q_{ei})/C_h - Q_{d,max}/C_{in} + 2\phi_F, \qquad (7.10b)$$

where $Q_{d,max}$ is the semiconductor charge when the semiconductor surface potential becomes $2\phi_F$. $Q_{d,max} = -\sqrt{qN_A\epsilon_{si}(2\phi_F)}$ for the case of N channel ISFET.

The sheet charge, Q_{ei}, may be dependent on the potential at the interface, ψ_0. In this respect, the charges in the Helmholtz layer are much like the surface state charge, Q_{ss}, at the insulator-semiconductor interface, because Q_{ss} is also dependent on the semiconductor surface potential ψ_s.

The EDL theory to be developed in the next section is for two relationships; $Q_{dif}(\psi_{dif})$ for the diffusion layer and $Q_{ei}(\psi_0)$ for the Helmholtz layer.

7.3.2.1 Theory of electric double layer: the diffusion layer

First, we consider the diffusion layer. Suppose that the ion concentration of the electrolyte of an EISFET structure such as that shown in Fig. 7.6 is n_0. The ions may be the ions intentionally introduced or naturally contained in the electrolyte solution to be tested. The charge density, ρ, in the electrolyte can be written as

$$\rho_0 = e\left([H^+] - [OH^-] + n_0^+ - n_0^-\right) \qquad (7.11)$$

$[H^+]$ and $[OH^-]$ can be neglected for the case when n_0 is much bigger than $[H^+]$ and $[OH^-]$. From the Poisson equation,

$$\frac{d^2\psi}{dy^2} = -\frac{e}{\varepsilon_w}\left[n_0 \exp\left(-\frac{\psi}{V_t}\right) - n_0 \exp\left(\frac{\psi}{V_t}\right)\right], \qquad (7.12)$$

where ε_w is the permittivity of the electrolyte solution, V_t the thermal voltage, $k_B T/e$, and the Boltzmann statistics has been used for n_0^+ and n_0^-. The solution to Eq. (7.13) gives for the relationship between Q_{dif} and ψ_{dif} as

$$Q_{dif} = -\frac{2\varepsilon_w V_t}{L_{d,sol}} \sinh\left(\frac{\psi_{dif}}{2V_t}\right),\qquad(7.13)$$

where $L_{d,sol}$ is the Debye length in given electrolyte solution, written by $\sqrt{\frac{\varepsilon_w V_t}{2en_0}}$. See Problem 7 for the derivation of Eq. (7.13).

The theory for the diffusion layer is called the Gouy–Chapman theory [26]. It is interesting to notice that the situation is very similar to the charge potential relationship at the surface of the intrinsic semiconductor where electron and hole concentrations are modulated by the surface potential of the semiconductor (see Problem 8).

7.3.2.2 The Helmholtz layer

Between the diffusion layer and solid surface (gate insulator in the EISFET structure), there exists the Helmholtz layer made of the polarized water molecules between the solid and ions in the electrolyte. Probably, the Helmholtz layer thickness may be regarded as the closed distance for ions to the solid surface. There may exist a fixed sheet charge Q_{ei} with the negligible thickness at the solid-Helmholtz layer interface.

7.3.2.3 Relationship between Q_{ei} and [H$^+$] in the electrolyte

In this section, we will derive the relationship between the Q_{ei} and ψ_0, called the interface potential at the EI interface. Q_{ei} originates from the charge states of the unsaturated bond at the insulator surface. We consider the SiO$_2$ layer as the gate insulator, and the theory can be easily extended to other insulators, such as Si$_3$N$_4$, with proper modifications in the surface parameters to be explained later. At the surface of the SiO$_2$ layer, the unsaturated SiO site, which is negatively charged, is saturated to SiOH and becomes neutral. The SiOH bond becomes positively charged if the SiOH site is bonded with another H$^+$ ion. In Fig. 7.9, the three states are shown according to the association and dissociation of the H ion to the SiOH bond.

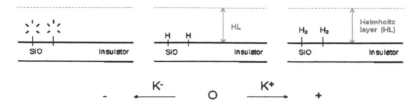

Figure 7.9. The surface of the insulator (SiO$_2$ as the example) is neutral when the SiOH bond is formed (SiO site is saturated with H). If H is dissociated, it becomes negatively charged. If H is additionally attached, it becomes positively charged. In this case, the surface site is called amphoteric. The chemical reaction constants from SiOH to SiO$^-$ and SiOH$_2^+$ are K_- and K_+, respectively.

$$\text{SiOH}_2^+ \overset{k_+}{\rightleftharpoons} \text{SiOH} + \text{H}_s^+ \tag{7.14a}$$

$$\text{SiOH} \overset{k_-}{\rightleftharpoons} \text{SiO}^- + \text{H}_s^+ \tag{7.14b}$$

Out of the total available sites N_s per unit area at the EI interface, a certain proportion of the states turns to one of the states [SiOH], [SiO$^-$], and [SiOH$_2^+$], which is represented by the areal density:

$$N_s = [\text{SiOH}] + \left[\text{SiO}^-\right] + \left[\text{SiOH}_2^+\right] \tag{7.15}$$

and

$$Q_{ei} = e\left(\left[\text{SiOH}_2^+\right] - \left[\text{SiO}^-\right]\right). \tag{7.16}$$

In equilibrium, the chemical constants, K_+ and K_- can be represented by the surface concentration of each site as

$$K_+ = \frac{[SiOH][\text{H}^+]_s}{[\text{SiOH}_2^+]} \tag{7.17}$$

$$K_- = \frac{[\text{SiO}^-][\text{H}^+]_s}{[\text{SiOH}]} \tag{7.18}$$

for the amphoteric (bidirectional) reaction for H and SiO.

Notice that the two constants are represented by the ionic concentrations at the interface, [H$^+$]$_s$, where the chemical reactions take place. Our goal is to relate those with the ionic concentrations at the bulk of the electrolyte, which are the quantities to be probed by the sensor. In order to relate the bulk and surface concentrations, we use the Boltzmann statistics and for K_+ and K_-,

$$K_+ = \left[\frac{[\text{SiOH}][\text{H}^+]_b}{\text{SiOH}_2^+} \exp\left(-\psi_0/V_t\right) \right] \tag{7.19}$$

$$K_- = \left[\frac{[\text{SiO}^-][\text{H}^+]_b}{\text{SiOH}} \exp\left(-\psi_0/V_t\right) \right]. \tag{7.20}$$

With K_+ and K_-, the constants containing the information on the surface kinetics between the $[\text{H}]_s$ and the surface sites on the insulator, the surface charge density, Q_{ei}, can be written as, from Eq. (7.16),

$$Q_{ei} = e\,[\text{SiOH}] \left\{ \frac{[\text{H}^+]_b}{K_+} \exp\left(-\frac{\psi_0}{V_t}\right) - \frac{K_-}{[\text{H}^+]_b} \exp\left(\frac{\psi_0}{V_t}\right) \right\}. \tag{7.21}$$

Note that Q_{ei} is a strong function of the EI interface potential, ψ_0 as the case for Q_{dif}.

Q_{ei} may originate from the surface, which may be charged due to the secondary ions, if they exist, in the electrolyte. The secondary ions may be Na^+, Cl^-, Mg^{++}, or K^+ ions depending on the electrolytes. For simplicity, we have not dealt with the effects of the secondary ions even though the theory explained in the book may be directly applied to the case of the secondary ions [25].

So far, we have found the relationship between the potential drop across the EDL, ψ_0, and the charges at the EI interface, Q_{ei} (Eqs. (7.16), (7.19), and (7.20)) for given parameters such as K_+ and K_-, and N_s(Eq. 7.15). In order to definitely determine the ψ_s and V_G relationship in (Eq. 7.10), we need to understand the semiconductor surface theory described in the following section.

7.3.2.4 Theory of semiconductor surface

From Eq. (3.26) in Chapter 3, the areal density of the surface charges, Q_s, can be rewritten as

$$Q_s = -\frac{\sqrt{2}\varepsilon_s V_t}{L_{d,s}} \left\{ \left[\exp\left(-\frac{\psi_s}{V_t}\right) + \frac{\psi_s}{V_t} - 1 \right] \right.$$
$$\left. + \frac{n_i^2}{N_a} \left[\exp\left(-\frac{\psi_s}{V_t}\right) - \frac{\psi_s}{V_t} - 1 \right] \right\}, \tag{7.22}$$

where $L_{d,s}$ is the Debye length of the semiconductor, expressed by $\sqrt{\varepsilon_s V_t / e N_a}$, and N_a and ψ_s are the channel doping concentration and semiconductor surface potential, respectively.

7.3.2.5 Total charges in the system and equivalent circuit

Before we move on to finding the relationship between the FET current and pH in the bulk of electrolyte, we consider the total charges in the system. Assuming that the system from the reference (or gate electrode) electrode to the semiconductor (see Fig. 7.10(a)) is a one-dimensional system, the space charge distribution is replotted in Fig. 7.10. We have neglected that the voltage across the reference

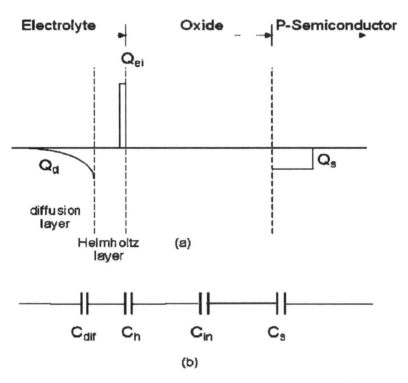

Figure 7.10. (a) The charge distribution in each region of the EIS system. Charges are represented by the charges/unit area. (b) The equivalent circuit represented by the equivalent capacitance to represent each region of the EIS system.

electrode (see Fig. 7.7(b)) and the electrolyte, meaning that there is not a chemical potential difference between the reference electrode and the electrolyte system. The system can be represented by a combination of capacitors in series, C_{dif}, C_h, and C_{in}, representing the capacitance component of the diffusion layer, Helmholtz layer, and the insulator, respectively. Now at the EI interface, there are Q_{dif}, Q_{ei}, and Q_s, which are the charge density per unit area in the diffusion layer, the Helmholtz, and semiconductor surface, respectively.

Now in the total system, since the charges in the reference electrode are balanced out by the charges in the electrolyte electrode interface, the charges shown in Fig. 7.10 satisfy the charge neutrality condition,

$$Q_{dif} + Q_{ei} + Q_s = 0 \qquad (7.23)$$

In order to relate the charges and potentials in each region (see Fig. 7.7(a)), the capacitances of the diffusion layer, Helmholtz layer, and the insulator are defined as C_{dif}, C_h, and C_{in}, respectively. Referring to the equivalent circuit in Fig. 7.10(b), the potential difference across each capacitor can be related to the charges as

$$\psi_o - \psi_s = -Q_s/C_{in} \text{ (semicondutor surface)} \qquad (7.24a)$$

$$\psi_0 - \psi_{dif} = -Q_{dif}/C_h \text{ (EI surface)} \qquad (7.24b)$$

7.3.2.6 ISFET current equation relating pH and current

It should be remembered that the goal of EISFET theory is to obtain the FET current for a certain $[H^+]$ in the bulk of the electrolyte with the secondary ions contained in the electrolyte. In the example considered below, we consider $[Na^+]$ and $[Cl^-]$ as the secondary ions.

It is quite unfortunate that we cannot obtain a simple analytical equation to relate the pH of the bulk electrolyte with the EISFET current. However, we can obtain the EISFET current in the following procedure using the equations derived in the previous section:

(i) For a given $[H^+]$ in the bulk of the electrolyte, Eqs. (7.19) and (7.20) are the governing equations for the relationship between pH in the bulk and ψ_0 and Q_{ei}. From Eqs. (7.19) and (7.20),

$$-2.303 \times \Delta\text{pH} = \frac{e\psi_0}{kT} + \frac{1}{2}\ln\left(\frac{[\text{SiOH}_2^+]}{[\text{SiO}^-]}\right)$$

$$\frac{K_-}{K_+} = \frac{[\text{SiOH}_2^+][\text{SiO}^-]}{[\text{SiOH}]^2}. \tag{7.25}$$

Notice that K_+ and K_- are constants determined by the surface kinetics for the chemical reactions between different sites in Fig. 7.9. Usually, the numbers are obtained by the kinetics of the different sites with the ionic molecules. Also, the N_s values are also constant, determined by the ionic concentrations of the electrolyte and the unsaturated sites at the insulator surface [27, 28]. In equation (7.25),

$$\Delta\text{pH} = \text{pH} - \text{pH}_{pzc}, \tag{7.26}$$

where pH_{pzc} corresponds to the pH at the point of zero charge on the surface, i.e., $[\text{SiOH}^+] = [\text{SiO}^-]$ (see Problem 9 for the expression of pH_{pzc}).

(ii) The current modulation of ISFET by $[\text{H}^+]$ in the electrolyte can be easily predicted by using the equation for the threshold voltage of ISFET in Eq. (7.10b),

$$V_{TH} = V_{FB} + \psi_{dif} - (Q_{d,\max} + Q_{ei})/C_h - Q_{d,\max}/C_{in} + 2\phi_F.$$

For a given $2\phi_F$ and $Q_{d,\max}$, V_{TH} can be obtained from the self consistent solution of $\psi_{dif} = Q_{dif}^{-1}(Q_{dif})$ in Eq. (7.10b), $Q_{dif} + Q_{ei} = -Q_{d,\max}$, and Eq. (7.21) and (7.25) for Q_{ei} and ψ_0 relationship. Once V_{TH} is obtained, drain current of the EISFET can be obtained from Eq. (7.7a) and (7.7b).

Eq. (7.25) can be regarded as the set of equations relating pH of the bulk solution with the interface potential, ψ_0, with rest of the equations as the auxiliary equations. It can be inferred from Eq. (7.25) that the best sensitivity of the ISFET is the case when the logarithmic term becomes minimum.

In Fig. 7.11, the measurement results for the relationship between the relative gate voltage and pH value. Also the theoretical calculation is plotted with the fitting parameters shown in the figures [24]. The theory fits well with the measurements and may provide a set of equations that may provide guidelines for the design of the ISFET devices.

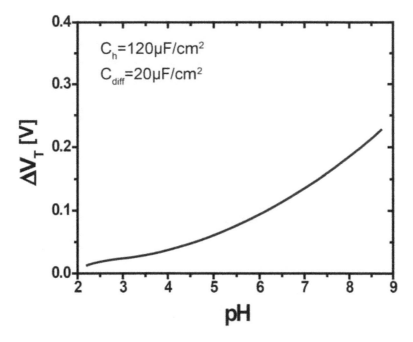

Figure 7.11. A typical relationship between pH and the change in the threshold voltage. Approximately, log-linear relationship is found, which is due to the charge and potential relationship in Eqs. (7.13) and (7.21). Redrawn from Ref. [24].

7.4 Biomolecule Sensors Based on the FET Principle

7.4.1 *Some Historical Background*

After the EISFET device was introduced, the EISFET structure has been extended to its utilization from pH sensing to sensing of various biological molecules contained in the electrolyte. One of the typical applications is DNA sensing, where single strand (ss) DNA, called the probe DNA, is immobilized onto the insulator, and the change in the charge due to hybridization with other oligomer DNA strands in the electrolyte sample is sensed by the change in the EISFET current [1]. The structure with the probe DNA is called the DNAFET. The molecular structure of the probe DNA should be

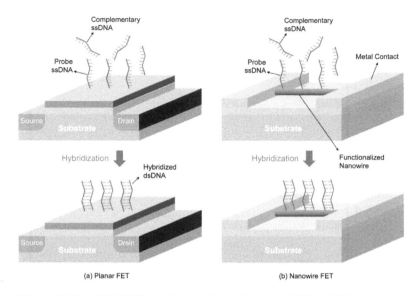

(a) Planar FET   (b) Nanowire FET

Figure 7.12. DNAFET based on a silicon (a) planar FET and (b) nanowire FET. Notice that both the probe DNA and the target DNA have negative charges in the electrolyte solution. See also Color Insert.

designed and fabricated in such a way that it is matched with the target DNA to be sensed.

Instead of the conventional planar type ISFET structure, silicon nanowires (Si-NWs) have been introduced as FET devices [29–32]. The principle of operation of the silicon nanowire DNAFET is similar to the operation of the conventional DNAFET. Fig. 7.12 shows a DNAFET based on the (a) planar FET, (b) nanowire FET, both of which share similar operating principles; the binding of the target complementary ssDNA with the probe DNA attached on the gate insulator modulates the NW surface potential through a field effect, which in turn produces a change in the current of the FET. The motivation for using the NW FET instead of the planar FET structure is that it may achieve better sensitivity because the surface-to-volume ratio of the device is bigger for the nanowire FET. Thus, if the same change in the surface charges occurs in the EI interface, the change in the FET current is bigger.

In this section, the operational principle of the DNAFET will be explained using the planar N-type FET as an example. First, change in the interface charge due to the DNA hybridization event will be described. Then, the DNA layer will be modeled as the sheet charge located at the interface between the Helmholtz layer and the diffusion layer. Then, more realistic treatment will be followed where the DNA layer is considered as an insulator layer having a finite thickness.

7.4.2 Theory of DNA FET

7.4.2.1 Binding events

The equation for the binding event between the single strand (ss) probe DNA and target DNA can be written as

$$\frac{dN_{ds}}{dt} = k_F \left(N_p - N_{ds} \right) C_t - k_R N_{ds},\qquad(7.27)$$

where N_{ds} and N_p are the areal densities of the hybridized double-strand DNA and the probe DNA, respectively. In Eq. (7.27), C_t is the volume density of the target DNA molecules at the surface and k_F and k_R are the surface kinetic constant of the binding event and dissociation event, respectively.

After the steady state is reached, N_{ds} can written as, by letting Eq. (7.27) to be zero.

$$N_{ds} = \frac{N_p}{1 + \left(\frac{k_R}{k_F} \frac{1}{C_t} \right)}\qquad(7.28)$$

If the distribution of the probe DNA on the surface of the gate insulator is not uniform either due to non-uniform attachment of the probe DNA or the non-uniform orientation of the probe DNA toward the electrolyte solution, more general expression for N_{ds} can be written [33] as

$$N_{ds} = \frac{N_p}{1 + \left(\frac{k_R}{k_F} \frac{1}{C_t} \right)^{-\gamma}},\qquad(7.29)$$

where γ is a parameter indicating the uniformity of the probe DNA molecules.

7.4.2.2 N_{ds} is considered the sheet charge layer

For simplicity, we consider an extreme case when the DNA film thickness is much smaller than the thickness of the diffusion layer as shown in Fig. 7.13(b). In the case, the DNA charge may be considered to locate between the Helmholtz layer and the diffusion layer as

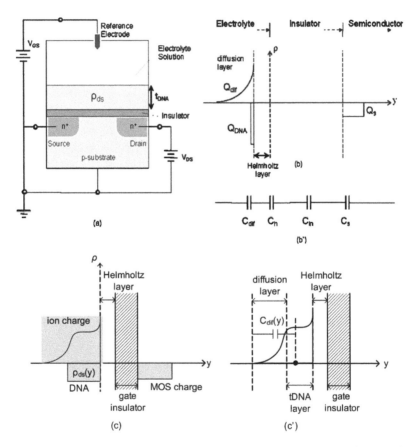

Figure 7.13. A schematic diagram of the DNAFET. (a) The double-strand DNA after the hybridization event is considered an insulator film containing a space charge, ρ_{ds}. (b) One-dimensional charge distribution and (b′) the equivalent circuit of the DNAFET immersed in an electrolyte solution with DNA sheet charge. (c) Schematic diagram of the DNAFET with the DNA layer as an insulator film and (c′) the equivalent capacitance from a point (y) of the DNA space charge to the electrolyte is denoted as $C_{dif}(y)$.

a delta function in space. The DNA charge gives rise to the induced charges. Referring to Fig. 7.13, the shift in the flat-band voltage due the DNA charge is (see Problem 5)

$$\Delta V_{FB} = -Q_{DNA}/C_{dif} \tag{7.30}$$

In Eq. (7.30), C_{dif} is the equivalent capacitance seen from DNA charge to the gate electrode. The larger C_{dif} is, the smaller the sensitivity of the sensor becomes. The change in the flat-band voltage induces change in the threshold voltage as shown in Eq. (7.10b). Once the threshold voltage is known, the modulation in the FET current can be predicted from the general equation for the MOSFET current in Eqs. (7.7a) and (7.7b).

Unlike the case for the fixed charge in the gate insulator, C_{dif} is not constant but a function of EDL potential as shown in Eq. (7.13). In the case, the Eq. (7.30) cannot be solved analytically and require a numerical solution using a computer.

7.4.2.3 N_{ds} has a finite thickness

More generally, the DNA layer is considered an insulator layer with permittivity ε_{DNA}, a thickness of t_{DNA}, and a volume charge density ρ_{ss} and ρ_{ds}, each standing for the negative charge density originated by the probe DNA and the hybridized molecules with the target DNA molecules. The thickness of the DNA layer is dependent on the length of the DNA molecules (usually having the number of bases of 20–30) and N_p, which is the surface density of the probe DNA molecules attached to the insulator surface. The charge densities are, of course, dependent on N_p and C_t, which are the concentration of the target DNA molecules contained in the electrolyte and the affinity between the probe and the target DNA molecules. The change in the electrical current in the DNAFET due to the difference in ρ_{ss} and ρ_{ds} gives information on C_{tDNA}, the concentration of the target DNA in the solution.

Figure 7.13(a) shows a schematic diagram of the DNAFET with the DNA layer as an insulator film. The distribution of the charges in one dimension and the equivalent circuit are shown in Fig. 7.13 (b) and (b'). Notice that the DNA film also includes the electrolyte so that the system may be regarded as a diffusion layer of the EDL with the dielectric permittivity modified by the DNA molecules and with the fixed space charge originated by the DNA molecules. Mathematically, the change in the flat-band voltage (so the threshold voltage) due to $\rho(y)$ can be written as

$$\Delta V_{FB} = -\int \rho(y)/C_{dif}(y)dy, \qquad (7.31)$$

where $C_{dif}(y)$ is the equivalent capacitance seen from $\rho(y)$ toward the electrolyte as shown in Fig. 7.13(c').

The threshold voltage of the n-type FET is increased by the negative charges in the DNA film. The change in the charge in the DNA layer induces counter charges in the diffusion layer and the semiconductor surface. Even though the Eq. (7.31) looks simple, the solution requires a computer simulation because ρ and C_{dif} are coupled [34, 35].

7.4.3 Some Limitations on DNA FET

The limitations on the DNA-type bio FET structure come from the fact that change in the charges caused by the binding events between the probe DNA and target DNA is screened out by the counter ions existing in the electrolyte. The Debye length determines the scale over which mobile charges screen out electric fields generated by the charges in the DNA molecules to be sensed. As the Debye length becomes shorter, the extent over which the change in the charges gives an effect becomes smaller. If the Debye length is outside the EDL, the field effect will be nullified before it reaches to the EI interface, giving effect to the surface potential of the FET.

The best way to understand the effects of the screening is to refer to Fig. 7.14. In Fig. 7.13(b), we assumed that Q_{DNA} is located between the Helmholtz layer and the diffusion layer in a delta function. However, if the DNA charge is distributed within the diffusion layer, the situation may be depicted as the equivalent circuit shown in Fig. 7.13(c). As the DNA charge and its extent of its effect (the Debye length) get close to the diffusion layer boundary with the electrolyte or outside of the diffusion layer, the effect to the threshold voltage becomes negligible. Fig. 7.14(b) is an extreme case; then the biomolecules are located outside the EDL layer and the mobile charges in the electrolyte screen out the change in the charges from

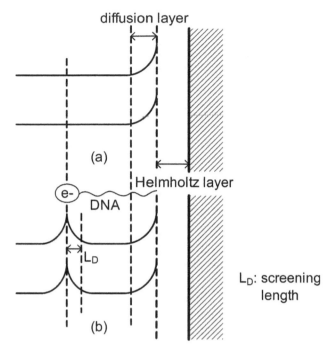

diffusion layer

(a)

Helmholtz layer

e-

DNA

L_D

L_D: screening length

(b)

Figure 7.14. The schematic diagram of the screening effect. (a)The energy diagram without the DNA charge . The effect of the DNA charge outside of the diffusion layer is considered in (b). If the Debye length is shorter than the distance from the diffusion layer, the effect of the DNA charge is screened.

the biomolecules. It should be noticed that the Debye length becomes shorter as the ionic concentration increases (see Problem 11) so that the screening effects may be more serious as the ionic concentration of the electrolyte increases.

In the explanation of the DNAFET, we neglected to consider the charges at the insulator surface. Since the isoelectric points of silicon oxide and nitride are quite far from the pH values of pure water and serum, the insulator film may be charged as is the case with an EISFET. In this case, the situation is far more complicated than the case we considered in the previous section. The detailed understanding of the Q_{DNA} and its interaction with the EDL, and thereby the modulation of the surface potential of the FET, is by no means fully understood and will be topic of the further research.

7.5 Summary

In this chapter, we studied the EISFET structure composed of the gate, electrolyte solution, gate dielectric film and the semiconductor FET. The structure can be viewed as the gate dielectric film sandwiched by two materials: electrolyte and semiconductor. As there exists an interface region between the gate dielectric film and the semiconductor, there is the interface layer called the EDL, consisting of the Helmholtz layer and the diffusion layer. The Helmholtz layer is made of the thin layer of water molecules attached to the solid surface. Also, we derived the diffusion layer equation relating the potential drop and the ion concentration of the electrolyte. In this respect, the EISFET system is similar to the conventional MOSFET with the polysilicon as the gate material. The diffusion layer of the EDL corresponds to the poly depletion layer.

As an application example of the EISFET structure, we studied ISFET and DNAFET. Both applications are based on the common operation principle; the threshold voltage of the FET, and so the drain current, is modified by charges at the interface between the gate dielectric material and the electrolyte. In the case of the ISFET with the SiO_2 as the gate insulator, the charges are originated by the protonation-deprotonation of SiOH, which saturated the bonding of the SiO_2 film at the interface. The charge states are a function of the proton concentration of the electrolyte solution.

In the case of the DNAFET, the charges at the interface are from the phosphorus backbone of the DNA molecules attached on the gate dielectric film. If the hybridization occurs with the target DNA molecules, the charge density doubles because the target DNA molecule has the same negative charges. The hybridization event will be manifested as the change in the threshold voltage of the FET, thereby causing the modification in the drain current. Even though the operation principle of the ISFET and DNAFET as application examples of the EISFET structure is straightforward, extensive scientific and engineering researches and developments are required before the devices have wide applications to molecular sensing. Seeing the history of the MOSFET researches to understand the dielectric–semiconductor interface and to make the devices as the core components of modern VLSI (very large scale integration), we can easily anticipate that exciting research areas are awaiting us, among which are better understanding and control of the interface between gate dielectric material and molecules, charge states, interaction with the electrolytes containing lots of ions, and unwanted impurity molecules, to mention a few.

PROBLEMS

1. Study the operational principle of the reference electrode made of Ag/AgCl, which is used as a reference electrode in many electrochemical measurements.

2. Explain step by step the glycolysis cycle taking place in the cell and explain how H^+ is generated or consumed in the cycle. Read, for example, Ref. 7.

3. Verify that the convenient conversion constant between "mol/L" and "#/cm^3" is 6×10^{20}.

4. Show that for $[HA]_i >> x$, $pH = 0.5(pK_a - \log_{10}[HA]_i)$.

5. **Change in the threshold voltage due to sheet charge**: In an ideal MOSFET, V_{FB} is applied to the gate so that the surface potential of semiconductor is zero and the surface charge is also zero. Consider that there is Q_{ox} at x_{ox} from the Si–SiO$_2$ interface. Verify that the new V'_{FB} becomes $V_{FB} - Q_{ox}/C_{eq}$, where $C_{eq} = \varepsilon_{ox}/(T_{ox} - x_{ox})$. (Hint: Draw the energy band diagram when V'_{FB} is applied. In the case, the semiconductor charge is zero, and $-Q_{ox}$ is induced at the gate–insulator interface so that additional voltage to V_{FB} is simply the voltage between Q_{ox} and the gate.) You can apply the similar process to obtain the change in V_{FB} in the EISFET when there exists Q_{ei}.

6. Consider the EISFET system shown in Fig. 7.6. Writing the Gauss equation with the Gaussian surface in the gate dielectric material, show that the electric field in the gate oxide is $E_{ox} = -Q_s/\varepsilon_{ox}$. Also, E_h in the Helmholtz layer can be written as $E_h = -(Q_s + Q_{ei})/\varepsilon_h$. Then show that $V_{ox} = -Q_s/C_{ox}$ and $V_h = -(Q_s + Q_{ei})/C_h$.

7. Verify Eq. (7.13) for the relationship between the potential drop across the diffusion layer and the areal charge. You can start with Eq. (7.12) by converting the left hand side to $-\frac{dE}{dy} = E\frac{dE}{d\psi}$. By multiplying both side by $d\psi$, you can integrate the left hand side with respect to E and the right hand side with ψ. Once E_{ei} at the interface is obtained, the areal density can be obtained by E_{ei} to give Eq. (7.13).

8. Derive the semiconductor surface potential vs. surface charge when the semiconductor is intrinsic. Compare the result with Eq. (7.13).

9. Verify that the pH_{pzc} value is an intrinsic quantity determined by the kinetic constants, K_+ and K_- as $(-1/2)\log_{10}(K_+K_-)$.

10. The sensitivity of the ISFET is defined as $\Delta I_D/\Delta pH$. Derive the maximum sensitivity achieved by the ISFET. (Hint: In Eq. (7.25), assume that the logarithmic term is zero.)

11. Obtain the screening length coming from the EDL layer for the solution whose Na^+ (Cl^-) concentration is 1mmol/L. Compare this with the probe DNA having 40 mers. (Hint: 1 mer of DNA is approximately 3.) Compare the results with the case for de-ionized water. (Hint: In this case, $[H^+] = [OH^-] = 10^{-7}$ mol/L.)

References

1. I. Willner, E. Katz, and B. Willner, *Electroanalysis*, **9**, 965–977 (1997).

2. M.K. Patra, K. Manzoor, M. Manoth, S.C. Negi, S.R. Vadera, and N. Kumar, *Defence Science Journal*, **58**, 650–654 (2008).

3. A. L. Sharma, V. Saxena, S. Annapoorni, B. D. Malhotra, *J. Appl. Polym. Sci.*, **81**, 1460–1466 (2001).

4. P. Bergveld., *IEEE Trans. Bio-Med.*, **15**, 102–105 (1968).

5. P. Bergveld., *IEEE Trans. Bio-Med.*, **17**, 70–71 (1970).

6. Y.-C. Wang, A.L. Stevens, and J. Han, *Anal. Chem.*, **77**, 4293–4299 (2005).

7. W. Purves, D. Sadava, G. Orians, and H. Craig Heller, *The Life*, 6th edn, www.whfreeman.com/Catalog/static/whf/thelifewire6e.

8. "Antibody" is a term used to describe the biomolecules that react with the antigen (usually causing disease) and disable the function of the antigen.

9. DNA (or RNA), called aptamer.

10. Antibody-antigen binding target protein by way of the "protein binding" for the case of antibody–antigen binding.

11. DNA (or RNA)-protein binding in the case of the Aptamer-protein binding.

12. T. Hill, *J. Am. Chem. Soc.*, **78**, 1577–1580 (1956).

13. D. van der Spoel, E. Lindahl, B. Hess, A. van Buuren, E. Apol, P. Meulen-hoff, *et al.*, *Gromacs User Manual*, version 3.2, http://www.gromacs.org (2004).

14. D. Eisenberg and A. McLachlan, *Nature*, **319**, 199–203 (1986).

15. P. Bergveld, *IEEE Trans. Bio-Med.*, **19**, 342–351 (1972).

16. S.D. Moss, J. Janata, and C.C. Johnson, *Anal. Chem.*, **47**, 2238–2242 (1975).

17. R.G. Kelly, *Electrochemica Acta*, **22**, 1–8 (1977).

18. S. Caras, and J. Janata, *Anal. Chem.*, **52**, 1935–1937, (1980).

19. J. Janata and R. J. Huber, in *Ion Selective Electrodes in Analytical Chemistry*, **2**, (ed. H. Freiser), New York: Plenum, pp. 107 (1980).

20. T. Matsuo and K. D. Wise, *IEEE Trans. Bio-Med.*, **21**, 485 (1974).

21. T. Matsuo and M. Esashi, *Sens. Actuators*, **1**, 77–96, (1981).

22. D. Harame, J. Shott, J. Plummer, and J. D. Meindl, in *IEDM Tech. Dig.*, p. 467 (1981).

23. C.D. Fung, P.W. Cheung, and W.H. Ko, *IEEE Trans. Electron. Devices.*, **33**, 8–18 (1986).

24. J.A. Davis, R.O. James, and J.O. Leckie, *J. Colloid Interface Sci.*, **63**, 480–499 (1978).

25. D. E. Yates, S. Levine, and T. W. Healy, *J. Chem. Soc., Faraday Trans.* 1, **70**, 1807–1818 (1974).

26. D.L. Dugger, J.H. Staton, B.N. Irby, B.L. McConell, W.W. Cummings, and R.W. Maatmen, *J. Phys. Chem.*, **68**, 757–760 (1964).

27. S. M. Ahmed, *J. Phys. Chem.*, **73**, 3546–3555 (1969).

28. J. Hahm and C. Lieber, *Nano Lett.*, **4**, pp. 51–54 (2004).

29. Z. Li, Y. Chen, X. Li, T.I. Kamins, K. Nauka, and R.S. Williams, *Nano Lett.*, **4**, 245–247 (2004).

30. M.M. Cheung, G. Cuda, Y.L. Bunimovich, M. Gaspari, J.R. Heath, H.D. Hill, *et al.*, *Curr. Opin. Chem. Biol.*, **10**, 11 (2006).

31. E. Stern, J.F. Klemic, D.A. Routenberg, P.N. Wyrembak, D.B. Turner-Evans, A.D. Hamilton, *et al.*, *Nature*, **445**, 519 (2007).

32. I. Quiñones and G. Guiochon, *J. Colloid. Interface Sci.*, **183**, 57–67 (1996).

33. C. Heitzinger and G. Klimeck, *J. Comput. Electron.*, **6**, 387 (2007).

34. Y. Liu, K. Lilja, C. Heitzinger, and R. Dutton, in *IEDM Tech. Dig.*, 491 (2008).

Appendix I

Physical Constants

Electronic charge: $e = 1.60 \times 10^{-19}$ C
Free electron mass: $m_e = 9.11 \times 10^{-31}$ kg
Plank constant: $h = 6.63 \times 10^{-34}$ Js
Reduced Plank constant: $\hbar = 1.06 \times 10^{-34}$ Js
Boltzmann constant: $k_B = 1.38 \times 10^{-23}$ J/K
Permittivity of vacuum: $\varepsilon_o = 8.85 \times 10^{-14}$ F/cm
Permittivity of silicon dioxide: $\varepsilon_{ox} = 3.45 \times 10^{-13}$ F/cm
Permittivity of silicon: $\varepsilon_{Si} = 1.04 \times 10^{-12}$ F/cm

Appendix II

Dirac Delta Function

The Dirac delta function, $\delta(x)$, is not strictly a function, but a generalized function that is zero everywhere except at $x = 0$. We can define the Dirac delta function as a limit of a sequence of functions. The simplest way of defining it is to start with a rectangular pulse function with a unit area as shown in Fig. A-II.1. This function can be described as

$$\delta_a(x) = \begin{cases} 1/a & \text{for } |x| \le a/2 \\ 0 & \text{for } |x| > a/2 \end{cases} \qquad \text{(A-II.1)}$$

If we take the limiting case where the width of the pulse becomes indefinitely small ($a \to 0$) while the area of the pulse is kept constant, we obtain the Dirac delta function.

$$\delta(x) = \lim_{a \to 0} \delta_a(x) \qquad \text{(A-II.2)}$$

Thus, the Dirac delta function, $(E_B d)\delta(x)$, can be used as a limiting case of the square barrier potential with height E_B and width d.

Although the above definition is very simple and easy to understand, it has one drawback: The rectangular pulse function is a discontinuous function. In some mathematical operations such as differentiation, a continuous and smooth function is much more useful. For this purpose, a Gaussian function defined as follows can be used.

$$\delta_b(x) = \frac{1}{b\sqrt{\pi}} \exp\left(-x^2/b^2\right) \qquad \text{(A-II.3)}$$

The shape of this function is shown in Fig. A-II.2

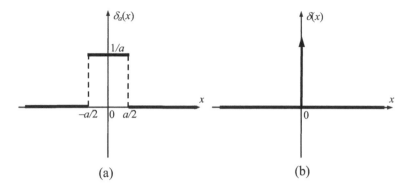

Figure A-II.1. (a) A rectangular pulse function with a unit area. (b) If $a \rightarrow 0$, this function becomes the Dirac delta function.

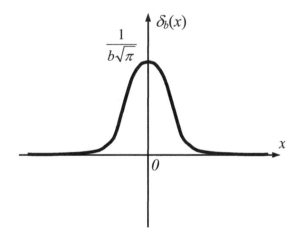

Figure A-II.2. Gaussian function with a unit area. If $b \rightarrow 0$, this function becomes the Dirac delta function.

One of the most useful properties of the Dirac delta function is its sampling characteristic as given in the following equation:

$$\int_{-\infty}^{+\infty} f(x)\delta(x)dx = f(0) \qquad \text{(A-II.4)}$$

The function $f(x)$ should be a continuous function. If we consider a constant function, $f(x) = 1$, we can confirm that the area under the delta function is always 1.

$$\int_{-\infty}^{+\infty} \delta(x)dx = 1 \qquad \text{(A-II.5)}$$

If we consider the Fourier transform of the Dirac delta function, we obtain

$$\int_{-\infty}^{+\infty} \delta(x) \exp(-j\omega x)dx = 1. \qquad \text{(A-II.6)}$$

Using the inverse Fourier transform, we obtain

$$\delta(x) = \frac{1}{2\pi} \int_{-\infty}^{+\infty} \exp(j\omega x)d\omega. \qquad \text{(A-II.7)}$$

This relation can be used as a very useful definition of the Dirac delta function.

Appendix III

Gamma Function

If z is a complex number with a positive real part, the gamma function, $\Gamma(z)$, is defined as

$$\Gamma(z) = \int_0^\infty t^{z-1} \exp(-t)dt. \qquad \text{(A-III.1)}$$

Integration by parts gives an important functional relation of the Gamma function

$$\Gamma(z+1) = z\Gamma(z). \qquad \text{(A-III.2)}$$

If z is a positive integer, n, then

$$\Gamma(n) = (n-1)!, \qquad \text{(A-III.3)}$$

That is, the Gamma function is a generalization of the factorial function to real and complex numbers.

It would be useful to evaluate the Gamma function for $z = 3/2$, since the Fermi integral can be approximated as

$$F_{1/2}(\eta_F) \cong \frac{2}{\sqrt{\pi}}\exp(\eta_F)\Gamma(3/2). \qquad \text{(A-III.4)}$$

According to the definition of the Gamma function (Eq. (A-III.1)),

$$\Gamma(3/2) = \int_0^\infty t^{1/2} \exp(-t)dt = \frac{1}{2}\int_0^\infty t^{-1/2} \exp(-t)dt. \qquad \text{(A-III.5)}$$

If we put $t = x^2$, we obtain

$$\Gamma(3/2) = \int_0^\infty \exp(-x^2)dx. \qquad \text{(A-III.6)}$$

If we square it,

$$[\Gamma(3/2)]^2 = \left[\int_0^\infty \exp(-x^2)dx\right]^2$$

$$= \frac{1}{4}\int_{-\infty}^\infty \int_{-\infty}^\infty \exp[-(x^2 + y^2)]dxdy. \quad \text{(A-III.7)}$$

Transforming the Cartesian coordinate system to the polar coordinate system, we obtain

$$[\Gamma(3/2)]^2 = \frac{1}{4}\int_0^\infty \int_0^{2\pi} r\exp[-r^2]drd\theta = \frac{\pi}{4}. \quad \text{(A-III.8)}$$

Thus,

$$\Gamma(3/2) = \frac{\sqrt{\pi}}{2}. \quad \text{(A-III.9)}$$

Appendix IV

Effective Mass Matrix

In order to deal with the acceleration of electrons in a three-dimensional (3D) space, we have to extend the concept of one-dimensional effective mass in Section 1.3.9 to that of 3D effective mass. In a 3D space, both the acceleration and the force are expressed as a vector with three components (a_x, a_y, and a_z for acceleration, and F_x, F_y, and F_z for force). Since each component of the acceleration vector may be a function of all three components of the force vector, we expect that a matrix is required to express the relationship between the acceleration and the force.

Let us consider the acceleration in a 3D space. The acceleration vector is given as

$$\vec{a} = \frac{d\vec{v}}{dt}. \tag{A-IV.1}$$

The velocity, $\vec{v}$, in turn, is related to the energy and the wave vector by Eq. (1.75).

$$\vec{v} = \frac{1}{\hbar} \nabla_k E(\vec{k}) \tag{A-IV.2}$$

Finally, the relationship between the wave vector, $\vec{k}$, and the force, $\vec{F}$, is given by Eq. (1.78).

$$\hbar \frac{d\vec{k}}{dt} = \vec{F}. \tag{A-IV.3}$$

Using these equations, we can derive an explicit form of the acceleration–force relationship. In the Cartesian coordinate system,

Eq. (A-IV.2) can be expressed as

$$\vec{v} = \frac{1}{\hbar}\left(\frac{\partial E}{\partial k_x}\hat{x} + \frac{\partial E}{\partial k_y}\hat{y} + \frac{\partial E}{\partial k_z}\hat{z}\right). \tag{A-IV.4}$$

Then, Eq. (A-IV.1) becomes

$$\begin{aligned}
\vec{a} &= \frac{d}{dt}\left[\frac{1}{\hbar}\left(\frac{\partial E}{\partial k_x}\hat{x} + \frac{\partial E}{\partial k_y}\hat{y} + \frac{\partial E}{\partial k_z}\hat{z}\right)\right] \\
&= \frac{1}{\hbar}\left[\left(\frac{\partial^2 E}{\partial k_x^2}\frac{dk_x}{dt} + \frac{\partial^2 E}{\partial k_x\partial k_y}\frac{dk_y}{dt} + \frac{\partial^2 E}{\partial k_x\partial k_z}\frac{dk_z}{dt}\right)\hat{x}\right. \\
&\quad + \left(\frac{\partial^2 E}{\partial k_x\partial k_y}\frac{dk_x}{dt} + \frac{\partial^2 E}{\partial k_y^2}\frac{dk_y}{dt} + \frac{\partial^2 E}{\partial k_y\partial k_z}\frac{dk_z}{dt}\right)\hat{y} \\
&\quad + \left.\left(\frac{\partial^2 E}{\partial k_x\partial k_z}\frac{dk_x}{dt} + \frac{\partial^2 E}{\partial k_y\partial k_z}\frac{dk_y}{dt} + \frac{\partial^2 E}{\partial k_z^2}\frac{dk_z}{dt}\right)\hat{z}\right] \tag{A-IV.5}
\end{aligned}$$

Therefore, we can obtain the x-, y-, z- components of acceleration as follows:

$$a_i = \sum_j \frac{1}{\hbar}\frac{\partial^2 E}{\partial k_i \partial k_j}\frac{dk_j}{dt} = \sum_j \frac{1}{\hbar^2}\frac{\partial^2 E}{\partial k_i \partial k_j}F_j, \qquad i, j = x, y, z \tag{A-IV.6}$$

From this equation, the ij^{th} element of the inverse effective mass matrix can be obtained.

$$\left(\frac{1}{m^*}\right)_{ij} = \frac{1}{\hbar^2}\frac{\partial^2 E}{\partial k_i \partial k_j}, \qquad i, j = x, y, z \tag{A-IV.7}$$

The full matrix is written as

$$\left[\frac{1}{m^*}\right] = \begin{bmatrix} \dfrac{1}{\hbar^2}\dfrac{\partial^2 E}{\partial k_x^2} & \dfrac{1}{\hbar^2}\dfrac{\partial^2 E}{\partial k_x\partial k_y} & \dfrac{1}{\hbar^2}\dfrac{\partial^2 E}{\partial k_x\partial k_z} \\[2ex] \dfrac{1}{\hbar^2}\dfrac{\partial^2 E}{\partial k_x\partial k_y} & \dfrac{1}{\hbar^2}\dfrac{\partial^2 E}{\partial k_y^2} & \dfrac{1}{\hbar^2}\dfrac{\partial^2 E}{\partial k_y\partial k_z} \\[2ex] \dfrac{1}{\hbar^2}\dfrac{\partial^2 E}{\partial k_x\partial k_z} & \dfrac{1}{\hbar^2}\dfrac{\partial^2 E}{\partial k_y\partial k_z} & \dfrac{1}{\hbar^2}\dfrac{\partial^2 E}{\partial k_z^2} \end{bmatrix}. \tag{A-IV.8}$$

The effective mass matrix can be obtained by the matrix inversion.

$$[m^*] = \left[\frac{1}{m^*}\right]^{-1} \tag{A-IV.9}$$

In the conduction and valence band edges of semiconductors, the symmetry in the band structure often makes the off-diagonal elements zero, so that the (inverse) effective mass matrix becomes a

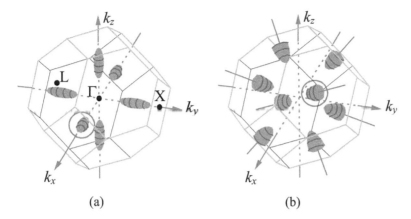

Figure A-IV.1. (a) Equal energy surfaces of silicon near the minimum energy points of conduction band, and (b) equal energy surfaces of germanium near the minimum energy points of conduction band. See also Color Insert.

diagonal matrix. For example, the conduction band of silicon has six minimum energy points in k-space and all the minimum energy points are located near the X-points and inside the first Brillouin zone (Fig. A-IV.1(a)). Three important symmetry points are indicated in the same figure. The Γ-point is the center of the Brillouin zone (or origin of the k-space), the X-point is the center of the square surface of the Brillouin zone boundary, and the L-point is the center of the hexagonal surface of the Brillouin zone boundary. There are six X-points and eight L-points on the Brillouin zone boundary. The region of k-space near the minimum energy point is often called a "valley." Near each minimum point (in other words, in each valley), the equal energy surface can be approximated by an ellipsoidal surface as shown in the figure.

Now, let us consider the effective mass matrix for the minimum energy point on the positive k_x-axis (as indicated by a circle in the figure). The ellipsoidal surface can be expressed as

$$E = \frac{\hbar^2 (k_x - k_{x0})^2}{2m_l^*} + \frac{\hbar^2 k_y^2}{2m_t^*} + \frac{\hbar^2 k_z^2}{2m_t^*}. \qquad \text{(A-IV.10)}$$

It can be obtained by rotating an ellipse on the k_x–k_y plane with the k_x-axis as the axis of rotation. $(k_{x0}, 0, 0)$ is the point where the energy is minimum, and $\sqrt{2m_l^*}/\hbar$ and $\sqrt{2m_t^*}/\hbar$ are the major and minor radius of the ellipse in the k_x–k_y plane, respectively.

For the point $(k_{x0}, 0, 0)$, we can easily calculate the elements of the inverse effective mass matrix using Eq. (A-IV.8). The result is

$$\left[\frac{1}{m^*}\right] = \begin{bmatrix} \frac{1}{m_l^*} & 0 & 0 \\ 0 & \frac{1}{m_t^*} & 0 \\ 0 & 0 & \frac{1}{m_t^*} \end{bmatrix}. \qquad \text{(A-IV.11)}$$

The diagonality of this matrix is originated from the rotational symmetry of the equal energy surface. The effective mass matrix is

$$[m^*] = \begin{bmatrix} m_l^* & 0 & 0 \\ 0 & m_t^* & 0 \\ 0 & 0 & m_t^* \end{bmatrix}. \qquad \text{(A-IV.12)}$$

As expected, the effective mass matrix is diagonal. m_l^* is called the longitudinal effective mass and m_t^* is called the transverse effective mass. The longitudinal effective mass of silicon is larger than the transverse effective mass, since the former is related with the major radius while the latter is related with the minor radius. The minimum energy point on the negative k_x-axis has exactly the same effective mass matrix.

The effective mass matrix for the other four energy minimum points can be obtained similarly. For the two minima on the k_y-axis, the effective mass matrix is

$$[m^*] = \begin{bmatrix} m_t^* & 0 & 0 \\ 0 & m_l^* & 0 \\ 0 & 0 & m_t^* \end{bmatrix}. \qquad \text{(A-IV.13)}$$

For the two minima on the k_z-axis, the effective mass matrix is

$$[m^*] = \begin{bmatrix} m_t^* & 0 & 0 \\ 0 & m_t^* & 0 \\ 0 & 0 & m_l^* \end{bmatrix}. \qquad \text{(A-IV.14)}$$

In case of germanium, there are eight energy minimum points in the conduction band, and they are located at L-points (Fig. A-IV.1(b)). Since the major axis of the equal energy ellipsoid is not aligned with any of the k_x-, k_y-, or k_z- axis, the equation for an equal energy surface is somewhat complicated. Fortunately, the same form as Eq. (A-IV.10) (with different values of m_l^* and m_t^*, of course) can be obtained for the equal energy surface of the L-valley in the [111] direction, if we use a rotated coordinate system, k'_x, k'_y, k'_z. The rotated coordinate system has the following relationship with the original coordinate system, k_x, k_y, k_z:

$$k'_x = \frac{1}{\sqrt{3}}k_x + \frac{1}{\sqrt{3}}k_y + \frac{1}{\sqrt{3}}k_z$$

$$k'_y = -\frac{1}{\sqrt{2}}k_x + \frac{1}{\sqrt{2}}k_y \qquad \text{(A-IV.15)}$$

$$k'_z = -\frac{1}{\sqrt{6}}k_x - \frac{1}{\sqrt{6}}k_y + \frac{\sqrt{2}}{\sqrt{3}}k_z$$

In the rotated coordinate system, the energy can be expressed as

$$E = \frac{\hbar^2(k'_x - k'_{x0})^2}{2m^*_l} + \frac{\hbar^2 k'^2_y}{2m^*_t} + \frac{\hbar^2 k'^2_z}{2m^*_t}. \qquad \text{(A-IV.16)}$$

Thus, in the new coordinate system, we can calculate the effective mass matrix for the (111) minimum point of germanium conduction band in the same way as the case of silicon conduction band. If we want to know the effective mass matrix in the original coordinate system, we have to substitute Eq. (A-IV.15) into (A-IV.16). Then, the effective mass matrix will not be diagonal anymore. Even though this calculation might be somewhat complicated, it would be unavoidable if we want to evaluate the combined effect of the multiple valleys.

Let us calculate the effective mass matrix for the L-valley in the [111] direction. Before we start the calculation, we can translate the energy minimum point from $(k'_{x0}, 0, 0)$ to the origin without altering the result of calculation. This invariance of mass under translation originates from the fact that the mass is related only with the curvature (second derivative), not with the slope (first derivative). In a quadratic polynomial, all the first- and lower-order terms will disappear after double differentiation! After translation, we obtain the following equation:

$$E = \frac{\hbar^2 k'^2_x}{2m^*_l} + \frac{\hbar^2 k'^2_y}{2m^*_t} + \frac{\hbar^2 k'^2_z}{2m^*_t} \qquad \text{(A-IV.17)}$$

Substituting Eq. (A-IV.15) into Eq. (A-IV.17), we obtain

$$E = \left(\frac{1}{6m^*_l} + \frac{1}{3m^*_t} \right) \hbar^2 (k^2_x + k^2_y + k^2_z)$$

$$+ \left(\frac{1}{3m^*_l} - \frac{1}{3m^*_t} \right) \hbar^2 (k_x k_y + k_y k_z + k_z k_x) \qquad \text{(A-IV.18)}$$

Using Eqs. (A-IV.8) and (A-IV.18), we can calculate the inverse effective matrix for the L-valley in the [111] direction.

$$\left[\frac{1}{m^*}\right] = \begin{bmatrix} \frac{1}{6m_l^*} + \frac{1}{3m_t^*} & \frac{1}{3}\left(\frac{1}{m_l^*} - \frac{1}{m_t^*}\right) & \frac{1}{3}\left(\frac{1}{m_l^*} - \frac{1}{m_t^*}\right) \\ \frac{1}{3}\left(\frac{1}{m_l^*} - \frac{1}{m_t^*}\right) & \frac{1}{6m_l^*} + \frac{1}{3m_t^*} & \frac{1}{3}\left(\frac{1}{m_l^*} - \frac{1}{m_t^*}\right) \\ \frac{1}{3}\left(\frac{1}{m_l^*} - \frac{1}{m_t^*}\right) & \frac{1}{3}\left(\frac{1}{m_l^*} - \frac{1}{m_t^*}\right) & \frac{1}{6m_l^*} + \frac{1}{3m_t^*} \end{bmatrix}$$

(A-IV.19)

The effective mass matrix can be obtained by the matrix inversion of this equation. However, Eq. (A-IV.19) will be more useful for the calculation of conductivity effective mass of germanium.

Besides the energy minimum point whose inverse matrix is calculated above, there are seven more in the conduction band of germanium. Fortunately, we need to consider only three more valleys due to the translational invariance of the effective mass in k-space. For the valley in the [$1\bar{1}1$] direction, we can choose a new rotated coordinate system as follows:

$$k_x' = \frac{1}{\sqrt{3}}k_x - \frac{1}{\sqrt{3}}k_y + \frac{1}{\sqrt{3}}k_z$$

$$k_y' = \frac{1}{\sqrt{2}}k_x + \frac{1}{\sqrt{2}}k_y$$

(A-IV.20)

$$k_z' = -\frac{1}{\sqrt{6}}k_x + \frac{1}{\sqrt{6}}k_y + \frac{\sqrt{2}}{\sqrt{3}}k_z$$

Once this relationship is established, we can follow the same procedure as in the case of the valley in the [111] direction. For the valley in the [$\bar{1}11$] direction, we can choose the following rotated coordinate system:

$$k_x' = -\frac{1}{\sqrt{3}}k_x + \frac{1}{\sqrt{3}}k_y + \frac{1}{\sqrt{3}}k_z$$

$$k_y' = -\frac{1}{\sqrt{2}}k_x - \frac{1}{\sqrt{2}}k_y$$

(A-IV.21)

$$k_z' = \frac{1}{\sqrt{6}}k_x - \frac{1}{\sqrt{6}}k_y + \frac{\sqrt{2}}{\sqrt{3}}k_z$$

For the valley in the $[\bar{1}\bar{1}1]$ direction, we have to choose the following rotated coordinate system:

$$k'_x = -\frac{1}{\sqrt{3}}k_x - \frac{1}{\sqrt{3}}k_y + \frac{1}{\sqrt{3}}k_z$$

$$k'_y = \frac{1}{\sqrt{2}}k_x - \frac{1}{\sqrt{2}}k_y \qquad\qquad \text{(A-IV.22)}$$

$$k'_z = \frac{1}{\sqrt{6}}k_x + \frac{1}{\sqrt{6}}k_y + \frac{\sqrt{2}}{\sqrt{3}}k_z$$

The remaining steps for the calculation of effective mass matrices are left for the readers, since it would be just a routine.

Appendix V

Density-Of-States (DOS) and Conductivity Effective Masses

In Appendix IV, we calculated the effective mass matrix for one minimum energy point (or valley). As mentioned in the same appendix, however, there can be multiple valleys or degenerate bands in the band structure. In such cases, the effect of all valleys or degenerate bands should be considered. When we calculate the effect of multiple valleys or degenerate bands, we have to exercise caution in combining the values of the effective mass matrix elements. There are two distinct categories of effective masses: one is the density-of-states (DOS) effective mass, and the other is the conductivity effective mass.

The DOS effective mass is introduced to express the density of states with a single number for the effective mass, instead of using a complicated formula describing the effect of all the relevant valleys and the non-spherical nature of the equal energy surfaces. Since we assumed the spherical symmetry of the equal energy surface (i.e., the effective mass is assumed to be independent of direction) when we calculated the density of states (Section 1.4.1), we have to develop a general method to deal with the non-sphericity of the equal energy surface. A non-spherical equal energy surface can be expressed as

$$E = \frac{\hbar^2}{2} \left(\frac{k_x^2}{m_x} + \frac{k_y^2}{m_y} + \frac{k_z^2}{m_z} \right) \tag{A-V.1}$$

If we use the following coordinate transformation, we can restore the spherical symmetry in the new coordinate system.

$$k'_x = \sqrt{\frac{m_e}{m_x}} k_x , \quad k'_y = \sqrt{\frac{m_e}{m_y}} k_y , \quad k'_z = \sqrt{\frac{m_e}{m_z}} k_z \qquad \text{(A-V.2)}$$

The energy is given by

$$E = \frac{\hbar^2}{2} \left(\frac{k'^2_x}{m_e} + \frac{k'^2_y}{m_e} + \frac{k'^2_z}{m_e} \right) = \frac{\hbar^2}{2m_e} (\mathbf{k'})^2 . \qquad \text{(A-V.3)}$$

In this coordinate system, the volume occupied by a lattice point is

$$\left(\frac{2\pi}{L_x} \sqrt{\frac{m_e}{m_x}} \right) \left(\frac{2\pi}{L_y} \sqrt{\frac{m_e}{m_y}} \right) \left(\frac{2\pi}{L_z} \sqrt{\frac{m_e}{m_z}} \right) . \qquad \text{(A-V.4)}$$

Following the same procedure as used in Section 1.4.1, we can calculate the density of states for a single valley.

$$g(E) = \frac{8\pi \sqrt{2 m_x m_y m_z}}{h^3} E^{1/2} . \qquad \text{(A-V.5)}$$

We can notice that, in this equation, the electron mass, m_e, is replaced by $(m_x m_y m_z)^{1/3}$. Thus, we can define the DOS effective mass, $m^*_{de,i}$, for a single valley designated by an index, i, as

$$m^*_{de,i} = (m_x m_y m_z)^{1/3} . \qquad \text{(A-V.6)}$$

If there are multiple valleys, we have to add the DOS of all valleys. Now, the total DOS effective mass is given as

$$(m^*_{de})^{3/2} = \sum_i (m^*_{de,i})^{3/2} . \qquad \text{(A-V.7)}$$

For N equivalent non-spherical valleys, we obtain

$$m^*_{de} = N^{2/3} (m_x m_y m_z)^{1/3} . \qquad \text{(A-V.8)}$$

Based on this result, we can evaluate the DOS effective mass of silicon. The conduction band of silicon has six energy minima that lie on the line segment between Γ- and X- point and occur at about 85% of the way to the zone boundary (Fig. A-IV.1(a)). The equal energy surfaces near the minima are ellipsoidal and the longitudinal and transverse effective masses are given as

$$m^*_l = 0.97 m_e, \qquad m^*_t = 0.19 m_e, \qquad \text{(A-V.9)}$$

where m_e is the free electron mass. By substituting these numbers into Eq. (A-V.9), we obtain the DOS effective mass of silicon.

$$m_{de}^* = 6^{2/3} \left[m_l^* \left(m_t^* \right)^2 \right]^{1/3} \approx 1.08 m_e \qquad \text{(A-V.10)}$$

The reason behind this relatively large value of DOS effective mass is the large number (six) of equivalent energy minima. In case of germanium, there are eight energy minima located at L-points (Fig. A-IV.1(b)). Since we have to count the number of quantum states that are inside the first Brillouin zone, we should consider only a half of each ellipsoid. Thus, the effective number of energy minima (or valleys) is four, not eight.

The conductivity effective mass should be expressed as a matrix, in general, since effective masses often have a matrix form and the conductivity is proportional to the effective mass. In case of a free electron, however, its mass is just a number, so that we can have a number, not a matrix, for the conductivity.

$$\sigma = \frac{n e^2 \tau}{m_e} \qquad \text{(A-V.11)}$$

Here, n is the electron concentration and τ is the scattering time. For a conduction band minimum with non-spherical equal energy surfaces around it, we have to use the following matrix form for conductivity.

$$\sigma^{(i)} = n^{(i)} e^2 \tau \left[\frac{1}{m^*} \right]$$

$$= \frac{n^{(i)} e^2 \tau}{\hbar^2} \begin{bmatrix} \dfrac{\partial^2 E}{\partial k_x^2} & \dfrac{\partial^2 E}{\partial k_x \partial k_y} & \dfrac{\partial^2 E}{\partial k_x \partial k_z} \\[2mm] \dfrac{\partial^2 E}{\partial k_x \partial k_y} & \dfrac{\partial^2 E}{\partial k_y^2} & \dfrac{\partial^2 E}{\partial k_y \partial k_z} \\[2mm] \dfrac{\partial^2 E}{\partial k_x \partial k_z} & \dfrac{\partial^2 E}{\partial k_y \partial k_z} & \dfrac{\partial^2 E}{\partial k_z^2} \end{bmatrix} \qquad \text{(A-V.12)}$$

If there are multiple valleys, we need to consider the contribution from all the valleys.

$$\sigma = \sum_i \sigma^{(i)} \qquad \text{(A-V.13)}$$

In case of silicon, electrons are evenly distributed to six valleys in the conduction band, and there are two equivalent valleys in k_x-, k_y-, and k_z-directions.

$$\sigma = 2 \cdot \left(\frac{n}{6}e^2\tau\right) \begin{bmatrix} \frac{1}{m_l^*} & 0 & 0 \\ 0 & \frac{1}{m_t^*} & 0 \\ 0 & 0 & \frac{1}{m_t^*} \end{bmatrix} + 2 \cdot \left(\frac{n}{6}e^2\tau\right) \begin{bmatrix} \frac{1}{m_t^*} & 0 & 0 \\ 0 & \frac{1}{m_l^*} & 0 \\ 0 & 0 & \frac{1}{m_t^*} \end{bmatrix}$$

$$+ 2 \cdot \left(\frac{n}{6}e^2\tau\right) \begin{bmatrix} \frac{1}{m_t^*} & 0 & 0 \\ 0 & \frac{1}{m_t^*} & 0 \\ 0 & 0 & \frac{1}{m_l^*} \end{bmatrix} \tag{A-V.14}$$

If we carry out the matrix summation, we obtain a surprisingly simple result.

$$\sigma = \frac{ne^2\tau}{3}\left(\frac{1}{m_l^*} + \frac{2}{m_t^*}\right) \begin{bmatrix} 1 & 0 & 0 \\ 0 & 1 & 0 \\ 0 & 0 & 1 \end{bmatrix} = \frac{ne^2\tau}{3}\left(\frac{1}{m_l^*} + \frac{2}{m_t^*}\right) \tag{A-V.15}$$

We can treat the identity matrix just like an ordinary number, 1, since the matrix multiplication of the identity matrix and an arbitrary matrix, **A**, results in the same matrix, **A**. Thus, even though the effective mass of each valley depends on the direction, the combined effect is independent of the direction. Finally, the inverse of the conductivity effective mass of silicon can be written as

$$\frac{1}{m_{ce}^*} = \frac{1}{3}\left(\frac{1}{m_l^*} + \frac{2}{m_t^*}\right) \tag{A-V.16}$$

By substituting the effective masses given by Eq. (A-V.9), we obtain the conductivity effective mass of silicon as

$$m_{ce}^* \cong 0.26m_e. \tag{A-V.17}$$

Note that the conductivity effective mass is much smaller than the DOS effective mass. This is because the degeneracy factor (six) does not appear in the conductivity effective mass, while it does in the DOS effective mass. We take the average value in the calculation of conductivity, while we take the sum in the calculation of DOS.

If we consider the conductivity effective mass of germanium, the calculation can be quite complicated. In order to take advantage of the symmetry in the eight valleys (half-valleys, to be accurate) of germanium conduction band, we have to use the original coordinate system shown in Fig. A-IV.1(b). In the original coordinate system, however, the off-diagonal elements of the effective mass matrix are not zero. After all the calculation is done (Problem 3, Chapter 2), however, the equation for the effective mass of germanium should have the same form as Eq. (A-V.16), as long as the longitudinal and transverse effective masses defined in the rotated coordinated system (Eq. (A-IV.16)) are used. This fascinating simplification occurs due to the symmetry in the three-dimensional structure of the valleys. During the averaging process, all the off-diagonal elements cancel themselves out!

In most of the semiconductors, the valence band edge has a very similar structure. The maximum energy points of the valence bands are located at the origin (Γ-point), and the band edge maintains spherical symmetry. In addition, the energy bands are degenerate as shown in Fig. A-V.1. That is, two energy bands with different curvatures are located at the same position in k-space. The one with a larger curvature is called a light hole band, while the one with a smaller curvature is called a heavy hole band. Obviously, these names are originated from the mass of holes residing in each band.

The DOS for holes can be calculated by adding up the heavy hole DOS and the light hole DOS. Since the DOS is proportional to $\left(m_h^*\right)^{3/2}$,

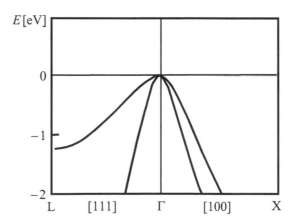

Figure A-V.1. Energy vs. k relationship for the valence band of silicon.

Table A-V.1. Effective masses of electrons and holes for silicon, germanium, and gallium arsenide (unit: free electron mass, $m_e = 9.11 \times 10^{-31}$ kg)

(a)

	Electrons			
	Longitudinal (m_l^*)	Transverse (m_t^*)	DOS (m_{de}^*)	Conductivity (m_{ce}^*)
Silicon	0.97	0.19	1.08	0.26
Germanium	1.64	0.082	0.56	0.12
Gallium arsenide	0.067	0.067	0.067	0.067

(b)

	Holes			
	Heavy (m_{hh}^*)	Light (m_{lh}^*)	DOS (m_{dh}^*)	Conductivity (m_{ch}^*)
Silicon	0.49	0.16	0.55	0.37
Germanium	0.28	0.044	0.29	0.21
Gallium arsenide	0.45	0.082	0.47	0.34

we have the following relationship for the DOS effective mass.

$$\left(m_{dh}^*\right)^{3/2} = \left(m_{hh}^*\right)^{3/2} + \left(m_{lh}^*\right)^{3/2} \tag{A-V.18}$$

The conductivity effective mass can be obtained by evaluating the weighted average of the heavy hole and light hole inverse masses. The weights should be the density of states.

Table A-V.1 summarizes the effective masses of electrons and holes for various materials: silicon (Si), germanium (Ge), gallium arsenide (GaAs). The minimum energy of GaAs conduction band is located at the origin (Γ-point) and the equal energy surfaces near this point has spherical symmetry. Due to the spherical symmetry, GaAs has the same DOS and conductivity masses.

Index

Color Insert

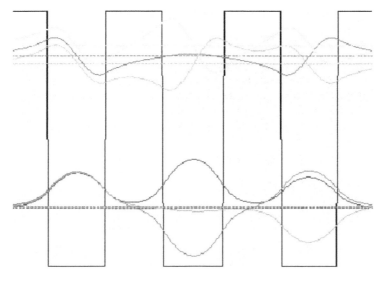

Figure 1.11. Bound state energy levels and the shape of the wave function for three square wells. The well depth (barrier height) is 3 eV, and the well width is 0.5 nm. The spacing between wells is 0.5 nm.

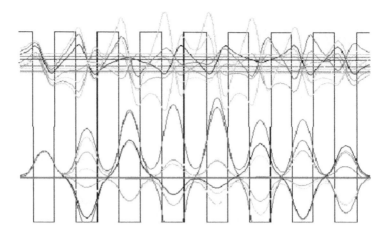

Figure 1.12. Bound state energy levels and the shape of the wave function for eight square wells. The well depth (barrier height) is 3 eV, and the well width is 0.5 nm. The spacing between wells is 0.5 nm.

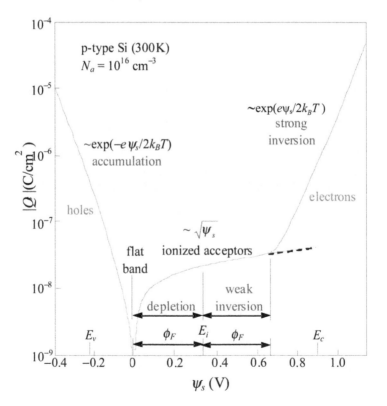

Figure 3.13 Semiconductor charge per unit area as a function of the surface potential.

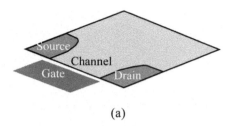

(a)

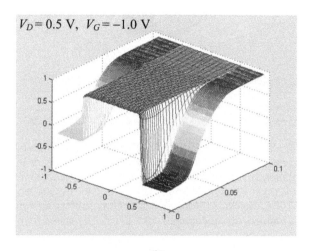

(b)

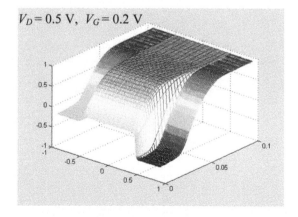

(c)

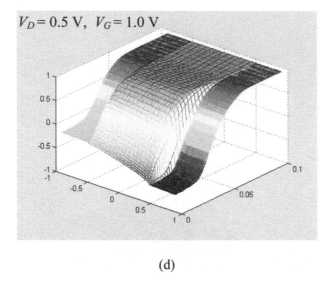

(d)

Figure 3.27 Simulated MOSFET band diagrams for various bias conditions: (a) cross-section of the MOSFET shown as it appears in a three-dimensional plot, (b) conduction band minimum as a function of location at $V_G = -1$ V, (c) at $V_G = 0.2$ V, and (d) at $V_G = 1$ V. The gate length of the MOSFET is 1 μm, the oxide thickness is 2 nm, the substrate doping is 10^{18} cm^{-3}, and the source/drain doping is 10^{20} cm^{-3}. The substrate material is silicon.

$V_G = 0.8$ V, $V_D = 0.4$ V

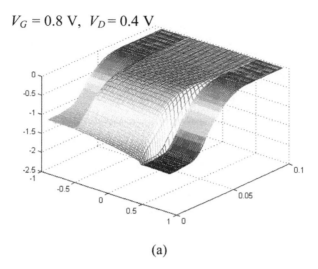

(a)

$V_G = 0.8$ V, $V_D = 1.0$ V

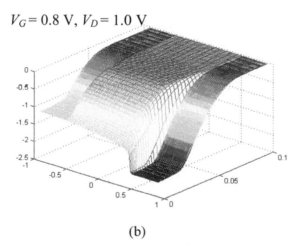

(b)

Figure 3.28 Simulated MOSFET band diagrams for various bias conditions: (a) conduction band minimum as a function of location at $V_G = 0.8$ V, $V_D = 0.4$ V and (b) at $V_G = 0.8$ V, $V_D = 1$ V. All the other device parameters are the same as those for Fig. 3.27.

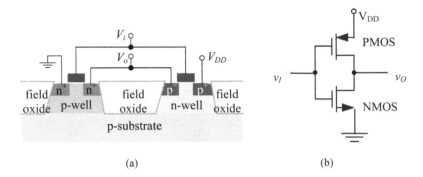

Figure 3.40 CMOS inverter: (a) cross section and (b) circuit diagram.

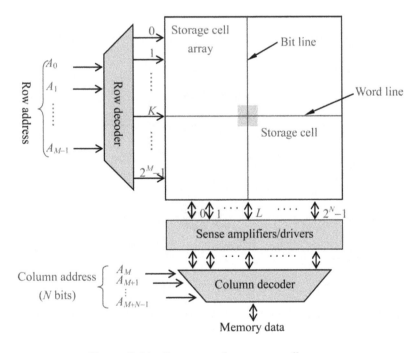

Figure 3.44 Structure of a memory cell array.

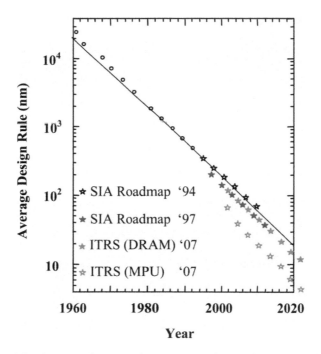

Figure 4.1 Average design rule vs. year. The predictions of the SIA roadmap and the ITRS are also plotted.

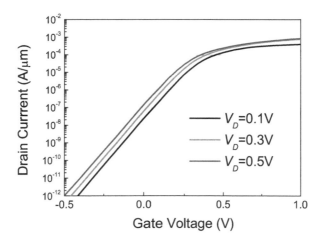

Figure 4.7 Subthreshold characteristics of a MOSFET showing DIBL. The channel length and the doping are the same as those of Figure 4.6.

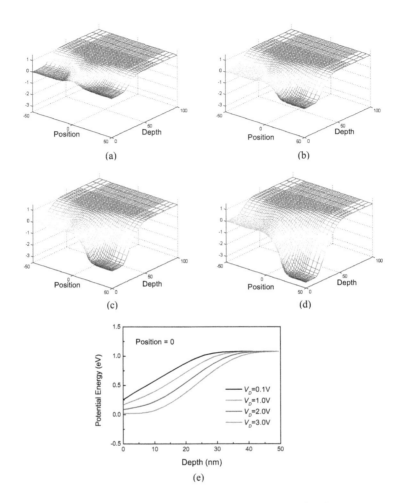

Figure 4.8 Progress of punch-through as we increase the drain voltage: (a) $V_D = 0.1$ V, (b) $V_D = 1$ V, (c) $V_D = 2$ V, (d) $V_D = 3$ V, and (e) potential barrier height as a function of depth from the oxide-silicon interface (position $= 0$). The channel length and the doping are the same as those of Figure 4.6.

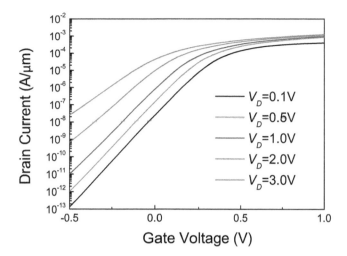

Figure 4.9 Subthreshold characteristics showing the impact of punch-through on the drain current. The channel length and the doping are the same as those of Figure 4.6.

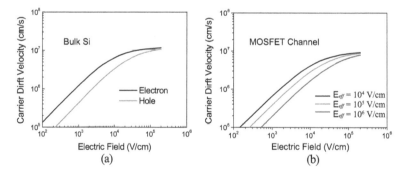

Figure 4.10 Carrier velocity vs. electric field: (a) electron and hole velocity in bulk silicon and (b) electron velocity in a silicon MOSFET channel.

(a) (x, y, z)

(b)

Figure 4.12 Position and shape of charge fluctuation: (a) the infinitesimal volume element where the charge fluctuation occurs, and (b) simplifying assumption of the charge fluctuation. In (b), the effect of the dopant fluctuation in (a) is assumed to be equivalent to that of a uniform delta function implant of dose ΔD and depth x. The top plate represents the semiconductor surface.

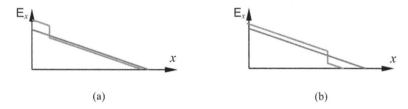

(a) (b)

Figure 4.13 Dependence of the electric field profile on the position of charge fluctuation: (a) electric field profile when x is small and (b) electric field profile when x is close to W_{dm}.

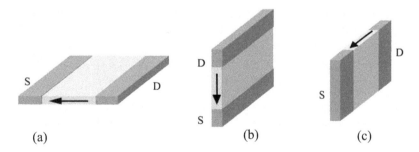

(a) (b) (c)

Figure 4.18 Three types of DG structures: (a) Type 1: horizontal length (L), horizontal width (W), (b) Type 2: vertical L, horizontal W, and (c) Type 3: horizontal L, vertical W. The arrow indicates the direction of current. Gates are formed on both sides of the channel, but are omitted in this figure.

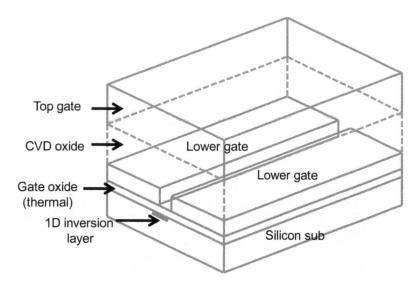

Top gate →

CVD oxide → Lower gate

Lower gate

Gate oxide (thermal) →

1D inversion layer →

Silicon sub

Figure 5.5 Formation of silicon 1DES utilizing gate depletion. The positively biased upper gate induces electrons in the silicon region, and the negatively biased lower gate depletes the electrons underneath to form a 1DES.

Figure 5.7 1DES fabricated from nanowires grown with a bottom-up method. The VLS-grown nanowires are spread over the SiO_2/Si substrate, and two metal contacts are deposited on both sides of the nanowire. The Si substrate acts as a back gate with the insulating layer of SiO_2.

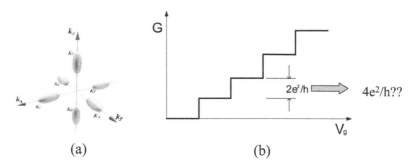

(a)   (b)

Figure 5.9 (a) Valley degeneracy of silicon. There are six oval shaped constant energy surfaces along the *k*-axes. (b) We can expect a different type of conductance quantization due to this valley degeneracy.

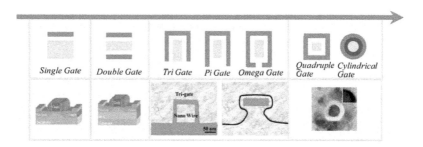

Figure 5.14 Evolution of silicon 3D transistors. The names of the devices reflect the number and shape of the gates. (Tri-gate and Omega-gate device in the lower row are schematically redrawn from Refs. 9 and 10. The photo of the cylindrical-gate device is from Ref. 2.)

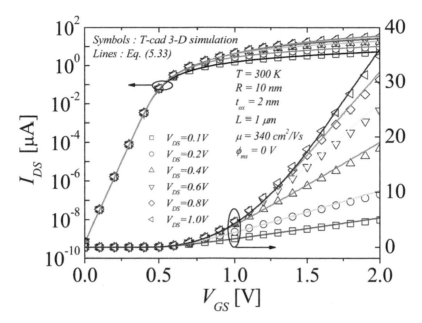

Figure 5.E5. I_{DS} as a function of V_{GS} at various values of V_{DS} in the linear regime. The results are calculated from Eq. (5.33). The results of numerical simulations are also shown.

(a) (b)

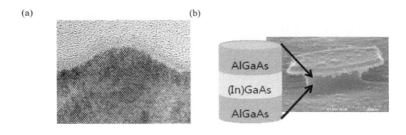

(c) (d)

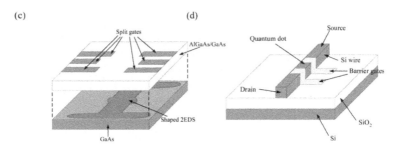

Figure 6.3 Examples of semiconductor 0DES: (a) a TEM photo of an InAs self-assembled quantum dot embedded in a GaAs matrix; (b) 0DES formed by the etching of a vertical pillar using a GaAs/AlGaAs quantum well wafer; (c) split gates on top of a GaAs/AlGaAs HEMT wafer; and (d) a 1D nanowire etched on an SOI wafer with two metal gates acting as barriers.

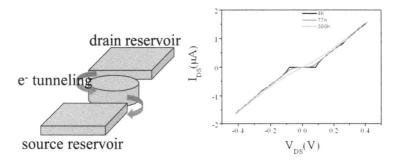

Figure 6.4 (Left) A schematic of the quantum dot connected to the source and the drain reservoir by tunnel barriers. (Right) A calculated current–voltage characteristic of such a quantum dot at three different temperatures. At the lowest temperature, a suppression of the current near zero bias can be observed. This is the manifestation of the Coulomb blockade phenomenon.

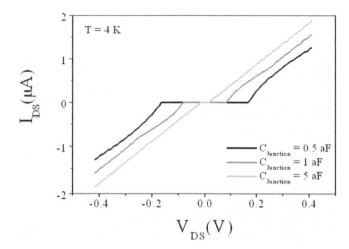

Figure 6.5 *I–V* characteristics of the quantum dot with values of $C = 0.5$, 1, and 5 aF and $T = 4.2$ K. All of them show clear gap structures near zero bias, suggesting that the single-electron charging energy is larger than $k_B T$ in all three quantum dots at this temperature.

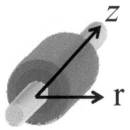

Figure 6.P1 A gate-all-around silicon nanowire field effect transistor. The electrons confined inside the nanowire feed an oval-shaped potential. The oval has the short axis in the r-direction of the cylindrical coordinate, and the long axis along the z-direction.

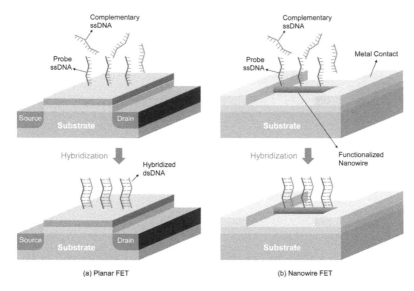

Figure 7.12 DNAFET based on a silicon (a) planar FET and (b) nanowire FET. Notice that both the probe DNA and the target DNA have negative charges in the electrolyte solution.

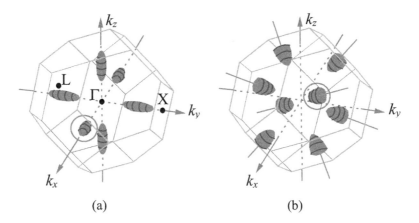

Figure A-IV.1. (a) Equal energy surfaces of silicon near the minimum energy points of conduction band, and (b) equal energy surfaces of germanium near the minimum energy points of conduction band.

Milton Keynes UK
Ingram Content Group UK Ltd.
UKHW031136141024
449569UK00006B/145